Metallography in Failure Analysis

Metallography in Failure Analysis

Edited by
James L. McCall
Battelle-Columbus Laboratories
Columbus, Ohio

and

P. M. French
Westinghouse Electric Corporation
Cheswick, Pennsylvania

Plenum Press · New York and London

Library of Congress Cataloging in Publication Data

Symposium on Metallography in Failure Analysis, Houston, Tex., 1977.
Metallography in failure analysis.

Includes indexes.
1. Metallography—Congresses. 2. Fracture mechanics—Congresses. I. McCall, James L. II. French, Peter Michael, 1935- III. American Society for Metals. IV. International Metallographic Society. V. Title.
TN689.2.S88 1977 620.1'6'3 78-7224
ISBN 0-306-40012-X

Proceedings of a Symposium on Metallography in Failure Analysis sponsored by the American Society for Metals and the International Metallographic Society held in Houston, Texas, July 17–18, 1977

A Division of Plenum Publishing Corporation
227 West 17th Street, New York, N.Y. 10011

Printed in the United States of America

PREFACE

Detailed analyses of failures of material components have proved to be valuable in many ways; by preventing further failures, by assessing the validity of designs and the selection of materials, by uncovering shortcomings in the processing of the materials involved through characterizations of defects, and by revealing problems introduced during the manufacture or fabrication of the component. Increased recognition of the value of performing failure analyses has caused the field to develop into a very active area of technical endeavor. Failure analysis has been employed in numerous different technical disciplines and has proven beneficial. The increased activity has caused many new and improved methods for performing these analyses to be developed. Among these are many methods which can be characterized as generally belonging to the field of metallography.

In recognition of the important role that metallography plays in the performance of failure analyses, the absence of a text that specifically discusses this subject, and the belief that communication of information on the subject would be of technical interest, The American Society for Metals and The International Metallographic Society co-sponsored a symposium. The intent was to bring together world-recognized authorities working in various aspects of the failure analysis and metallographic fields to share methods they use, results they have obtained, and the purposes to which they utilized these results. The symposium, entitled "Metallography in Failure Analysis", was held in Houston, Texas, USA, July 17–18, 1977. It followed three earlier symposiums co-sponsored by the same two societies on other subjects of interest to the metallographic community, *Microstructural Analysis – Tools and Techniques* (1972), *Metallographic Specimen Preparation – Optical and Electron Microscopy* (1973) and *Interpretive Techniques for Microstructural Analysis* (1975).

The wide-spread interest in the symposium, specifically shown by the large attendance and enthusiastic participation has encouraged us to publish all of the formally-presented papers. These papers comprise the current volume. Our hope is that these proceedings will serve as a useful reference for individuals active either full- or part-time in the field of failure analysis.

Organizing a symposium of the size that resulted would not have been possible without contributions from numerous individuals. To all we owe a deep debt of gratitude, but, especially, we want to mention Dr. L.R. Cornwell, General Chairman of the 1977 International Metallographic Convention of which the symposium was a part. The cooperation of both co-sponsoring societies was assured through several individuals, most directly Dr. E.J. Myers and Mr. Oren Huber of The American Society for Metals and Messrs. J.H. Richardson and R.J. Gray of the International Metallographic Society. We also give a special thanks to Connie McCall for putting the entire proceedings in a uniform format and typing them in camera-ready form. Finally, we thank all the authors and session chairmen without whose participation the symposium obviously would not have been possible.

We hope the combined efforts of these and many more unnamed individuals has resulted in a publication which will prove to be useful to the scientific community.

James L. McCall
Battelle—Columbus Laboratories

P.M. French
Westinghouse Electric Corporation

CONTENTS

FAILURE MECHANISMS AND METALLOGRAPHY: A REVIEW

I. Le May*

INTRODUCTION

In spite of the best efforts of design engineers and materials specialists, and in spite of well-thought-out procedures for quality assurance and control, engineering components still fail in service from time to time. In the majority of cases the failure does not lead to serious consequences and the component is merely replaced as having served its useful life and worn out, and in such cases optimization of the procedures of design and quality assurance have probably not been of unduly great concern, economic considerations dictating that avoidance of failure is not of overriding importance. However, in other cases failure may lead to serious consequences which can cause serious financial loss, environmental contamination or loss of life: in such situations and to minimize the risk of failure, the design and quality assurance procedures must have been very carefully studied and controlled, and in the event of such a failure occurring it is frequently necessary to establish the root causes in terms of design, choice and quality of material, fabrication procedure, and so on. Thus, the metallographer may be required to conduct an investigation to determine the mechanisms of failure, based on the appearance of the fracture surfaces and the microstructure of the components which have failed. It is the purpose of this paper to review the various mechanisms of fracture and to discuss the identifying morphological and microstructural features appropriate to each.

BASIC FRACTURE MECHANISMS

There are two basic and distinct mechanisms of fracture, namely cleavage and ductile fracture. The former occurs under tensile stress and involves separation along crystallographic planes with little or no plastic flow taking place, and leads to a brittle appearance of the fracture surface, with a small amount of energy being dissipated: the latter involves plastic deformation by slip and the energy dissipation involved is much greater, but depends on the extent of the plastic flow. It will be large when extensive shear, necking or void formation occurs, but relatively small when plastic flow is localized to the tip of a propagating crack, in which case the fracture may be termed brittle in an engineering sense, but is ductile on a strictly mechanistic basis.

However, it is insufficiently informative to classify fractures strictly in terms of the mechanisms of cleavage or ductile fracture, and we must look at the various factors which allow and cause cracks to propagate gradually or in a stepwise manner until they reach a critical length at which rapid failure takes place under load, either by cleavage, ductile fracture or some combination of the two. Indeed, it is relatively uncommon to find fractures of a simple overload nature outside of the laboratory, and much more

*University of Saskatchewan, Saskatoon, CANADA S7N OWO

common to find that failure in service has taken place over a period of time under the normal working conditions of loading. Thus, we shall identify fatigue crack growth, intergranular cracking and environmentally-assisted cracking as additional mechanisms leading to failure.

Fatigue crack growth occurs under repeated or cyclic tensile stress and is by far the commonest cause of failure in engineering components and structures. Intergranular cracking may arise in several ways: it may take place during creep deformation at high temperature either as a result of grain boundary sliding or from the growth of voids at grain boundaries as a result of vacancy deposition; or it may arise from segregation of specific elements at grain boundaries, so leading to an embrittled structure which can fracture under impact or monotonic loading. Environmentally-assisted cracking relates to fracture under the combined action of stress and an environmental effect such as corrosion, the presence of an embrittling fluid such as hydrogen, or the adsorption of a chemical species on to the fracture surface. Thus, environmentally-assisted cracking includes the phenomena of stress corrosion cracking, corrosion fatigue, hydrogen embrittlement and liquid metal embrittlement. In addition, we may extend it to include failure caused by radiation damage in conjunction with the service loading.

Two additional mechanisms of failure may be identified, corrosion and wear. These are both extremely important from a practical standpoint, although they are of a somewhat different nature to the mechanisms of *fracture* identified above.

INVESTIGATIONAL PROCEDURES

The determination of the failure mechanisms which were operative in causing a component to fracture or cease to function involves consideration of the design, specifications and service conditions for the component, the evidence of witnesses to the failure, and detailed examination of the fracture surfaces on both macro- and micro-scales, together with possible destructive tests to determine the microstructure and mechanical properties of the part. It is rare that evidence provided by the fracture surfaces alone will furnish sufficient information to determine unequivocally the cause(s) of failure, and such evidence may be quite misleading, for example when artifacts are present due to rubbing or subsequent damage inflicted on the fracture surfaces. The overall procedures involved in a failure analysis are discussed fully in appropriate reference texts [1,2].

It should be emphasized that it is absolutely essential that all non-destructive examinations and metallography are completed and documented before any destructive testing is begun.

Macro examination, both visual and under a low power microscope (preferably a stereo microscope), can be of great value in determining the overall nature of the fracture (e.g., whether it is essentially brittle or ductile on an engineering scale), the relationship of the fracture plane or planes to the operative directions of stressing, the point of initiation of the failure, and damage which may indicate an impact or other load. Thus, Fig. 1 illustrates a fatigue failure with the origin of the fracture pinpointed by means of the beach marks or arrest lines, while Fig. 2 shows clearly the path of a brittle fracture arising from impact loading. In many such cases it is unnecessary to conduct detailed examination of the fracture surfaces on a micro-scale, and one may move directly to metallographic study of the microstructure at the crack initiation point.

Detailed microscopic examinations, the results of which we shall relate to the specific mechanisms of fracture, include metallography using both replica techniques and transmission electron microscopy (TEM) and direct observation under the scanning electron microscope (SEM), as well as optical and electron microscopic examinations

(a)

(b)

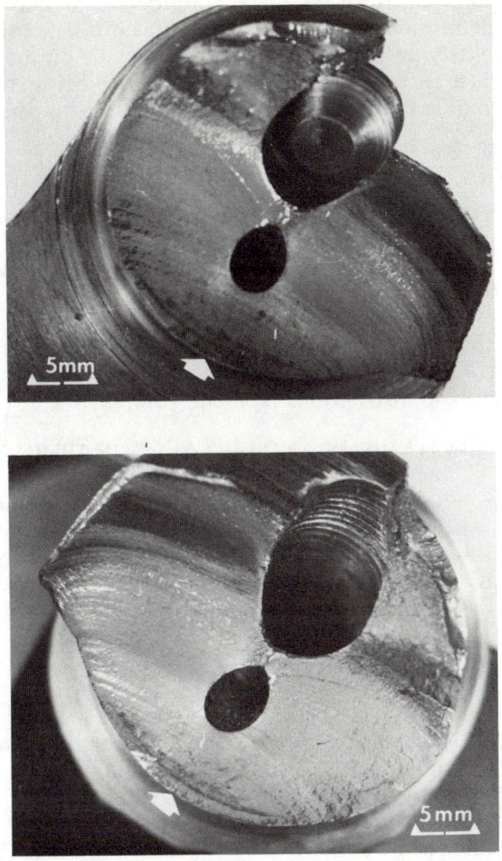

Fig. 1. Matching surfaces, (a) and (b), of fatigue failure in a crank-shaft. Origin of fracture is arrowed.

Fig. 2. Brittle fracture in connecting rod which broke suddenly following fatigue failure of bolts at big end bearing. Arrow indicates crack propagation direction.

of metallurgical specimens cut from the failed component. In addition, it may be necessary in some cases to determine the distribution of specific elements adjacent to the fracture surfaces, or to identify particular precipitates which lie on these, using more specialized analytical techniques such as Auger spectroscopy or electron microprobe analysis, these and related techniques being discussed in a previous symposium volume [3].

Analysis using the SEM has considerable advantages regarding simplicity and the avoidance of the somewhat tedious procedures involved in replication, but in practice failed components are generally much too large to be examined directly in the SEM, and so the component must be cut up for such analysis: consequently, examinations using replicas and the TEM may be more appropriate if the fracture surface and the component are not to be destroyed [4].

The techniques involved in preparing replicas for TEM are well documented and, rather than discuss them here, the interested reader is referred to the appropriate sources for detailed information and instructions (see e.g., [5,6,7]).

SPECIFIC FAILURE MECHANISMS AND METALLOGRAPHIC EVIDENCE

In examining the various failure mechanisms and metallographic evidence associated with each, we shall rely heavily on TEM studies using replicas, as these provide much finer detail than do optical microscopy or most SEM studies. Where appropriate, these will be supplemented by additional metallographic evidence based on sections through the microstructure.

Cleavage

Cleavage may occur in BCC and HCP metals and is associated with rapid loading and low temperature. It is facilitated by the presence of high triaxial stresses which allow a high tensile stress to exist for a given value of uniaxial yield stress. Cleavage can be a major problem in low strength structural steels, and has in the past led to the catastrophic failure of many large structures including the classic case of the tanker *Schenectady* in calm, cool weather while at her fitting-out dock moorings [8] (see Fig. 3).

Since the late 1940's extensive study has been made of the conditions leading to brittle fracture in steel, and our knowledge of the transition from transgranular cleavage to ductile fracture as temperature is raised and of the effect on this transition temperature of metallurgical and microstructural variables has become reasonably complete. Nonetheless, such failures still do occur today, and are of particular concern when dealing with structures for low temperature environments such as the Arctic.

Because of its crystallographic nature, cleavage leads to a fracture surface composed of flat, shiny facets. The crack is essentially flat and lies approximately normal to the operative tensile stress axis, but changes its orientation slightly when it crosses a grain boundary because of the differing orientations of adjacent grains and because the crack follows specific planes (e.g., $\{100\}$ in α-iron). If the boundary is a high angle or tilt one, the crack may restart in the adjacent grain as a single one; however, in the case of a twist boundary, many small parallel cracks may form, with cleavage steps between them (Fig. 4). These steps run together to form larger ones and lead to the characteristic river patterns seen in cleaved polycrystalline metals (Fig. 5). Cleavage steps may also form at the intersection of a cleavage crack and a screw dislocation [9], as shown in Fig. 6.

Another characteristic feature observed in cleaved iron is that of tongues [10], which are formed by local fracture along twin-matrix interfaces, the mechanism being

Fig. 3. The T–2 tanker *Schenectady* after sudden, brittle failure while in her fitting-out dock. (Courtesy of the U.S. Ship Structure Committee).

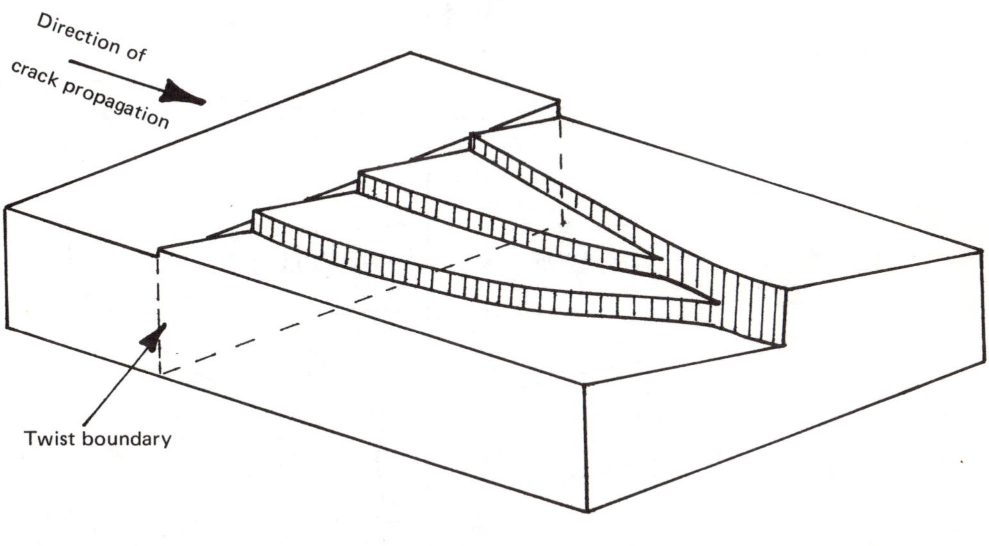

Fig. 4. Schematic showing cleavage steps produced when cleavage crack crosses a twist boundary.

illustrated in Fig. 7, while the appearance of the resulting tongues on a microfractograph is shown in Fig. 8. Tongue formation occurs as a result of the high velocity at which a cleavage crack propagates (it has a limiting velocity between 0.4 and 0.5 of the speed of sound) which produces a local strain rate which is too high for slip processes to provide all the accommodation required, and hence many twins are normally formed just ahead of the moving crack tip.

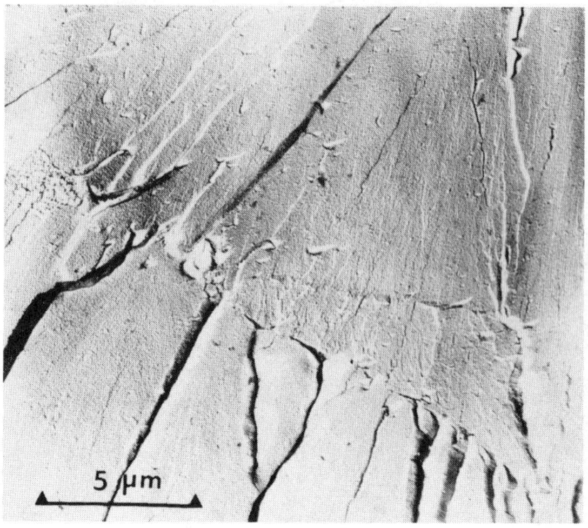

Fig. 5. River pattern on cleaved mild steel bolt.

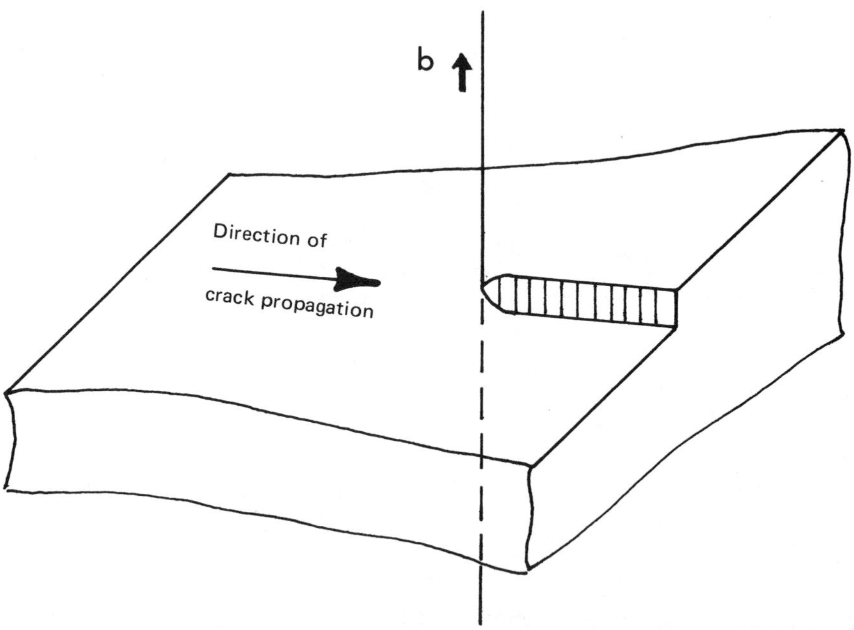

Fig. 6. Schematic showing a cleavage step initiated at a screw dislocation intersecting the cleavage plane.

Although cleavage is crystallographic in nature, in metals local plastic flow will begin when the local stress reaches the yield stress for shear, and a cleavage crack can initiate after some critical displacement has occurred at the crack tip. Above the ductile-brittle transition temperature (DBTT), yield will take place rather than cleavage.

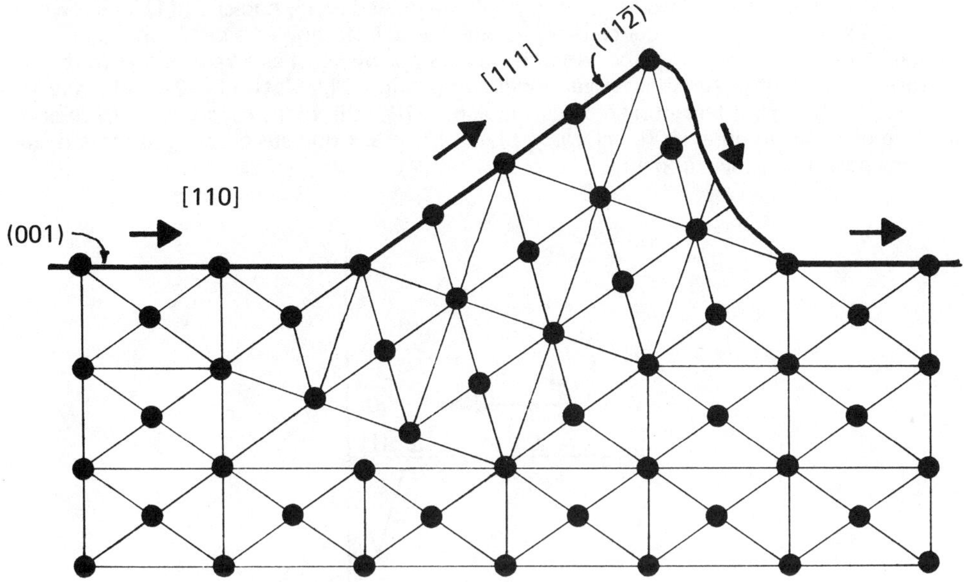

Fig. 7. The formation of a tongue at a twin-matrix interface in α-iron.

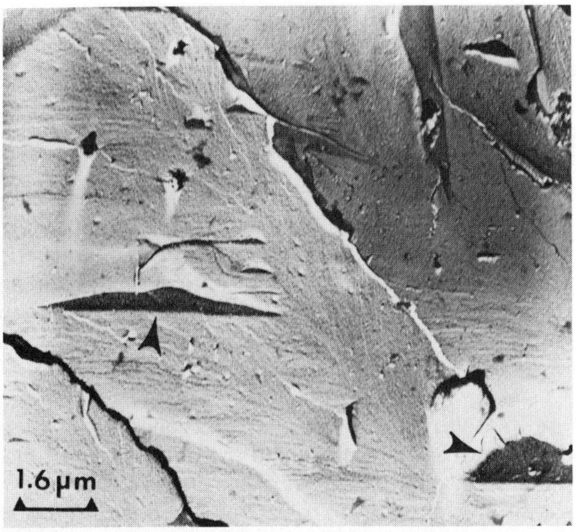

Fig. 8. Microfractograph showing tongues (arrowed) in cleaved mild steel.

During the propagation of a cleavage crack below DBTT, plastic flow always takes place at the crack tip where a high stress concentration exists. Thus, the energy which is required to create the new surfaces is greater than that strictly required for separation of atomic planes.

A number of models have been formulated to explain the nucleation of cleavage cracks on the basis of dislocation movement, the simplest one envisaging dislocations at the head of a pile-up coalescing to form a wedge-shaped crack nucleus [11] as shown in Fig. 9. The resulting stress concentration may cause initiation of a crack on a plane roughly normal to the slip plane, when the stress on this reaches a value equal to the theoretical strength. An alternative model was proposed by Cottrell [12] and is shown in Fig. 10. In this, intersecting dislocations on 110 slip planes coalesce to form an immobile dislocation on a 100 cleavage plane, the crack nucleus growing as more dislocations pass down the slip planes.

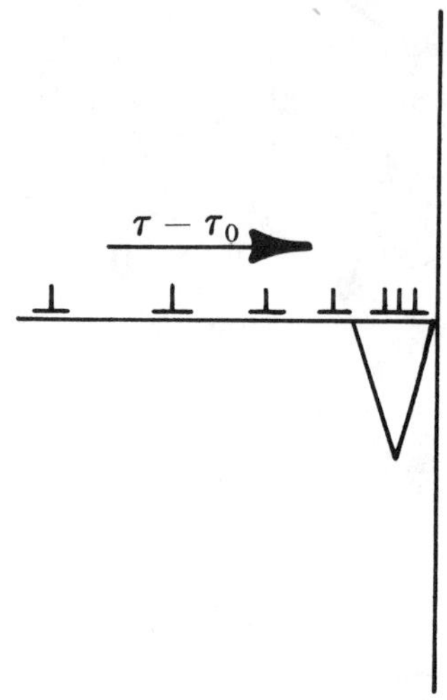

Fig. 9. Cleavage crack nucleated from coalescence of dislocations at head of a pile-up.

Cleavage cracks may also be initiated at twins [13] or in second phase particles such as carbides in steel. Smith [14] has discussed the case of cleavage being initiated in a grain boundary carbide by plastic deformation in the neighboring ferrite grain as shown in Fig. 11. This may, in turn, initiate cleavage in the next grain if the local grain boundary stress is sufficiently high and crack formation is energetically favorable. Smith's theory is of particular value as it allows us to examine the important effect of carbide thickness on the brittle fracture of steels in a semi-quantitative manner.

Ductile Fracture

Ductile fracture involves plastic flow, and the general term covers a wide range of failure modes. Under monotonic loading, and in the absence of cleavage, the ductile fracture may take place with a large degree of variation in the extent of plastic deformation, and when this is small the fracture may still be considered brittle in an engineering sense, with the crack propagating in a plane normal to the applied tensile stress. In such a case, however, the localized plastic deformation at the crack tip precludes the formation of cleavage facets and steps.

Various types of ductile fracture are illustrated schematically in Fig. 12, ranging from tensile fracture in pure, ductile metals which can result in necking down to a point, to creep fracture at high temperature by intergranular cavitation. In the present discussion, we shall concentrate on the rupture of engineering materials containing second phase particles or inclusions, and defer discussion of intergranular creep fracture to a later section. This restriction is hardly a severe one, as engineering materials always

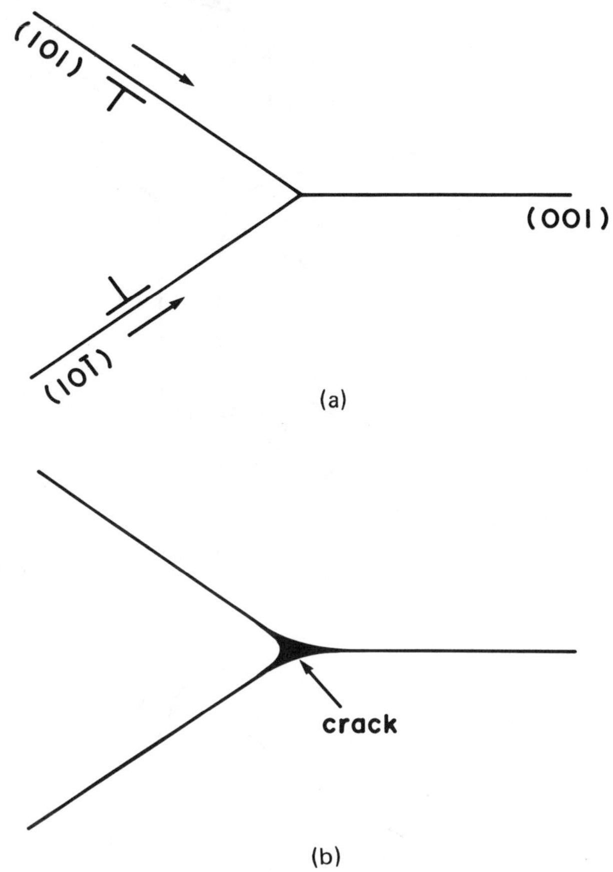

Fig. 10. Cottrell's model for cleavage crack initiation in BCC metals from intersecting dislocations forming immobile dislocation on the (001) cleavage plane. Additional dislocations moving down slip planes cause nucleus to grow as in (b).

contain large quantities of second phase particles, ranging from fine precipitates which are only detectable using thin foil TEM as in precipitation hardened alloys, to large particles visible under the optical microscope and consisting of various alloying elements which have been added to improve some properties, such as machinability.

The mechanism of ductile fracture is generally related to the initiation of microvoids at second phase particles, and their subsequent growth and coalescence. As a consequence of this process, the fracture surfaces contain dimples which show evidence of the initiation points at the particles (see Fig. 13). Depending on the mode of fracture, viz., tension, shearing or tearing, the dimples may have equiaxed or elongated shapes, and their formation is illustrated in Fig. 14, while the two different dimple shapes are apparent in Figs. 13 and 15 respectively. Obviously, in order to distinguish between shear dimples and tear dimples, it is necessary to prepare matching replicas from adjacent surface areas. When the particles are large and brittle, microvoid initiation may take place by the fracture of the particles rather than decohesion at the particle-matrix interface. Thus, cracked particles may be observed during metallographic examination of longitudinal sections through the material, and cleaved particles may be seen on the fracture surfaces adjacent to the dimpled surface of the matrix.

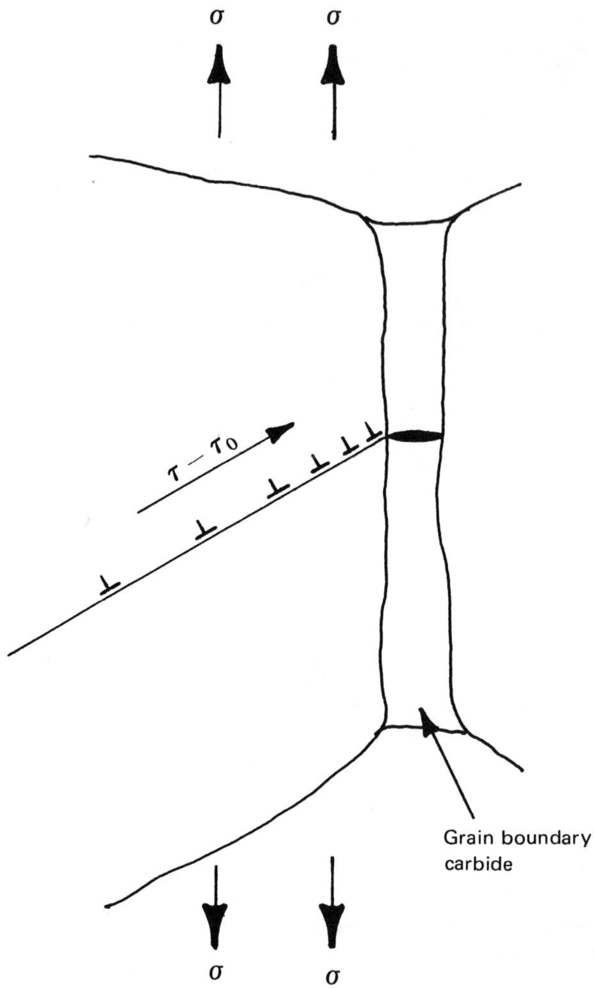

Fig. 11. Cleavage crack initiated in grain boundary carbide by plastic flow in neighboring ferrite grain.

A simple model for microvoid initiation by decohesion at the particle-matrix interface was proposed by Gurland and Plateau [15], being based on the concept that the strain energy relieved by void formation must be sufficient to produce the required energy for the newly created surfaces. This lead to an expression for the applied uniaxial tensile stress, σ, to produce decohesion as

$$\sigma = \frac{1}{q}\left(\frac{E\gamma}{a}\right)^{1/2} \tag{1}$$

where q is the stress concentration factor at the inclusion, γ the specific surface energy of the crack, E a weighted average of the elastic moduli of inclusion and matrix, and a is the particle size (diameter). This model has been extended by Lui and LeMay [16] to take into account the plastic work being dissipated around the particle, and the following relation is obtained

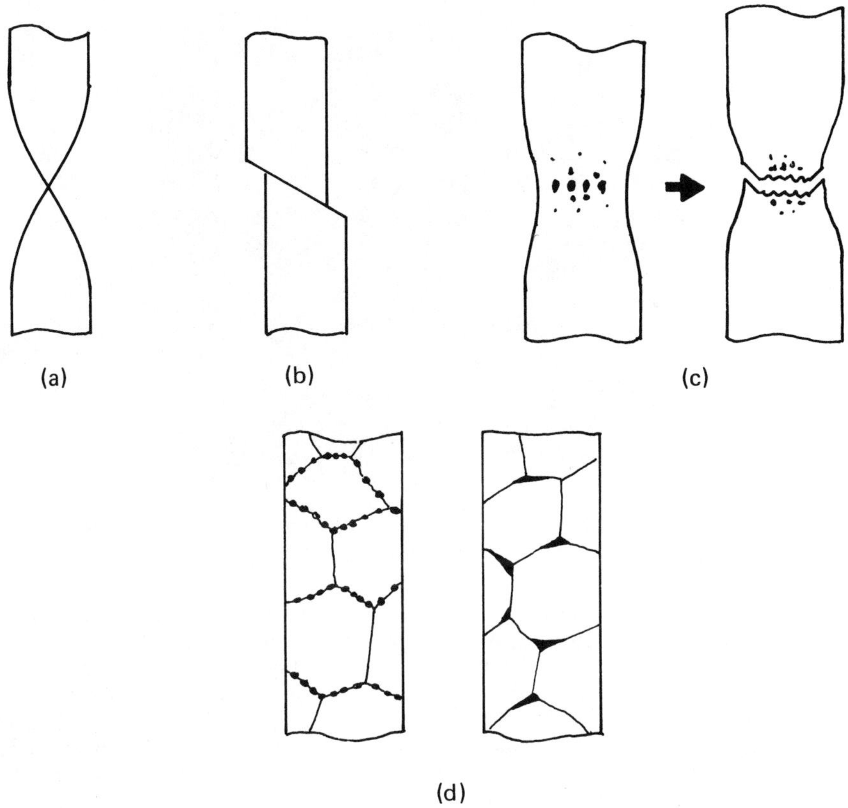

Fig. 12. Schematic illustration of different types of ductile fracture in tension: (a) Necking to a point in very pure metal; (b) Failure by shearing off; (c) Cup and cone fracture caused by void nucleation and coalescence with final shear; (d) Intergranular creep fracture showing different void nucleation points.

$$\sigma = \frac{1}{q}\left(\frac{E\gamma}{a}\right)^{1/2} + \frac{\sigma_\gamma}{q}\left(\frac{\Delta V}{V}\right)^{1/2} \qquad (2)$$

in which σ_γ is the average yield stress of the matrix for a given particle shape and volume fraction of precipitates, and is assumed to be independent of particle size and strain. V is the volume of a particle, while $\Delta V/V$ indicates the volume deformed around the particle in relation to the particle, and is regarded as being approximately constant for given particle shape.

Equation (2) indicates that the stress to initiate fracture is a function of $(1/a)^{1/2}$, and it has also been shown that the UTS varies in a like manner for some high strength quenched and tempered steels and for spheriodized carbon steels, for a given carbide shape [16]. Hence, it has been suggested that the controlling step in ductile fracture in such materials may be microvoid initiation, this occurring at a point close to the maximum on the tensile load-extension curve.

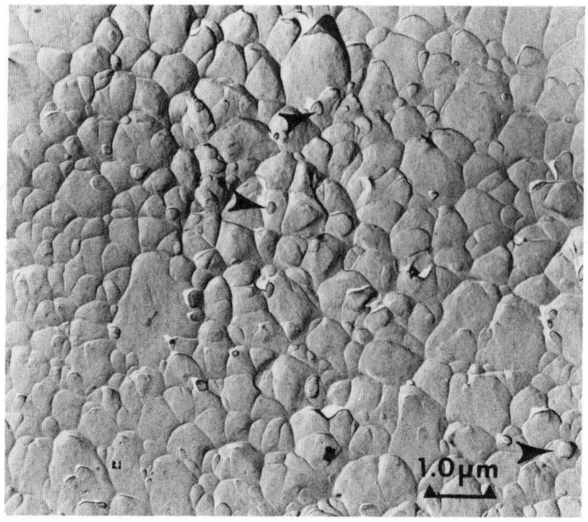

Fig. 13. Equiaxed dimples on fracture surface of quenched and tempered 4140 steel. Note the carbides at which local fracture initiated (arrowed).

Where extensive work hardening takes place, voids grow under the influence of a tensile stress as well as of a lateral stress, as occurs during necking, and it has been demonstrated that the fracture ductility can be calculated on the basis of the voids meeting each other, and is seen to be a function of the volume fraction of voids [17]. In high strength materials, where work hardening capacity is generally low, the microvoids may link up with a propagating crack by shear fracture along slip bands [18], as illustrated in Fig. 16. This can give rise to a distinctly jagged surface as may be seen on the weld metal shown in Fig. 17 which failed in a tensile stress-rupture test.

Fatigue

The great majority of engineering components fail by fatigue, i.e., under repeated or cyclic loading, but one cannot assume that fatigue is the cause of failure without the accumulation of appropriate evidence.

In examining the nature of fatigue fractured surfaces it is important to distinguish between the three stages of fatigue failure, viz., crack initiation, crack propagation, and final rupture which occurs when the crack reaches a critical length at which point it propagates rapidly to cause sudden failure during a single tensile load application. Fatigue is not in itself a *mechanism* of fracture, and we must look at the separate mechanisms of crack initiation and crack propagation.

Fatigue cracks generally initiate at a surface discontinuity or stress raiser, their formation being aided by the presence of a notch or inclusion; however, they may in some cases initiate at a subsurface defect such as a casting defect, inclusion, or quench or hydrogen-assisted crack. They are formed through the occurrence of local cyclic plastic deformation, which generates high local dislocation densities on slip planes, and can lead to the formation of small surface intrusions and extrusions which may be seen either by optical or replica electron microscopy. Figure 18 illustrates these schematically. The nominal stresses at which fatigue cracks can form may be well below the elastic limit of the material, but on a micro-scale the *local* stresses may be much higher as a result of the stress concentrations around defects and inclusions, leading to highly localized plastic deformation which is not apparent on a normal engineering scale.

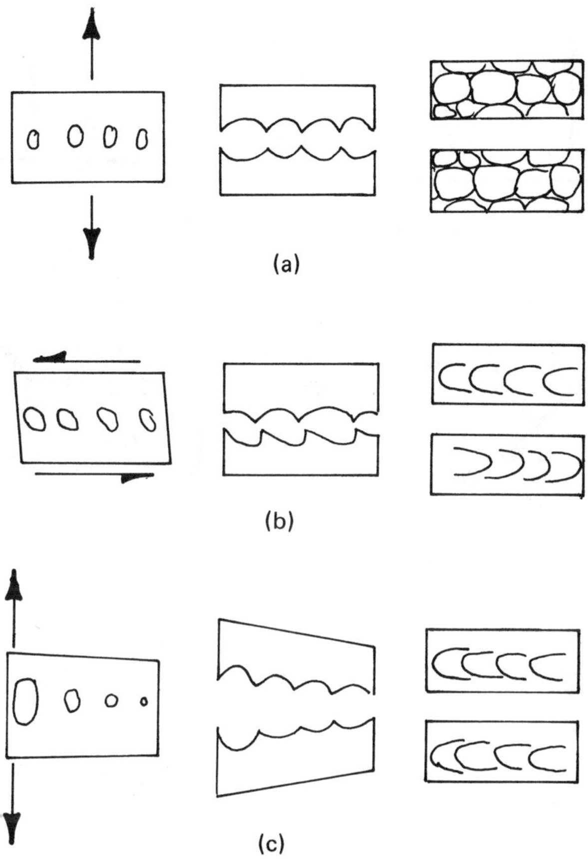

Fig. 14. Formation of dimples by (a) tension; (b) shear; (c) tearing. Variation in fracture morphology is shown at the right.

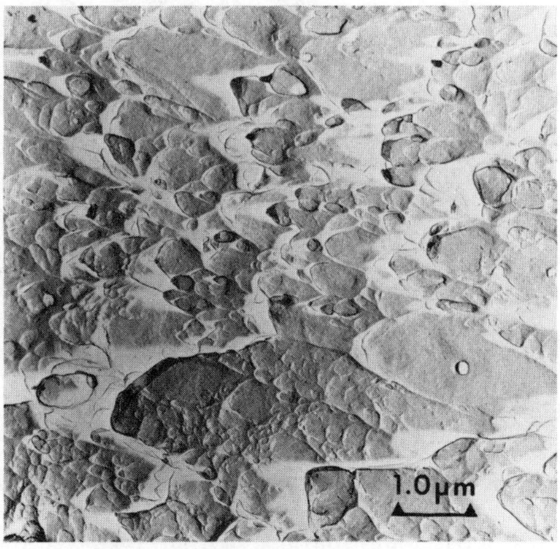

Fig. 15. Elongated dimples on fracture surface of quenched and tempered 4140 steel.

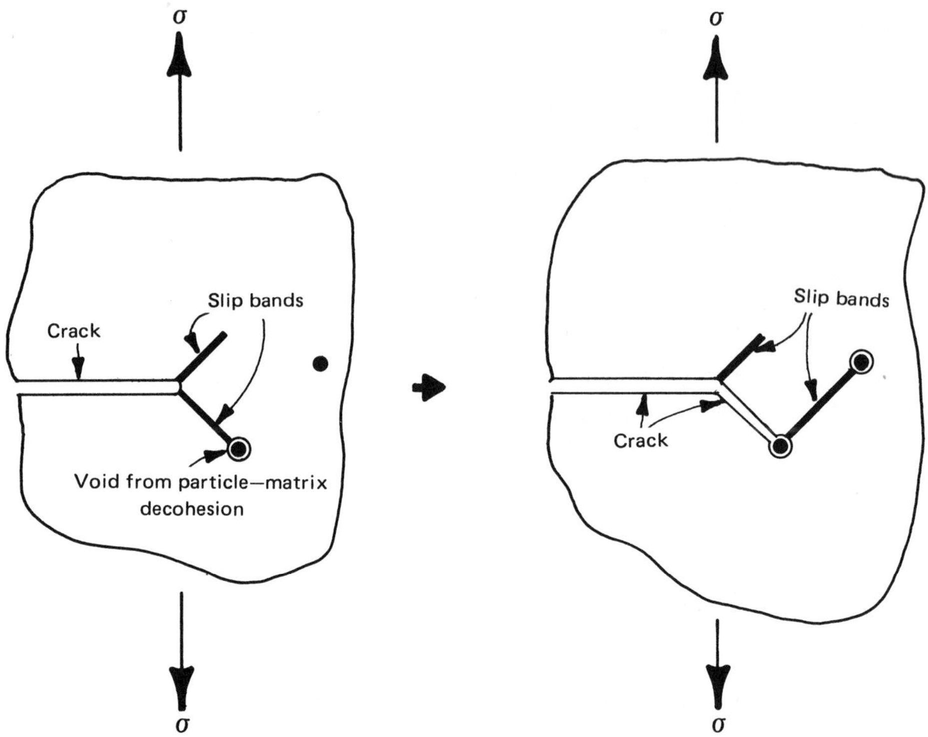

Fig. 16. Microvoids linking by shear along slip bands in a high strength alloy with limited work hardening capacity.

Fig. 17. Fracture surface of 316L weld metal specimen tested at high temperature in tension, showing jagged appearance from linking of microvoids by shear.

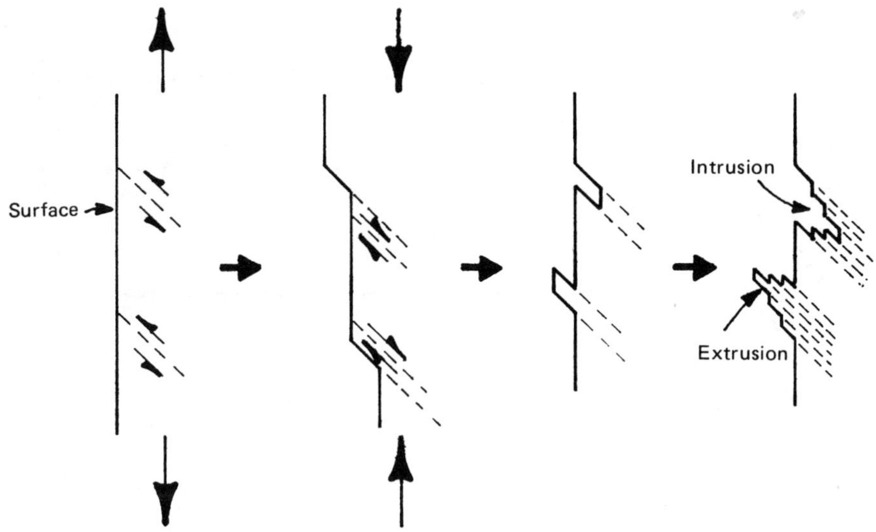

Fig. 18. Formation of intrusions and extrusion at surface during fatigue loading.

A fatigue crack propagates as a result of local plastic deformation at the crack tip, this occurring during each tensile loading period. Hence, we may note that fatigue cracks will not propagate under compressive loading conditions. The crack grows initially along a slip band where the initial nucleus was formed in the highly dislocated structure, and after some time it changes from this Stage I crack propagation mode to Stage II mode, which is roughly normal to the tensile axis, as indicated in Fig. 19. The length of the Stage I crack is generally small, and its surfaces are often essentially featureless; however, during Stage II crack propagation, which extends over a much greater distance, distinctive features termed fatigue striations (see Fig. 20) are frequently formed on it. The morphology of these striations varies widely with material and environment, and to understand the reasons for this, we shall look more closely at the mechanisms of fatigue crack propagation.

Laird and Smith [19] proposed their "plastic blunting model" for the formation of striations during stress reversal, and this is illustrated in Fig. 21. A number of other workers have proposed mechanisms for striation formation based both on a continuum mechanics approach and on consideration of slip processes at the crack tip. In particular, Neumann [20], Pelloux [21], and Broek and Bowles [22] have made detailed examination of the slip processes, and the main features of the coarse slip model which has been proposed as a result of their studies are shown in Fig. 22. It is seen that the crack propagates by slip occurring on intersecting slip planes meeting at the current crack tip, producing crack blunting. Upon load reversal, slip takes place in the reverse direction, but the crack does not reweld because of surface oxidation, and crack tip resharpening occurs. Hence, a series of fine slip steps appear on the fracture surface and a larger step or striation is produced during each stress cycle.

In real engineering materials slip will not always be possible on slip planes which exactly pass through the crack tip as they may be blocked by means of dislocations producing a back stress, second phase particles, and so on. Lui [23] has considered this situation for FCC metals, and the mechanisms involved and resulting fracture sur-

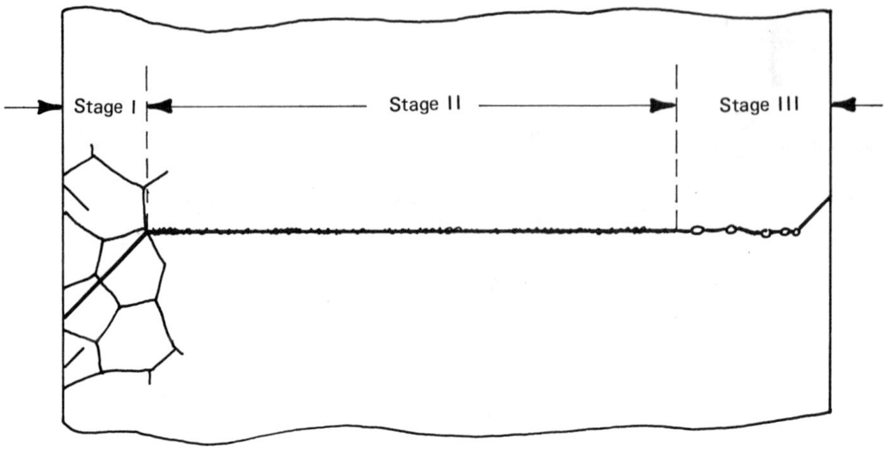

Fig. 19. The three stages of fatigue crack propagation.

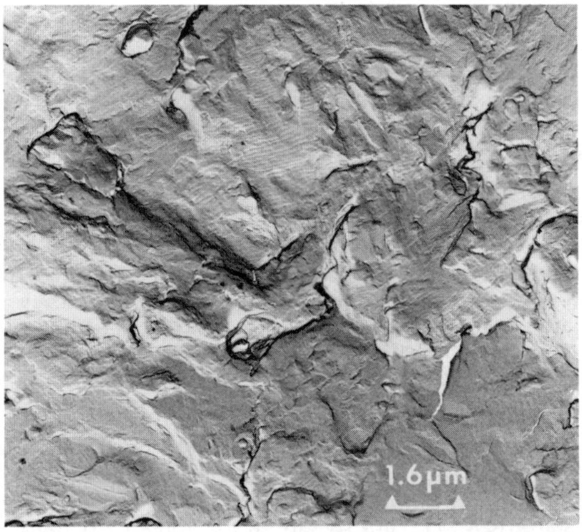

Fig. 20. Fatigue striations on quenched and tempered 4140 steel.

face morphology are similar to those proposed independently by Neumann [24]. Lui's model for crack growth is shown in Fig. 23, and it is capable of explaining the wide variety of striation shapes observed in practice, depending on the number of active forward and reverse slip planes. The model can also be considered to be applicable to BCC metals although the geometry of slip will be different. The formation of slip band cracks, which occur particularly in more ductile materials, and which are shown in Fig. 24, can be explained in terms of restraint imposed on reverse slip by large carbides or other second phase particles in conjunction with the large extent of forward slip taking place at high stress levels (see Fig. 25).

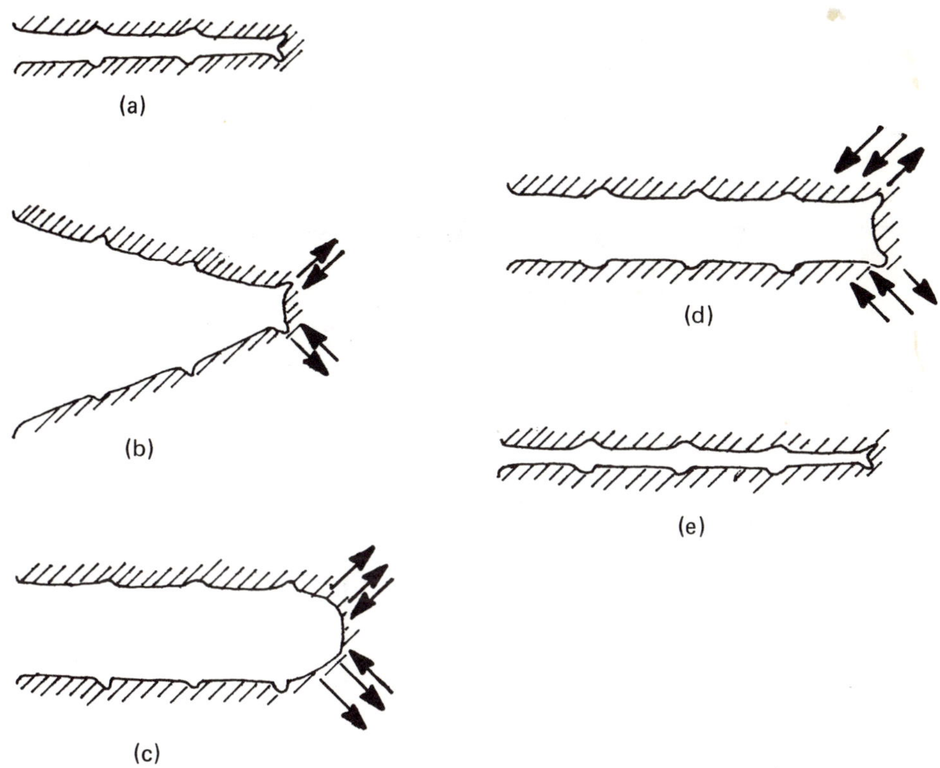

Fig. 21. The plastic blunting model of Laird and Smith [19] for the formation of fatigue striations. One complete stress cycle is illustrated.

Fatigue crack propagation in vacuum is characterized by a much decreased (by 5 to 10 times) crack growth rate and by an absence of striations as shown in Fig. 26 [21]. This may be explained by the fact that reverse slip can occur on the same slip planes as were involved in the crack advance, and rewelding can also take place, there being no surface oxide film. If rewelding were complete, no crack propagation would take place from cycle to cycle, and it has been suggested that rewelding is complete only during the first closure, becoming progressively poorer during subsequent compressive parts of the cycle [25]. Thus, a crack would first open as in air, take a series of cycles (5–10) before it ceased to be rewelded, and then move forward by another step. This process would give rise to periodic, but irregular, slip steps as have been observed in practice in vacuum, these having a spacing approximating that of striations formed in air [25]. Objections to the mechanisms may be raised on the basis that the crack should move forward during each cycle by the width of the area not rewelding, but this ignores the strain hardening occurring in the plastic zone ahead of the crack, and it seems more probable that the crack will progress very little after each partial rewelding on closure, until it jumps forward on breaking through the cyclicly hardened zone at its tip.

Striations are not always discernible on Stage II fatigue fracture surfaces formed in air, particularly in high strength ferrous alloys. The reasons may be that they are too fine (at very low crack propagation rates) or too poorly defined in terms of height

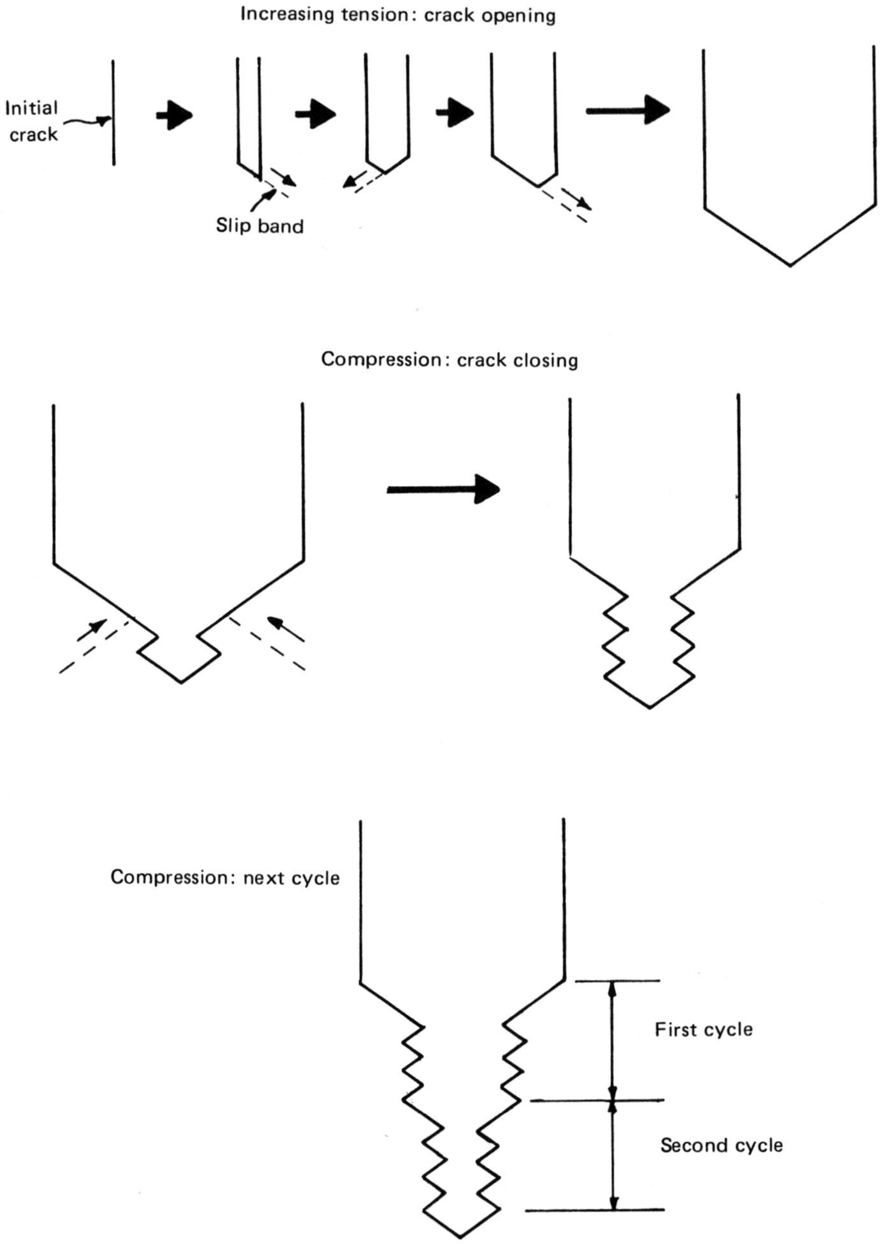

Fig. 22. The coarse slip model for fatigue crack growth.

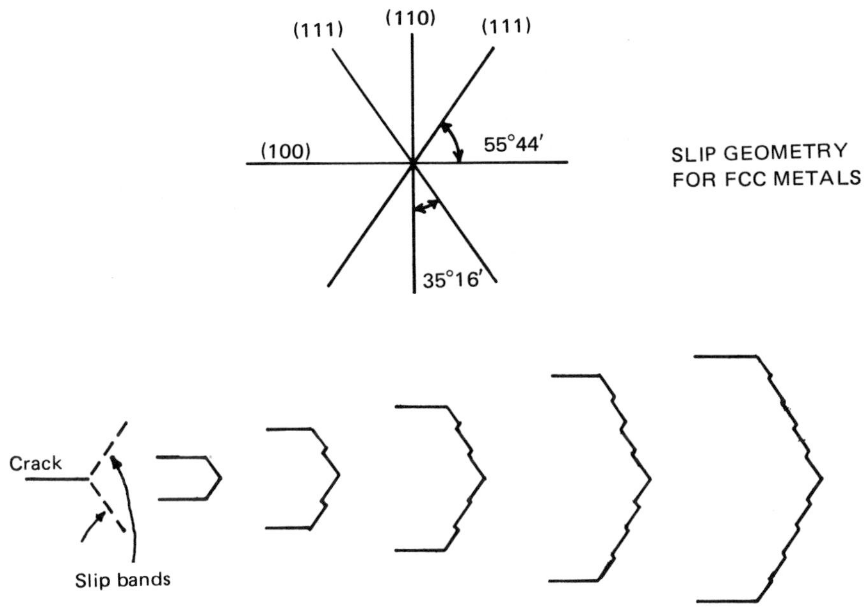

Fig. 23. The model of Lui [23] for fatigue crack growth on multiple slip planes.

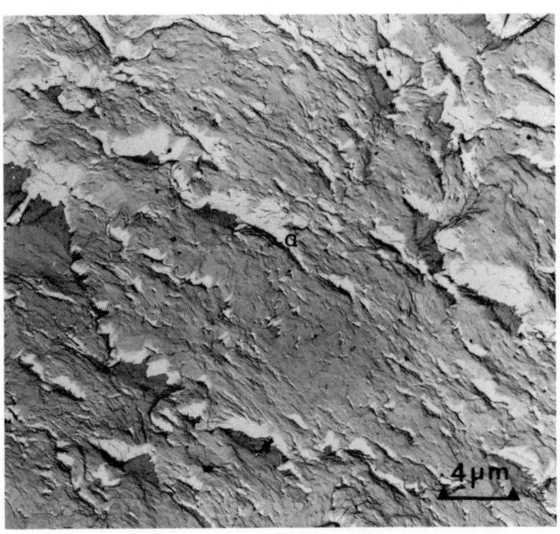

Fig. 24. Slip band cracks on quenched and tempered 4140 steel.

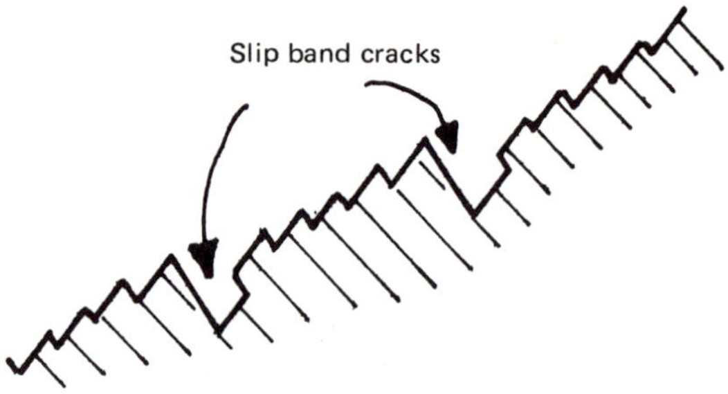

Fig. 25. Schematic of striation profile with slip band cracks formed by restraint of reverse slip from the model of Lui [23].

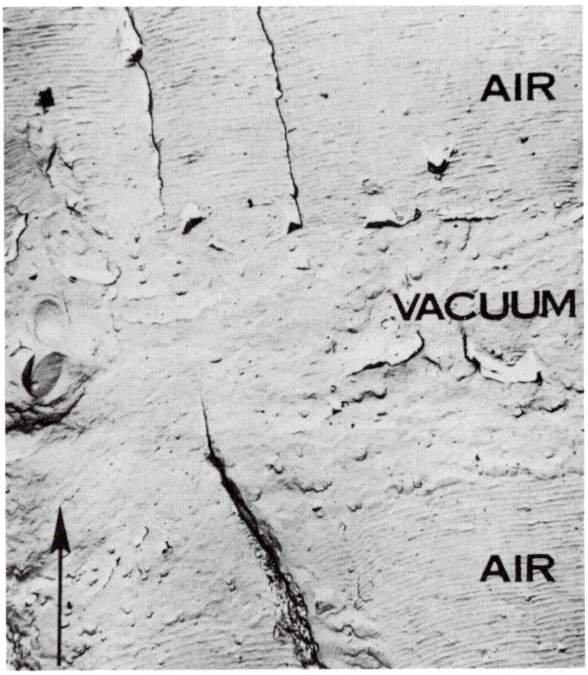

Fig. 26. Fatigue fracture surface of 2024–T3 aluminum tested successively in air, 216 cycles in vacuum of 5×10^{-6} torr, in air. From Ref. [21] by courtesy of R.M.N. Pelloux and the American Society for Metals. Copyright (1969) by American Society for Metals.

because of the very limited plastic flow occurring in high strength steels, or else the loading mode may not be conducive to their formation (e.g., tension-tension plane stress conditions [26]). They have, however, been observed even on fatigue crack surfaces which propagated along the prior austenite grain boundaries in as-quenched high strength steel at the interface with grain boundary carbides [27,28], as shown in Fig. 27.

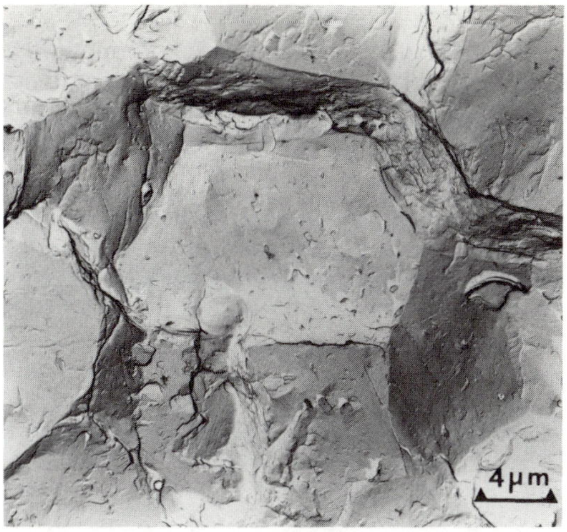

Fig. 27. Fatigue fracture along prior austenite boundaries in as-quenched 4140 steel, with striations apparent in places on the boundaries.

The presence of striations on the fracture surface is not unique to fatigue crack propagation, as will be emphasized later when discussing stress corrosion cracking. However, one feature which is sometimes observed on fracture surfaces and *is* unique to fatigue is tire track formation. Figure 28 shows some typical examples, and these features are formed when small second phase particles are either detached and embedded in a film of metal dust which moves forward as the crack propagates, or else by projecting particles from one surface which indent the other during crack closure, as the surfaces become offset during stress cycling after the crack tip has moved ahead by some distance [29]. Such features are very valuable in determining fatigue fracture to have been operative, as they are frequently observed when the crack propagation rate was high and few (or no) striations may be visible as in low-cycle fatigue.

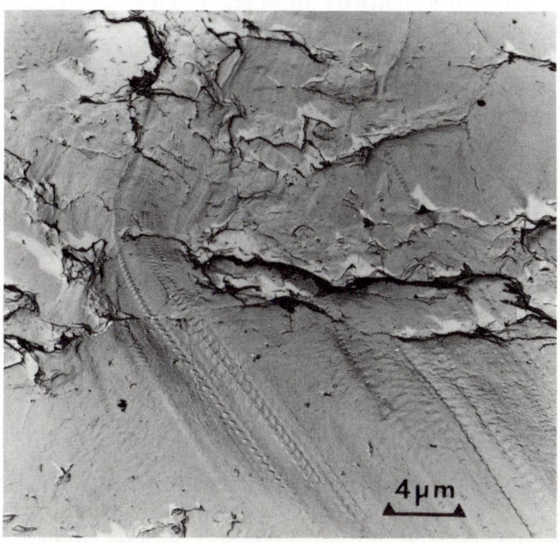

Fig. 28. Tire tracks on 4140 steel quenched and tempered at 700°C.

Stage III of the crack propagation process (see Fig. 19) is a stage characterized by rapid crack growth, microvoid coalescence, and dimple formation, and is a precursor to final rupture. Before this occurs, stretching may take place at the crack tip upon loading [31], leading to the formation of a stretched zone as shown in Fig. 29. This is generally similar in appearance to the stretched zone which may be seen on the surface of a fracture mechanics specimen between the prior fatigue crack and the fast fracture region [32].

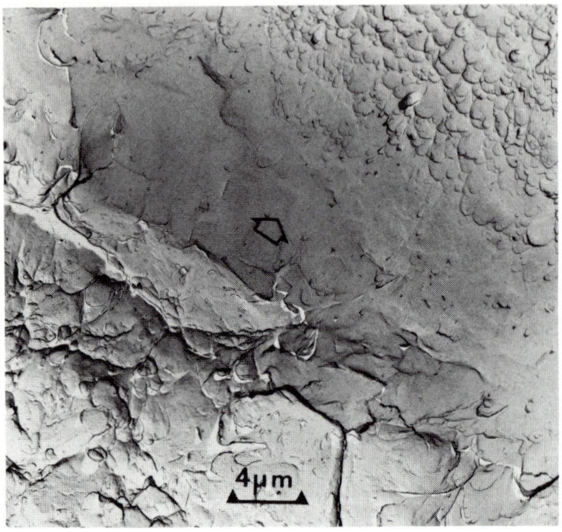

Fig. 29. Stretched zone just prior to final shear fracture in fatigued high strength steel.

Intergranular Fracture

As temperature is raised and rate of straining decreased, polycrystalline metals which normally fracture in a transcrystalline ductile manner undergo fracture of an intergranular nature. Thus, the majority of creep service failures are intergranular, and a typical fracture is shown in Fig. 30.

The development of intergranular creep fracture depends on the nucleation, growth and subsequent linking of voids on the grain boundaries, and Gifkins [33] provided a useful distinction between the two types of cavities seen on grain boundaries: the first is usually associated with cracking at triple-points and is designated as a "w" (for wedge) type cavity; the second type is the isolated rounded cavity, termed an "r" cavity. These are illustrated in Figs. 31 and 32 respectively.

Wedge cracks form at triple points due to grain boundary sliding, and may be promoted by decohesion at interfaces between grain boundary precipitates and the matrix as shown in Fig. 33. Fracture produces a rough surface, and grain boundary precipitates may be identified on this in appropriate circumstances as shown in Fig. 34.

The nucleation and growth of r-type voids at high temperature is strongly dependent upon the stress state. The effective stress may lead to their nucleation, while subsequent growth is dependent on the maximum tensile stress. The latter promotes the stress-induced flow of vacancies to voids [34], and these grow preferentially on boundaries having high tensile stress acting on them and, indeed, their formation can be suppressed by superposition of hydrostatic compressive stress [35]. At lower levels of

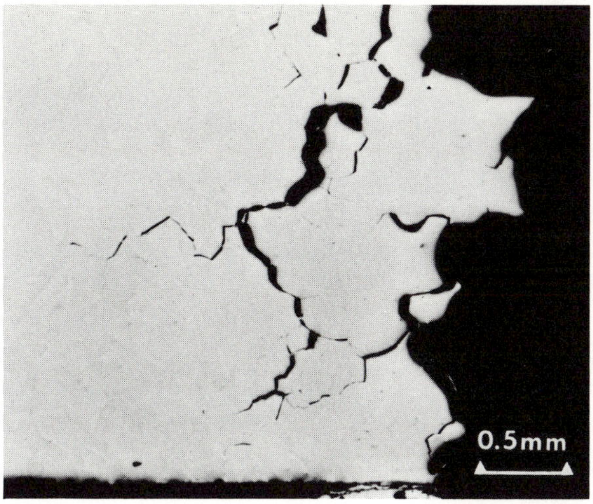

Fig. 30. Intergranular creep fracture in a nickel-base superalloy gas turbine blade.

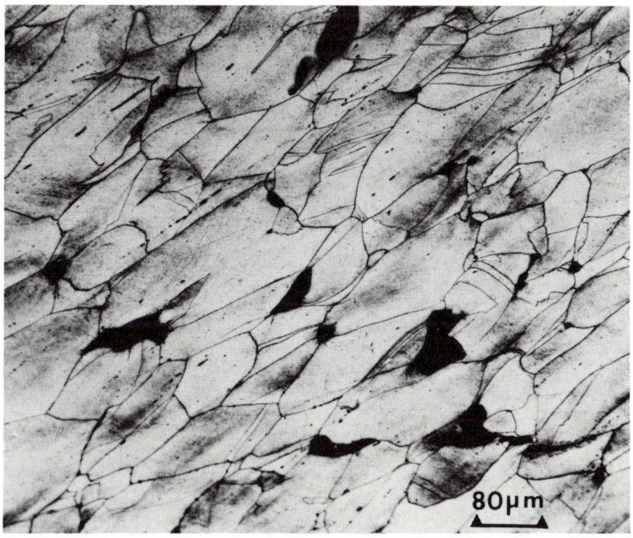

Fig. 31. Wedge-type cavities in 316 stainless steel which failed by creep.

stress or somewhat lower temperature, fracture may take place by the shear of material between voids, giving rise to a mixed intergranular-transgranular fracture surface [36].

Intergranular fracture may also take place under monotonic loading at ambient temperatures in embrittled materials. Embrittlement may arise from segregation of specific elements (from Groups IV to VI in the Periodic Table, in the case of steels) to grain boundaries, or from precipitation of brittle phases such as intermetallics or carbides on grain boundaries. In such cases the fracture has a faceted appearance with the crack following the grain boundaries, and detailed analysis of the surface and immediately underlying material may be necessary to prove conclusively the nature of the embrittling

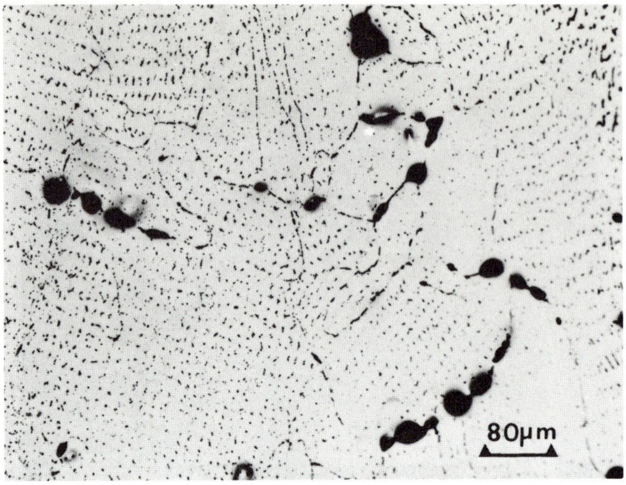

Fig. 32. 316L weld metal which failed by creep at high temperature, showing r–type voids.

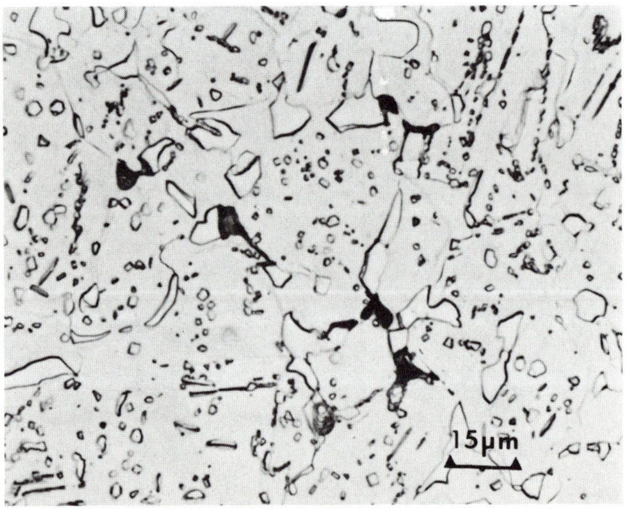

Fig. 33. Formation of w–type cavities by decohesion at precipitate-matrix interfaces in 316 stainless steel.

effect. Some examples of grain boundary precipitates are included in an earlier review paper prepared by the author [37].

Environmentally–Assisted Fracture

This, as indicated earlier, is a large and complex subject, and we may identify the following cases where the influence of environment coupled with stress may lead to failure: stress corrosion cracking (SCC); corrosion fatigue; hydrogen embrittlement; liquid metal embrittlement; and radiation damage. In the space available it is not possible to

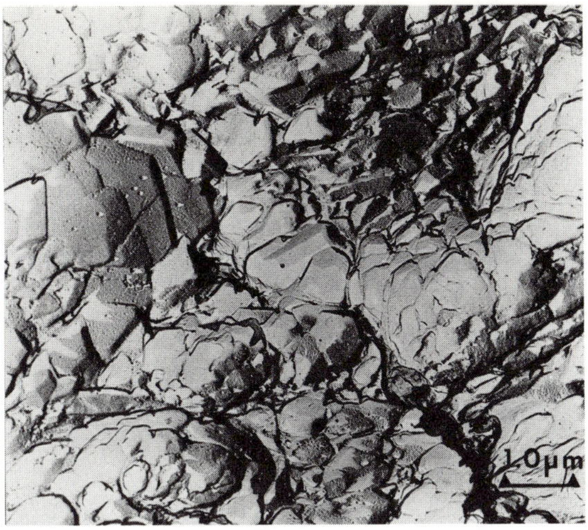

Fig. 34. Microfractograph of 316 stainless steel which failed in creep, showing carbides and σ–phase particles on fracture surface.

look at these in any great detail, so reference will be made, where appropriate, to recent reviews where more detailed discussion may be obtained.

The mechanism of stress corrosion cracking is illustrated in Fig. 35. Essentially, corrosion is localized at the tip of a propagating crack which may be either intergranular or transgranular, depending on the material and the corrosive environment. The surface of the crack becomes passivated or covered by a protective film, and this is ruptured when slip takes place at the crack tip, leaving a fresh surface. This in turn is attacked and subsequently either passivated or otherwise protected, the crack tip having moved forward. The process continues on a stepwise basis. The fracture morphology produced can vary widely, and may include striations as well as linear features running parallel to the direction of crack propagation, an example being shown in Fig. 36, which is taken from the fracture surface of an AISI 316L orthopaedic implant [38]. Stress corrosion fractures are characterized by little ductility, and in many cases an oxide coating may be seen on the crack surfaces. A typical photomicrograph of a crack tip is shown in Fig. 37.

Corrosion fatigue and SCC are closely allied, and the latter may be thought of as a special case in which the stress level is maintained constant. The fatigue limit of materials which have a poor resistance to corrosion may be very low indeed in an aqueous environment, perhaps less than one tenth of the ultimate tensile strength. Again, a wide range of fracture morphologies may result, and it should be noted that corrosion fatigue is not confined to aqueous or liquid environmental conditions. It may be the operative mechanism of failure in humid air, which can produce greatly reduced fatigue lives as compared with those obtained in dry air [39]. A recent review by Schijve [40] serves to emphasize the importance of the problem and to pinpoint some of the gaps in our knowledge. Similarly, two recent Conferences have attempted to review some of the areas of activity at the present time [41,42].

Hydrogen embrittlement can be a severe problem in ferrous materials in particular, because of its very high mobility in the BCC iron lattice at ambient temperatures. When present in sufficient quantity, hydrogen lowers the cohesive strength of the lattice [43], and it also tends to accumulate at defects, producing a high internal pressure

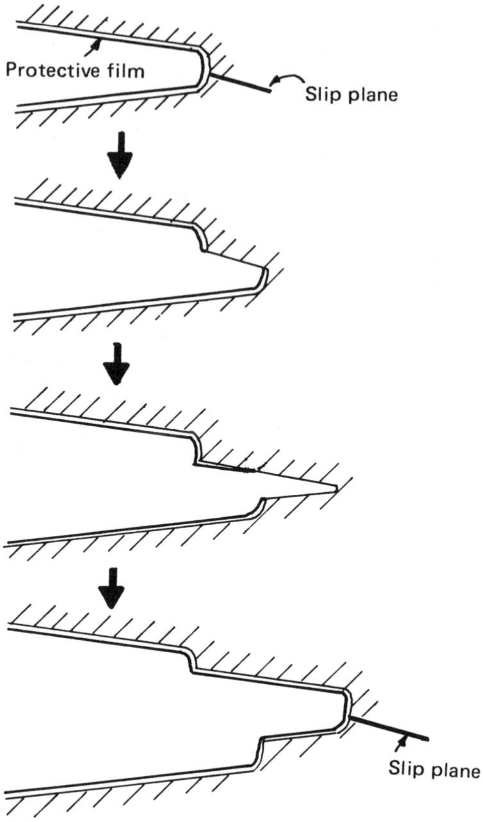

Fig. 35. Schematic illustration of stress corrosion crack propagating by rupture of the protective film on plastic deformation at the crack tip, subsequent corrosion, repassivation, and build up of stress to repeat the cycle.

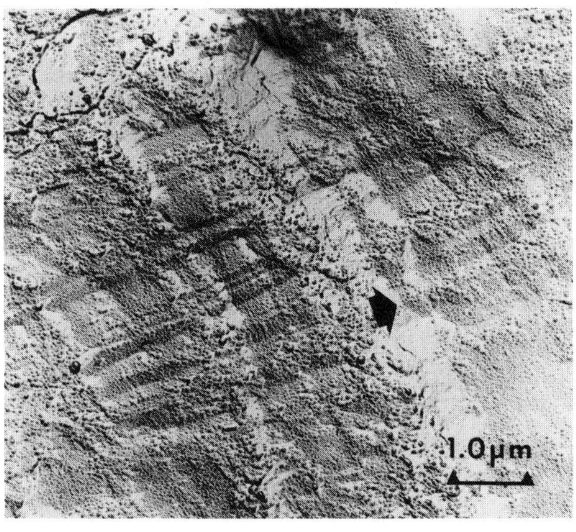

Fig. 36. Stress corrosion cracking striations on the fracture surface of 316L stainless steel, with linear features parallel to crack growth direction (arrowed), indicative of "tunnelling".

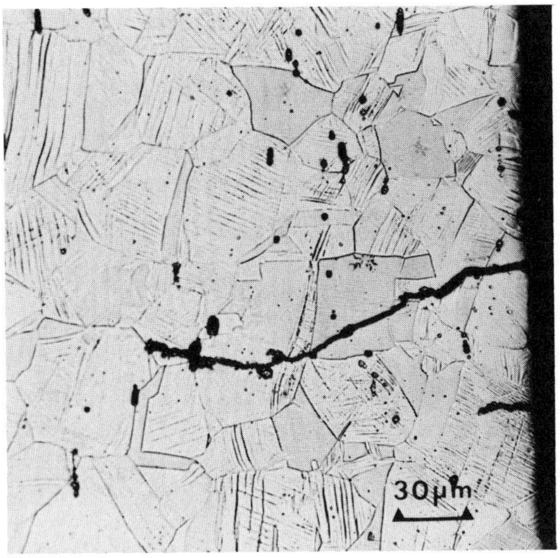

Fig. 37. Stress corrosion crack in 316 stainless steel.

which may cause rupture, particularly in high strength alloys. Hydrogen pickup may occur in several ways: it may take place from water in the scrap used in steelmaking, the gas becoming trapped during solidification; it may arise from pickling or plating operations; it may be picked up from the service environment; it may be a product when welding with electrodes having moist coating; or it may be produced at the tip of a propagating stress corrosion or corrosion fatigue crack. In the last mentioned case, it can be generated either from reaction of the corrosive medium with the freshly exposed metal surface or from dissociation of a gas (such as water vapor) on this surface: the subsequent absorption and embrittlement at the tip of the crack is considered to be a significant and frequently occurring mechanism in both SCC and corrosion fatigue.

Entrapped hydrogen can give rise to a number of features which may be detected metallographically. For example, on a macroscopic scale blisters may be produced in ferrite, probably being nucleated at sub-surface inclusions [44]. Flakes or small internal fissures may sometimes be detected by ultrasonic examination or, destructively, by optical metallography as shown in Fig. 38. These give rise to shiny facets on the fracture surface. When the fracture surface of hydrogen embrittled material is examined using electron microfractography, small microcracks may sometimes be seen.

Hydrogen embrittlement does occur in materials other than steel, such as titanium, zirconium and the refractory metals. Titanium and zirconium form brittle hydrides while the refractory metals are thought to become embrittled from supersaturation of hydrogen in solution [45].

Liquid metal embrittlement has become of increasing interest in the light of the developments around the world on liquid metal fast breeder reactor (LMFBR) technology, as normally ductile metals may fail in a catastrophic brittle manner by cleavage or, more commonly for polycrystalline materials, by rapid intergranular crack propagation. The mechanism postulated is that a liquid metal atom reduces the bond strength of the atoms of the solid material by chemisorption at the tip of a crack or at the end of a dislocation pile-up near a surface obstacle. For more detailed discussion of the mechanisms, see the reviews by Westwood *et al.* [46] and by Kamdar [47].

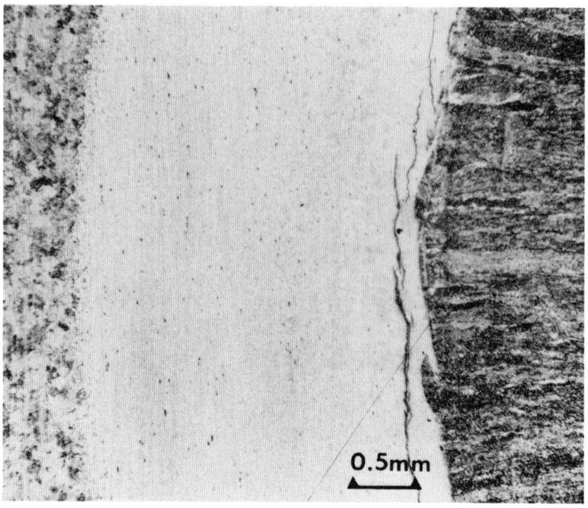

Fig. 38. Hydrogen cracking in HAZ of armour-plate welded with improperly dried electrodes.

Radiation damage may lead to void formation, locking of dislocations through interaction with radiation-induced defects, or excess precipitation of a second phase. Generally, the effect is to cause the material to lose ductility and to become more prone to brittle fracture. A comprehensive review of the relations between fracture mechanisms and irradiation damage has been given by Bement *et al.* [48], to which the interested reader is referred.

Corrosion and Wear

These two failure mechanisms are apparently straightforward and can be recognized with relatively little need for significant metallographic examination. However, it is often very difficult to provide solutions to corrosion problems except in straightforward cases where the wrong materials for the environment or dissimilar metals in contact have been chosen. Equally, while we can recognize wear, it is often difficult to provide a suitable, alternative choice of material to stand up to a particular abrasive service condition, and our knowledge of the detailed mechanisms of wear as they relate to microstructure leaves a considerable amount to be desired.

One particular situation which combines corrosion and wear is the insidious problem of fretting corrosion. In this regard, Waterhouse has prepared an excellent monograph of this subject [49], which discusses the mechanisms of, and metallographic evidence for, fretting corrosion, together with methods of combatting it.

CONCLUDING REMARKS

Hopefully, this paper has touched on the mechanisms of failure and the metallographic evidence for each, which are of relevance to practicing metallographers involved in, or interested in becoming involved in the analysis of service failures. Once again, however, it must be emphasized that *all* the evidence must be weighed and that metallographic examination is an ancillary tool which, when properly used, can provide definite confirmation or otherwise of mechanisms which may be postulated on the basis of the environmental conditions, overall appearance of the fracture and so on.

ACKNOWLEDGEMENTS

The author is grateful to his former graduate students and associates, many of whose names appear in the References, for stimulating discussions on the subject of failure mechanisms. In particular, it is appropriate to acknowledge the assistance provided by Dr. W.E. White, now of the University of Calgary in this respect and also in the preparation of this paper.

Acknowledgement is also made to the National Research Council of Canada, for support of the author's studies in the areas of deformation and fracture.

REFERENCES

1. "Failure Analysis and Prevention", *Metals Handbook,* 8th Ed., Vol. 10, ASM, Metals Park, Ohio, (1975).
2. Colangelo, V.J., and Heiser, F.A., *Analysis of Metallurgical Failures,* John Wiley, New York, (1974).
3. McCall, J.L., and Mueller, W.M. (eds.), *Microstructural Analysis: Tools and Techniques,* Plenum Press, New York – London, (1973).
4. Lui, M.–W., Le May, I., and Coote, R.I., "Nondestructive Metallurgical Examination of a Damaged 100 MW Generator Rotor", *Praktische Metallographie,* 12, pp. 319–326, (1975).
5. "Fractography and Atlas of Fractographs", *Metals Handbook,* 8th Ed., Vol. 9 ASM, Metals Park, Ohio (1974).
6. Warke, W.R., Nielsen, N.A., Hertzberg, R.W., Hunter, M.S., and Hill, M., "Techniques for Electron Microscopic Fractography", in *Electron Fractography,* STP 436, ASTM, Philadelphia, pp. 212–230 (1968).
7. Brammer, I.S., and Dewey, M.A.P., *Specimen Preparation for Electron Metallography,* Blackwell Scientific Publications, Oxford (1966).
8. Biggs, W.D., *The Brittle Fracture of Steel,* Macdonald and Evans, London, (1960).
9. Friedel, J., *Dislocations,* Pergamon Press, Oxford, p. 321 (1964).
10. Berry, J.M., "Cleavage Step Formation in Brittle Fracture Propagation", Trans. ASM, *51*, pp. 556-581 (1959). See also discussion of this paper by Crussard, C., Plateau, J., and Lean, J.B., on pp. 581–584.
11. Zener, C., "The Micro-Mechanism of Fracture" in *Fracturing of Metals,* ASM, Cleveland, pp. 3–31 (1948).
12. Cottrell, A.H., "Theory of Brittle Fracture in Steel and Similar Metals," *Trans AIME,* 212, pp. 192–203 (1958).
13. Hull, D., "Nucleation of Cracks by the Intersection of Twins in α–Iron", *Phil. Mag., Eighth Series,* 3, pp. 1468–1469 (1958).
14. Smith, E., "The Nucleation and Growth of Cleavage Microcracks in Mild Steel", in *Physical Basis of Yield and Fracture,* Inst. of Physics and the Physical Society, London, pp. 36–46 (1966).
15. Gurland, J., and Plateau, J., "The Mechanism of Ductile Rupture of Metals Containing Inclusions", *Trans. ASM,* 56, pp. 442–454 (1963).
16. Lui, M.–W., and Le May, I., "The Relation Between Ultimate Tensile Strength and Carbide Size in Steel", *J. of Eng. Matls. and Technology, Trans. ASME,* 98, pp. 173–179 (1976).
17. McClintock, F.A., "Local Criteria for Ductile Fracture", *Int. J. Fracture Mech.,* 4, pp. 101–129 (1968).
18. Knott, J.F., "Micro–Mechanisms of Fracture and the Fracture Toughness of Engineering Alloys", in *Fracture 1977,* D.M.R. Taplin (ed.), Vol. 1, Univ. of Waterloo Press, Waterloo, Ont., pp. 61–92 (1977).
19. Laird, C., and Smith, G.C., "Crack Propagation in High Stress Fatigue", *Phil. Mag., Eighth Series,* 7, pp. 847–857 (1962).

20. Neumann, P., "Bildung und Ausbreitung von Rissen bie Wechselverformung", *Z. Metallkunde,* 58, pp. 780–789 (1967).
21. Pelloux, R.M.N., "Mechanisms of Formation of Ductile Fatigue Striations", *Trans. ASM,* 62, pp. 281–285 (1969).
22. Broek, D., and Bowles, C.Q., "The Study of Fracture Surface Profiles in the Electron Microscope", *Int. J. Fracture Mech.,* 6, pp. 321–322 (1970).
23. Lui, M.–W., "Relations Between Microstructure, Mechanical Properties and Fracture Morphology of AISI 4140 Steel," Ph.D. Thesis, University of Saskatchewan, Saskatoon, (1973).
24. Neumann, P., "The Geometry of Slip Processes at a Propagating Fatigue Crack – II", *Acta Met.,* 22, pp. 1167–1178 (1974).
25. Broek, D., "Some Contributions of Electron Fractography to the Theory of Fracture", Review No. 185, *International Metallurgical Reviews,* 19, pp. 135–182 (1974).
26. Plumbridge, W.J., and Ryder, D.A., "The Influence of Specimen Geometry on the Mode of Fatigue Crack Growth in Aluminum", *Acta Met.,* 17, pp. 1449–1452 (1969).
27. Le May, I., and Lui, M.–W., "Fractographic Observations of Fatigue Fracture in High-Strength Steels", *Metallography,* 8, pp. 249–252 (1975).
28. Lui, M.–W., and Le May, I., "Effects of Grain Boundary Carbides on the Fatigue Fracture of an AISI 4140 Steel", in *Grain Boundaries in Engineering Materials,* J.L. Walter, J.H. Westbrook and D.A. Woodford (eds.), Claitor's, Baton Rouge, LA, pp. 397–407 (1975).
29. Beachem, C.D., "Microscopic Fatigue Fracture Surface Features in 2024–T3 Aluminum and the Influence of Crack Propagation Angle upon their Formation", *Trans. ASM,* 60, pp. 324–343 (1967).
30. Lui, M.–W., and Le May, I., "A Fractographic Study of Fatigue in AISI 4140 Steel", *IMS Proceedings 1971,* pp. 227–241 (1972).
31. Lui, M.–W., Shaw, W.J.D., Whelan, J.M., and Le May, I., "A Comparison of Transition Zones on the Fracture Surfaces of High Strength Steels", in *Microstructural Science,* Vol. 3A, P.M. French, R.J. Gray and J.L. McCall (eds.), Elsevier, New York, pp. 3–20 (1975).
32. Spitzig, W.A., Pellissier, G.E., Beachem, C.D., Brothers, A.J., Hill, M., and Warke, W.R., "A Fractographic Analysis of the Relationships Between Fracture Toughness and Surface Topography in Ultrahigh–Strength Steels", in *Electron Fractography,* STP 436, ASTM, Philadelphia, pp. 17–30 (1968).
33. Gifkins, R.C., "Mechanisms of Intergranular Fracture at Elevated Temperatures", in *Fracture,* B.L. Averbach, D.K. Felbeck, G.T. Hahn, and D.A. Thomas (eds.), MIT Press, Cambridge, Mass., pp. 673–623 (1959).
34. Hull, D., and Rimmer, D.E., "The Growth of Grain-Boundary Voids Under Stress", *Phil. Mag, Eighth Series,* 4, pp. 673–687 (1959).
35. Needham, N.G., Wheatley, J.E., and Greenwood, G.W., "The Creep Fracture of Copper and Magnesium", *Acta Met.,* 23, pp. 23–27 (1975).
36. White, W.E., "A Study of Creep in Stainless Steel Weldments", Ph.D. Thesis, University of Saskatchewan, Saskatoon, (1977).
37. Le May, I., "Microstructural Analysis and Fractography", in *Microstructural Analysis: Tools and Techniques,* J.L. McCall and W.M. Mueller (eds.), Plenum Press, New York – London, pp. 153–168 (1973).
38. White, W.E., Postlethwaite, J., and Le May, I., "On the Fracture of Orthopaedic Implants", in *Microstructural Science,* Vol. 4, E.W. Filer, J.M. Hoegfeldt and J.L. McCall (eds.), Elsevier, New York, pp. 145–158 (1976).
39. Shaw, W.J.D., and Le May, I., "Fatigue Crack Propagation in High Strength Steel Under Controlled Humidity Conditions", in *Fatigue Testing and Design,* Vol. 2, R.G. Bathgate (ed.), Soc. of Environmental Engineers, London, pp. 31.1–31.24 (1976).

40. Schijve, J., "Fundamental and Practical Aspects of Crack Growth Under Corrosion Fatigue Conditions", *Proc. Inst. Mech. Engrs.,* London, 191, pp. 107–114 (1977).
41. *Proceedings of the Conference on the Influence of Environment on Fatigue,* Inst. Mech. Engrs., London, (1977) (in press).
42. ASTM Symposium on Corrosion Fatigue, Denver, (November 1976) (to be published).
43. McMahon, C.J., Briant, C.L., and Banerji, S.K., "The Effects of Hydrogen and Impurities on Brittle Fracture in Steel", in *Fracture 1977,* Vol. 1, D.M.R. Taplin (ed.), University of Waterloo Press, Waterloo, Ont., pp. 363–385 (1977).
44. Evans, U.R., *The Corrosion and Oxidation of Metals,* Ed. Arnold, London, (1960).
45. Lawley, A., Liebmann, W., and Maddin, R., "Effect of Hydrogen on the Yielding Behavior of Molybdenum", *Acta Met.,* 9, pp. 841–850 (1961).
46. Westwood, A.R.C., Preece, C.M., and Kamdar, J.H., "Adsorption–Induced Brittle Fracture in Liquid–Metal Environments", in *Fracture: An Advanced Treatise,* Vol. III, H. Liebowitz (ed.), Academic Press, New York, pp. 589–644 (1971).
47. Kamdar, M.H., "Embrittlement by Liquid Metals", *Prog. Mat. Sci.,* 15, pp. 289–374 (1973).
48. Bement, A.L., Hoagland, R.G., and Smidt, F.A., "Fracture Mechanisms and Radiation Effects", in *Fracture: An Advanced Treatise*, Vol. III, H. Liebowitz (ed.), Academic Press, New York, pp. 535–587 (1971).
49. Waterhouse, R.B., *Fretting Corrosion,* Pergamon Press, Oxford (1972).

MACROSCOPIC EXAMINATION PROCEDURES FOR FAILURE ANALYSIS

G.F. Vander Voort *

The most important step in any failure analysis is the visual, or macroscopic, examination of the failed component. This step precedes all other examination techniques, including those directed at the microscopic level. An integral part of many macroscopic examinations is the use of macroetching to reveal or more clearly delineate features associated with the fracture including mechanical abuse marks, grinding scorch, decarburization, hardness or structure gradients, etc.

The extent of time required for the visual analysis varies with the type of failure. For a routine or simple component failure, such as a broken punch, the time required may be quite brief, whereas a complex problem, such as a failed structure, may require many hours. The approach used for each of these types of failures is similar in principle. The investigator visually (or with a simple hand lens) studies the fracture and its surrounding area and records pertinent features photographically or schematically. These features include:

- Origin of failure, or multiple origins
- Location of contributing stress concentrators
- Presence of relevant contaminants on the fracture, e.g., temper color, scale, paint
- Direction of crack propagation and sequence of failure
- Failure mode and mechanism
- Orientation and magnitude of stresses
- Imperfections contributing to the failure
- Sizes and other important physical data.

Obviously, as the magnitude of the failure increases (considering both size and legal ramifications) much more care must be taken in documenting the location of the wreckage. This is of great importance in reconstructing the sequence of the failure. Even when there are witnesses to the failure, the arrangement of the failed components should be carefully documented. Witnesses may only see the later stages of a failure and not the initiating features. Also, witnesses do not always agree with one another due to many factors: vantage point, surprise, panic, speed of incident, etc. The investigator must also inventory all the pieces at the accident site and compare these with what should be present. Observation of a missing critical component may explain the failure. In wreckage analysis one of the most difficult problems is separating those parts that broke as a result

---

*Metallurgical Service and Investigation Group, Product Research, Bethlehem Steel Corporation, Bethlehem, Pennsylvania, USA.

of the incident from those that broke causing the failure. Hence, determining the sequence of the failure is very important if the failure origin is to be determined. If the origin is not found, meaningful conclusions regarding the cause of failure usually cannot be made.

In many cases the investigator does not have the opportunity to go to the failure site but must rely on others to examine the failure and determine the origin. This situation can cause several problems. For example, rough handling or lack of protection can result in important fracture features being destroyed. Often a commercial photographer is hired to photograph the damage. While these people may take artistically and technically correct photographs, they may not know what should be included in the photographs, i.e., what is important and what is not. Most metallographers are excellent photographers and prefer to make their own documentation. The quality of these photographs, i.e., their ability to convey important technical information to the viewer, is of great importance if the failure analysis results are to be properly received and utilized.

BASIC PROCEDURES FOR EXAMINING FRACTURES

After a component or structure has catastrophically failed or a crack has been observed in a component, certain steps must be taken to ensure that the fracture features are not obliterated. In some cases, the fracture face will be destroyed in the incident, or a fire may occur after the fracture which may severely affect the fracture and will alter the apparent properties of the material. The failed component must be carefully protected from further damage. Rough handling or exposure may obscure important evidence.

There are two overpowering urges that all neophyte failure analysts seem to have. The first is to try to remate the fracture surfaces to see if they fit together. This seemingly innocent and harmless handling can severely alter the fine fracture features. Likewise, fingering of fracture surfaces can be most harmful. The second overpowering urge is to rush to the nearest machine shop and cut up the fracture before carefully gleaning all possible information and recording all necessary details photographically. The location and orientation of any samples cut from the fracture should be carefully recorded for future reference. Care should be taken during sectioning, grinding, polishing, and handling to avoid creating artifacts that would confuse or inhibit fracture interpretation. A common procedure is to spray the fracture surfaces with a clear enamel or acrylic before such procedures. The protective spray can be removed later using trichloroethylene or acetone.

If the fracture face must be removed from some large member so that it can be handled and examined in the laboratory, any saw cutting, radiac cutting, or burning should be done well away from the fracture initiation area. The fracture can be sprayed with clear enamel to protect it from any cutting fluids. For shipment, the fracture should be carefully packed so that it does not rub against other pieces. Such packages should be watertight, and dessicant should be added to absorb any moisture present. If the piece is very hard, such as a tool steel, it should not be tempered back to facilitate saw cutting, because such a procedure oxidizes the fracture face and alters the microstructure and properties.

Do not be too hasty in cleaning fractures. Note any deposits on the fracture as these may be important clues as to the life or origin of the fracture. Frequently, the fracture must be cleaned of dirt and grease in order to observe the macroscopic details.

Cleaning should be performed carefully, otherwise important information may be lost. For example, removal of a layer of rust stain on a fracture showed that the fracture was covered by paint to a depth of 5/16 inch. X-ray diffraction identified lead oxide (Pb_3O_4) confirming the identity of the paint. Knowledge of the date that the component was last painted gave useful information regarding the crack growth rate and life.

The gross fracture features can easily be examined after simple cleaning operations. Numerous procedures have been developed for cleaning fractures. In general, the mildest procedures should be employed first. If inadequate, stronger solutions can be used. The simplest method is to wash the fracture in alcohol in an ultrasonic cleaner. A soft toothbrush can be used to remove adherent dirt or debris. Fracture surfaces can also be cleaned by repeated application and stripping of cellulose acetate tape softened in acetone. The tape should be allowed to dry before stripping. Corrosion products can be removed by cleaning in a dilute nitric acid-ethanol solution. Alternatively, the sample could be immersed in an inhibited acid bath of 6N HCl with 2 g/l hexamethylene tetramine for several minutes. Heavily oxidized samples cleaned by such techniques may be difficult to interpret unambiguously when the fine features are examined [1]. Macroscopic interpretation will not be hindered. Lightly oxidized surfaces may be cleaned and viewed microscopically without trouble from interfering artifacts.

If the fracture face has been severely altered by corrosion, fire, or rubbing, the failure analysis may be severely impaired. Fire or tempering will destroy the fine fracture details but the macroscopic features will still be visible. Rubbing, of course, will destroy the fine features and may remove the macroscopic evidence as well.

Careful macroscopic examination is extremely important to the failure analysis. The origin, or origins, of failure usually must be determined if any conclusive results are to be obtained. When properly interpreted, fracture features describe the manner of loading, the level of the applied stress, the mechanisms involved, and the relative ductility or brittleness of the material. Other details may be revealed by the fracture, such as the presence of hardened cases, apparent grain size, and internal imperfections. Design-induced stress concentrators, material imperfections, fabrication imperfections, or machining imperfections which served to locate the failure should be detected during visual examination.

The usual sequence of operations in the examination of fractured components is:

1) Visually survey the entire component to obtain an overall understanding of the component.
2) Classify the fracture from a macroscopic viewpoint as ductile, brittle, fatigue, torsion, etc.
3) Determine the origin of failure by tracing the fracture back to its starting point or points.
4) Based on the fracture features, estimate the manner of loading (tension, compression, bending, etc.), the relative stress level (high, medium, or low), and the stress orientation.
5) If it appears necessary to examine the fine fracture features, use microfractographic techniques to determine the fracture mode (transgranular or intergranular) and to detect other characteristic features associated with the fracture. The fracture mechanism can be confirmed by observation of features such as dimples, cleavage, river-marks, or striations.

Frequently, the investigator is required to open up either the primary or secondary cracks for examination. The investigator must use his ingenuity in such cases so that the details are not destroyed while opening the cracks. For example, if the primary fracture path is heavily oxidized, tight secondary cracks may be opened and examined. In some cases the fracture has not severed the component completely and cutting must be done to permit observation of the primary fracture path.

If legal action is anticipated, the analyst must take extreme care to document all details of the wreckage and subsequent examination. Remember that others will want to examine the relevant specimens and they will be highly critical of any abuse or destruction of key evidence. In such cases the primary analyst should *personally* supervise all handling and testing steps.

The concept of macroscopic investigation cannot be narrowly viewed as pure and unadulterated visual examination unaided by any other procedures. As already noted in the introductory remarks, macroetching is typical of the procedures that can be employed to reveal or more sharply delineate features associated with the cracking. Additionally, the metallographer has many nondestructive tools available to help him detect surface or internal defects. Application of these methods is often quite important, not only in detecting the damage but in documenting it and in determining where components should be sectioned. Various types of liquid-penetrant, magnetic-particle, eddy-current, radiographic, ultrasonic and accoustical procedures may be applied depending on the nature of the problem and their availability to the failure analyst. As examples, Fig. 1 shows a cracked cutter die before and after magnetic particle inspection and Fig. 2 shows two examples of components before and after liquid-penetrant inspection.

Finding the origin (or origins) of the fracture is of great importance if the reason for crack initiation and contributing features are to be determined. The analysts must learn how to read the fracture surface markings. Fatigue fractures exhibiting "clam shell" marks and brittle fractures with chevrons are among the easiest fractures to interpret. However, not all fatigue fractures exhibit clam shells and such marks can be caused by other mechanisms. Hence, the analyst must supplement visual analysis with other tests, such as microfractography, to reach a definite conclusion. Usually, intergranular fractures cannot be detected by visual examination. Low magnification viewing is sometimes sufficient but generally microfractographic procedures are required. Intergranular fractures are characteristic of certain types of failure mechanisms: stress-corrosion cracking, hydrogen embrittlement, temper embrittlement, etc. Hence, when in doubt, the investigator should always examine the fine fracture details to either confirm or identify the fracture mode and mechanism. Figures 3 to 6 show examples of various types of failure origins.

MACROETCHING – AID TO VISUAL EXAMINATION

In addition to providing a basic description of the fracture surfaces, the macroscopic examination step in a failure analysis may be used to detect problems such as grinding abuse, "bruises", or other manifestations of surface damage. Indeed, the macroscopic examination cannot be artificially limited to gross fracture interpretation. Inasmuch as the gross fracture features are inevitably associated with microstructural features, macroetching can be employed to detect variations in such features which alter the fracture appearance and thereby provide: (a) a better understanding of the fracture and its possible causes, and (b) pinpoint those areas or features that may require further examination

Fig. 1. Magnetic particles were used to reveal crack network on this AISI S1 cutter die which cracked and spalled after redressing due to abusive grinding.

by one or more additional procedures. Failure analysis, as a total discipline, recognizes that the macroscopic and microscopic techniques cannot be viewed as rigidly separate and independent techniques, i.e., a dichotomy, but as overlapping, mutually supportive techniques.

Of course, not all failures involve fractures; some occur as a result of excessive wear or corrosion rate. Also, failures may occur as a result of excess plastic deformation which renders the part inoperative without fracture. Various macroetching and special print techniques can be applied to evaluate material homogeneity and quality and to detect some of the above problems.

The fracture appearance can be altered by the presence of different microstructures with different strengths. For example, Fig. 7 shows a carbon steel die that cracked during water quenching. Note that the cracking was located at the bottom of a cylindrical hole and there is a well defined rim of finer fracture around the periphery of the die where martensite was formed. Note that cracking began at the bore and proceeded outward as can be seen by the radial fracture marks. A disc was cut from behind the fracture of the half shown on the right side of Fig. 7 and surface ground. Cold etching in 10% nitric acid and water reveals an etch pattern (bottom, Fig. 7) identical to the fracture pattern. If the history of the part were unknown, hardness testing and microscopic examination would be required to determine the nature of the surface condition.

Fig. 2. Two examples of use of dye-penetrant to reveal tight cracks. Left—AISI S7 plastic mold sleeve, Right—AISI T1 part. Views above are prior to inspection, views below are after inspection.

Surface damage due to excessive frictional heat can also be revealed by macro-etching. For example, Fig. 8 shows a cutter blade after etching revealing the heat pattern due to the service conditions. Figure 9a shows a spall from a hardened steel roll. Dye-penetrant testing of the reverse side of the spall (Fig. 9b) brought out details of a crack pattern indicative of mechanical abuse that caused the spall. Details such as this can easily be missed unless the investigator carefully searches for such evidence. Improper grinding procedures can also cause cracking in high hardness steels. Two examples are shown in Fig. 10. Cold etching reveals the scorch pattern associated with the grinding cracks.

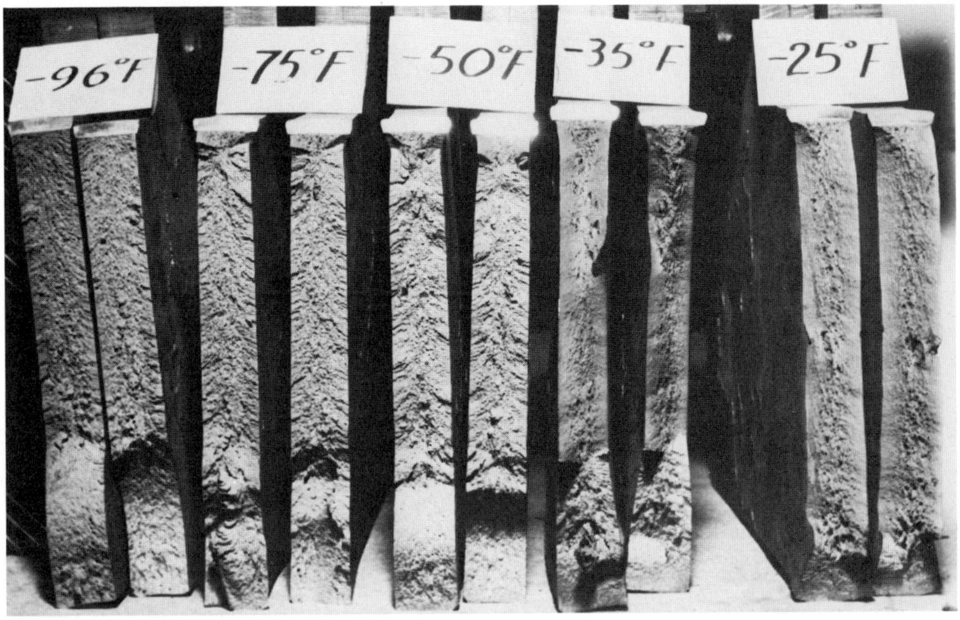

Fig. 3. Appearance of chevrons in ship steel samples broken over range of temperatures. Fractures began at notch (top). Mating fracture halves shown for each temperature.

Macroetching may also be used to reveal heat treatment irregularities. Figure 11 shows how macroetching can reveal the presence of decarburized surfaces. Figure 12 shows a cold-heading die that failed early in service. Cold etching revealed that the area around the central hole was hardened and only a shallow hardened case surrounded the working area. Note that the design incorporates a sharp corner (stress concentrator) at the base of the working face cavity. Figure 13 shows a wire-forming die that failed prematurely. Cold etching of a disc removed just behind the fracture revealed an unusual etch pattern. The dark area was about 8 Rc harder than the lighter etching area. Figure 14 shows a carbon steel part that cracked after quenching. Cold etching of the surface revealed soft spots. Figure 15 shows a ram that cracked during heat treatment after hardening. The part had been put through two carburizing and heat treatment cycles because the ram hardness was low after the first cycle. Cold etching a disc from the ram revealed a complex condition at the outside surface (and center hole). Microscopic examination is required in a case of this type to confirm the reason for the macroetch appearance. Figure 16 shows a macroetch of a disc through a cracked pump plunger. The macroetch showed that the crack was present prior to nitriding as was confirmed by microscopic examination.

Failures may also be caused by material problems and visual examination plus macroetching may be applied to reveal these problems. As an example, Fig. 17 shows chuck jaws that cracked during quenching due partly to the presence of an unconsolidated center condition. Hot acid etching (1:1 HCl in water at 160°F) was used to bring out the extent of the center condition. A second example is given in Fig. 18. A cold-header die cracked during water quenching due to an unconsolidated center (microporosity and inclusions). Hot acid etching was again used to reveal this condition.

Fig. 4. Examples of fatigue fractures. Crankshaft (upper right) has multiple origins. Note that the clam shell patterns are curved due to the rotational load. Other fatigue fractures have single origins.

CLASSIFICATION OF FRACTURES

A key aspect of the failure analysis is determination of the responsible failure mechanism. This involves classification of the fracture features, both macroscopic and microscopic, according to some logical scheme. Numerous systems have been devised but none are totally adequate. For example, the fracture can be classified according to its growth mechanism. This, however, ignores the crack initiation phase which may be quite different than the propagation mechanism.

The macroscopic features might be used to classify the fractures and such descriptions are quite useful. On a macroscopic level, the fracture may appear brittle but microscopically, the fracture may exhibit characteristics of ductile failures. Hence, in describing

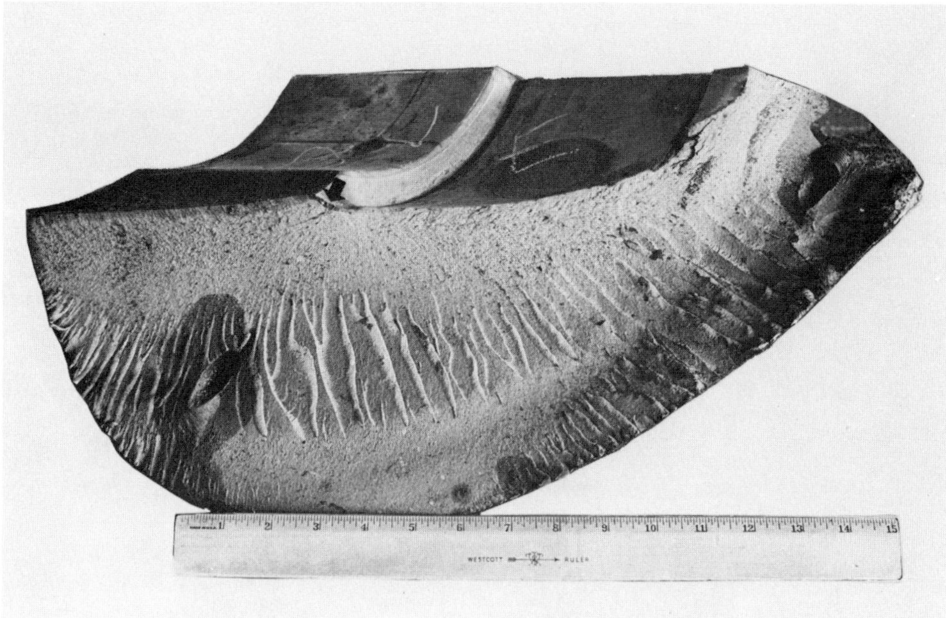

Fig. 5. Brittle fracture in large steel component. Note the pronounced radial marks.

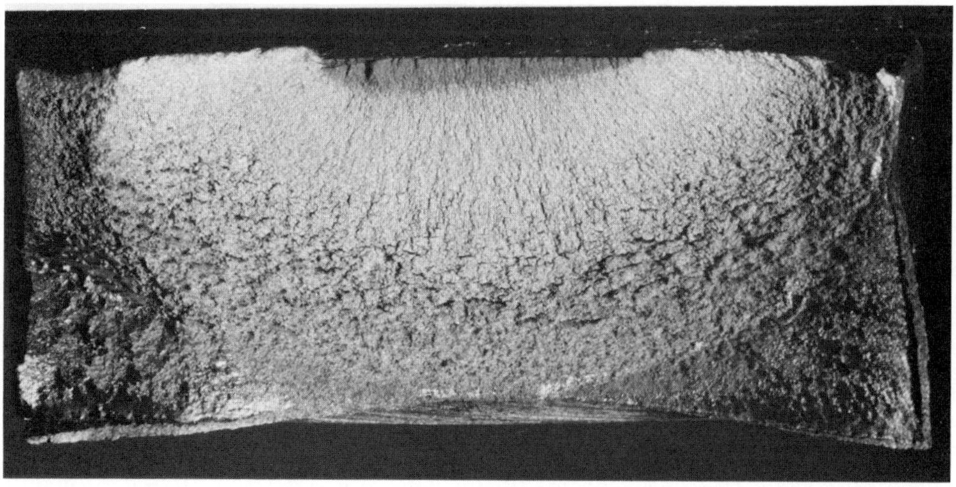

Fig. 6. Brittle fracture of steel sample began at discolored area. Outer, lower corners are shear lips.

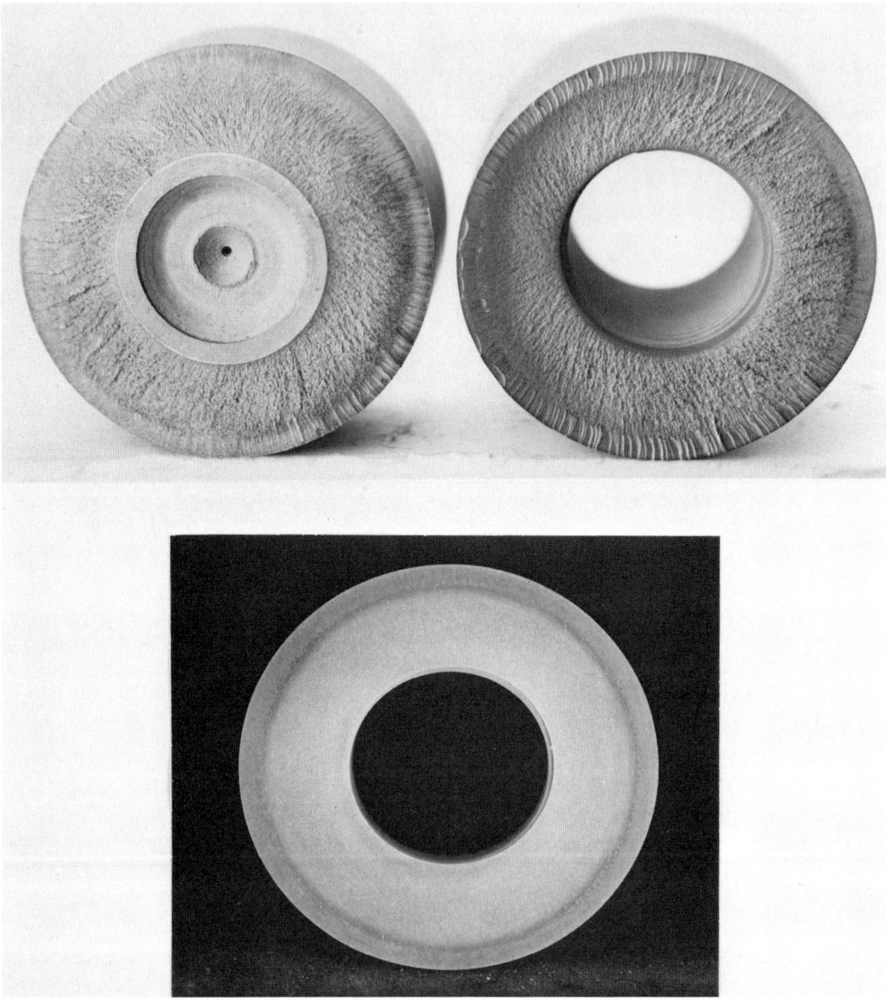

Fig. 7. AISI W2 die cracked during quenching. The fracture clearly shows the chill depth also shown in the cold etch disk (10% nitric acid in water)

a fracture it is important to describe and classify the fracture from both macroscopic and microscopic viewpoints. Classification of the fracture might be done as follows. The gross characteristics of the fracture should be used to classify it as either ductile, brittle, fatigue or creep. Then specific aspects of loading, such as torsion, tension, compression, etc. should be mentioned. Finally, microscopic procedures should be employed to detect features such as cleavage, dimples, striations, etc. and to determine if the crack path at the origin and during propagation was transgranular or intergranular. In this manner, the role of loading, temperature, material and environment can be assessed and utilized in establishing the failure mechanism. Factors responsible for crack initiation must also be discerned.

Fig. 8. Cold etching revealed localized heating during service of this AISI S7 cutter blade (10% nitric acid in water).

MACROSCOPIC CHARACTERISTICS OF FRACTURES

From a *macroscopic* standpoint, brittle fractures are characterized by the following:

- Little or no visible plastic deformation precedes the fracture.
- The fracture is generally flat and perpendicular to the surface of the components.
- The fracture may appear granular or crystalline, and is often highly reflective to light. Facets may also be observed, especially in coarse-grained steels.
- Herringbone, or chevron, patterns may be present.
- Cracks lengthen rapidly, often with a loud noise.

Figure 19 shows the classic chevron pattern that is frequently observed on brittle fractures of low-strength steels. The apexes of the chevrons point back toward the origin of the fracture. Note the wavy pattern along the surfaces of the fractured piece. In general, the more highly developed the chevron pattern, the greater the waviness of the edge. Materials that are still more brittle, such as cast iron or a high-alloy tool steel, will not exhibit chevrons. Figure 20 shows the crystalline appearance that is typical of a tensile fracture in such materials; no apparent necking is observable.

Chevron patterns are characteristic of many brittle fractures, not only in metals but in glass and plastics also. When such a pattern develops, it indicates that the fracture is discontinuous, that is, it proceeds first by initiation or initiations, followed by union of the initiation centers to form the fracture surface. Tipper [2] studied chevron patterns and showed that such fractures are caused by discontinuous regions of cleavage fracture joined by regions of shear and the chevrons featured are the ridges between the cleavage

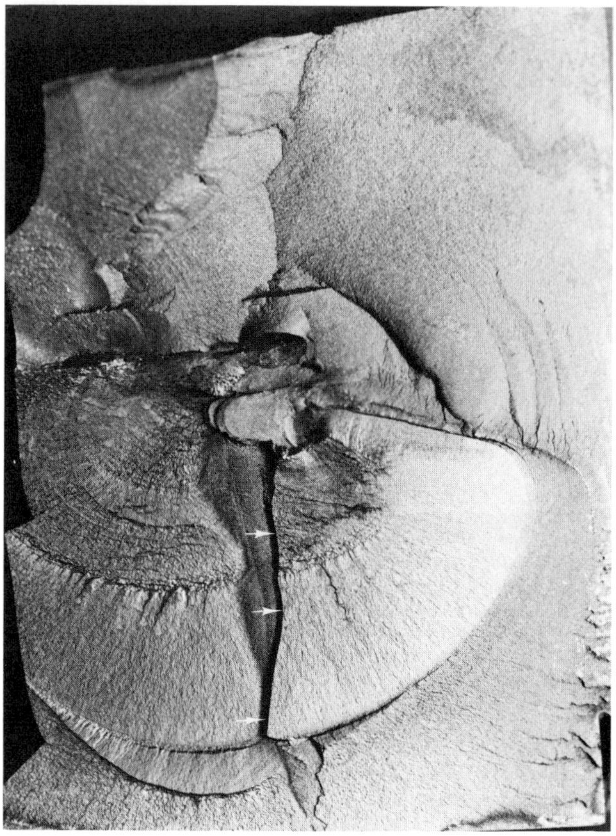

Fig. 9a. Circular spall from 16 in. diameter hardened steel roll. Original Mag. 2¼X. (Reduced approximately 58 percent for reproduction).

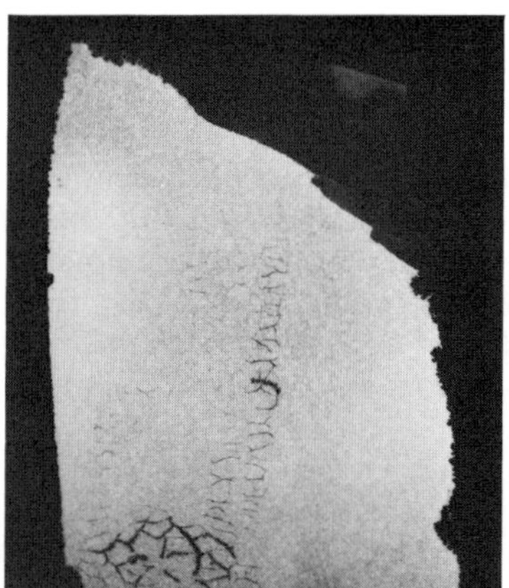

Fig. 9b. Reverse side of spall shown in Fig. 9a. Dye-penetrant testing revealed mechanical abuse on surface above spall. Abuse pattern is directly above arrows shown in Fig. 9a.

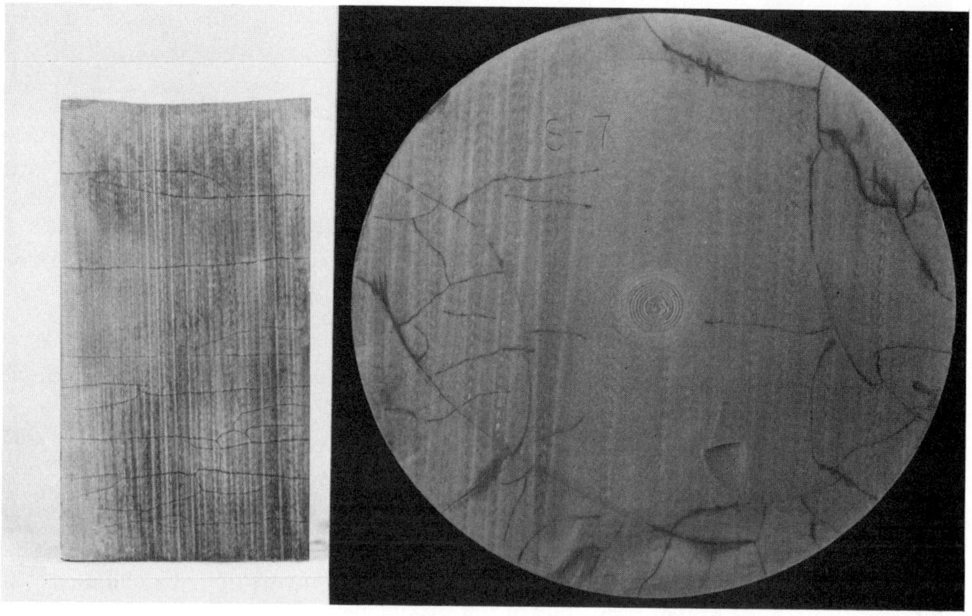

Fig. 10. Two examples of the use of cold etching to reveal abusive grinding. Left— An AISI 01 part; Right— An AISI S7 mult (10% nitric acid in water).

Fig. 11. Cold etching of AISI 01 die revealed extensive surface decarburization.

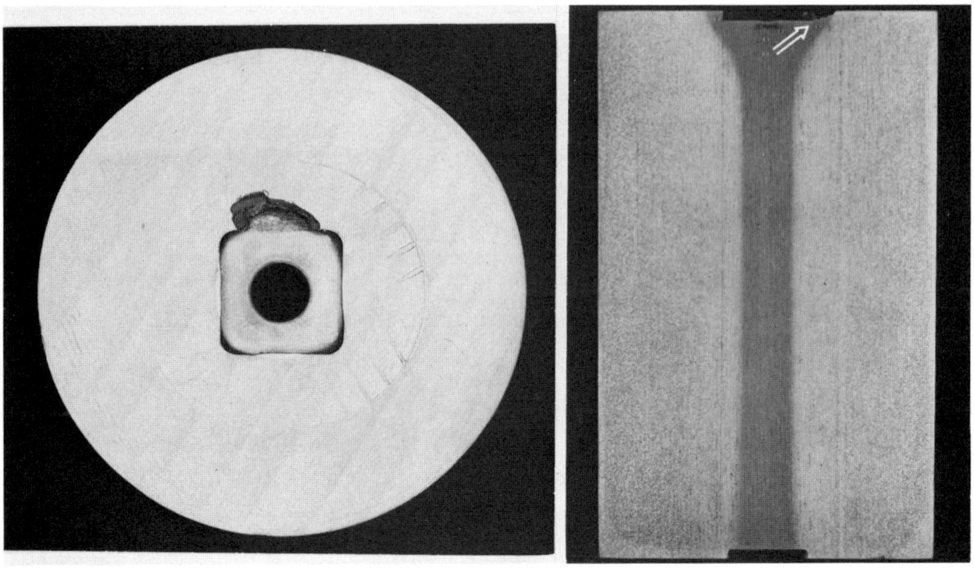

Fig. 12. AISI W1 (1% carbon) cold heading die that failed prematurely by chipping at the interface of the shallow hardened zone of the striking face. Disk cold etched with 10% nitric acid in water.

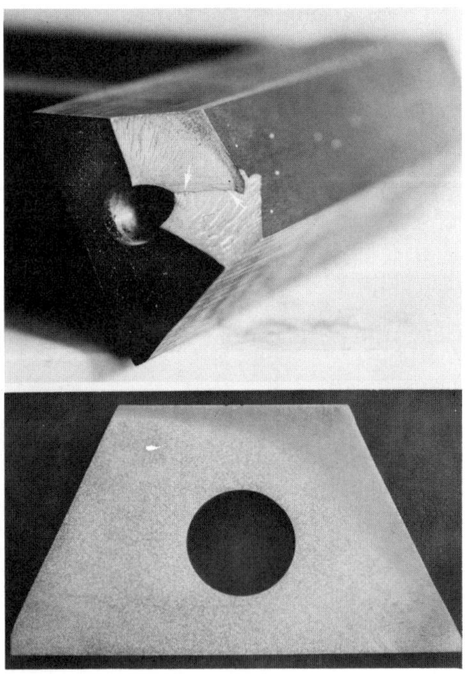

Fig. 13. AISI W1 wire forming die that failed prematurely due to a quenching irregularity. A chill was not developed around the bore working surface. The dark area was 38 HRC while the light area was 29–31 HRC (10% nitric acid in water).

Fig. 14. AISI W2 part that cracked during quenching due to soft spots. Surface etched with 10% nitric acid in water.

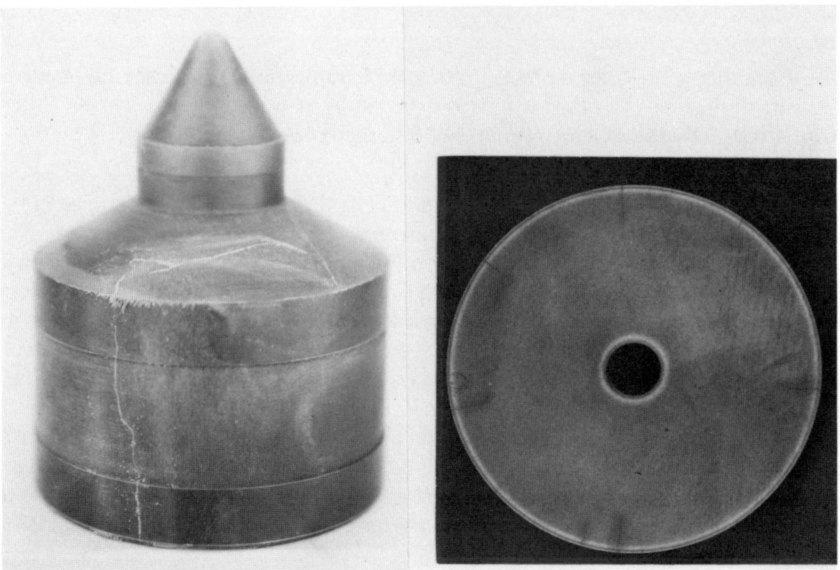

Fig. 15. AISI S7 ram which was improperly carburized and heat treated. Since the surface was soft, the ram was recarburized and again hardened. The cold etched disc reveals an outer decarburized zone, a carburized zone and an inner decarburized zone (10% nitric acid in water).

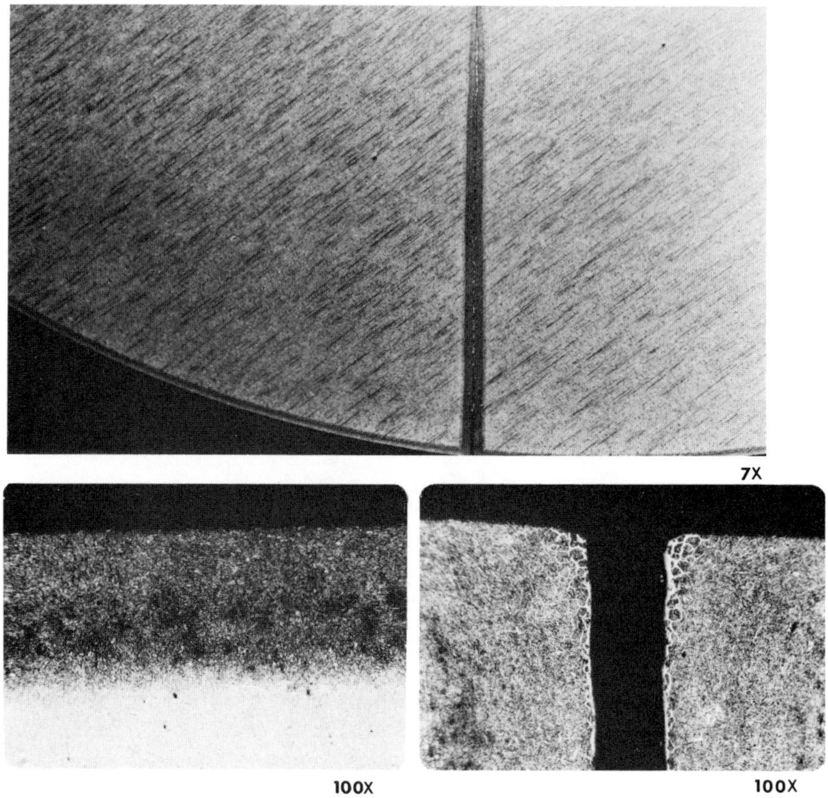

Fig. 16. Cold etch disk from a cracked AISI H11 pump plunger. Cracking occurred prior to nitriding since the crack wall was nitrided as well as the OD surface. (Micrographs at 100X, nital). (Reduced approximately 40 percent for reproduction).

and shear zones. The fracture tends to extend into the metal and creates cracks below the main fracture surface. This tendency is greatest in the crack growth direction. The spread of the crack towards the metal surfaces is not always equal, thus often producing asymmetrical chevron patterns. The front of the propagating crack tends to become curved and traces of this curved front can often be observed. Boyd [3] has shown that the crack front could be considered as a parabolic envelope enclosing a number of individually initiated cracks spreading radially. His model for chevron formation is shown in Fig. 21. His model predicts that the angle θ (see Fig. 21) will be 72 degrees and the angle is independent of the material thickness and properties. Boyd made 616 measurements on twenty fracture surfaces and observed a mean angle of 69 degrees which varied somewhat depending upon the presence and size of the shear lip. Since plastic deformation is required in the formation of chevron patterns, such patterns are not observed in "more brittle" fractures.

Boyd described "silky" fractures as the texture on fracture surfaces of maximum shear, such as the wall of a cup-and-cone tensile fracture. "Mat" fracture, however, is the dull fibrous texture in fractures that are predominantly flat and perpendicular to the

Fig. 17. Chuck jaws, made of AISI 4150 alloy steel, cracked during quenching due partly to an unconsolidated center (below, standard etch). Cracking was also promoted by heat treatment irregularities and was located by coarse machining marks.

tensile stress direction. The term "fibrous" fracture is often used in reference to either silky or mat fractures.

Structural fractures containing chevron markings were described by Boyd (Fig. 21) as:

1) The fracture has a substantially flat middle portion with either a mat or crystalline texture, or a mixture of both, which is marked by ridges and hollows forming the chevron pattern. Such chevrons are well developed in mat fractures but faint or absent on completely crystalline fractures.
2) Between the flat portion and either edge of the material are borders of silky texture which are at roughly 45 degrees to the plane of the flat portion of the fracture. The width of these "shear lips" vary appreciably and may not be visible in completely crystalline fractures.
3) The thickness along the fracture may be reduced, as in tensile specimens, and surface deformation may be present. Necking may be imperceptible in completely crystalline fractures.

Fig. 18. AISI W1 (0.90% carbon) cold header die which cracked during quenching due to an unconsolidated center. The dotted lines show the location of the etch disk (at right, standard etch).

Fig. 19. Typical appearance of a brittle fracture (note chevrons). Failure occurred in the web of a rail section.

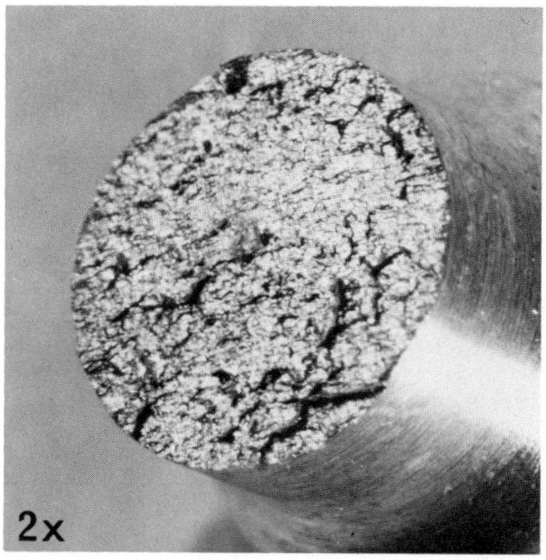

Fig. 20. Granular appearance of brittle fracture (large cleavage facets). (Reduced approximately 30 percent for reproduction).

4) The fracture front is curved with the "spearhead" near the middle of the plate thickness ahead of the visible crack at the plate surfaces.
5) Chevron marks appear to be orthogonal to the fracture front.
6) Chevron fractures occur by alternation of cleavage and shear fracturing. Small cleavage cracks occur in grains ahead of the main crack front. The walls between these cracks and the main crack front rupture by shearing thus producing ridges of fibrous texture separated by valleys of crystalline texture. The amount of fibrous or crystalline fracture will vary depending on the brittleness of the fracture.

Hence, we see that the clarity of the chevron pattern and the size of shear lips, if present, indicate the relative brittleness of the fracture.

Figure 21 (locations A to E) shows the close resemblance between fractures containing chevrons and the cup-and-cone tensile fracture. The fracture begins initially along the centerline and spreads as a disc-shaped crack towards the surfaces. However, before reaching the surface, the fracture mode changes and rupture occurs on the conical surfaces of maximum shear, approximately 45 degrees from the tensile axis (shown as dotted lines in sections A, B and C of Fig. 21).

As necking progresses, plastic deformation (shaded area in Fig. 21) occurs ahead of the crack front with lines of maximum shear strain approximately 45 degrees to the plate axis. If this plastic zone is large enough, fracture will progress by shear only. However, if necking is limited, thus limiting the extent of the plastic deformation, many small shear fractures of differing orientation will form resulting in a rough or serrated appearance. Cleavage is not an essential feature of the flat portion of the fracture although it can be present in varying amounts in fractures exhibiting chevrons.

The radial markings, sometimes visible, on tensile bar fractures can be explained in terms of Boyd's chevron fracture model. The straight radial lines are orthogonal to an expanding circular crack front. These radial lines start near the center (frequently the origin is visible) and progress to the shear lip region where the fracture mode changes.

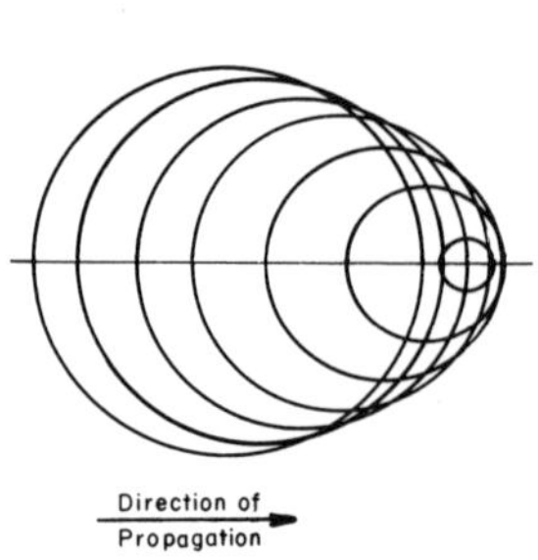

a) Crack front propagation model

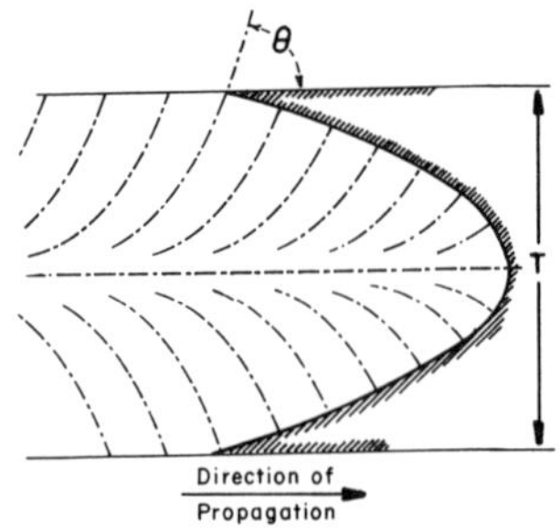

b) Crack front at given time

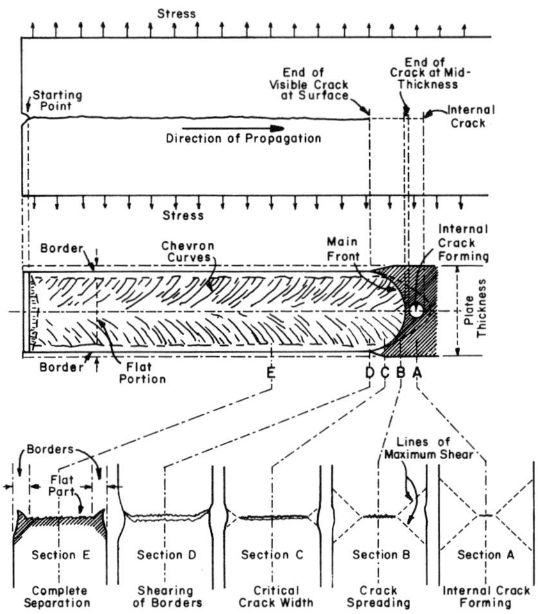

c) Analogy between chevron formation in plate and cup-and-cone fracture

Fig. 21. Boyd's model for chevron crack formation.

Unlike brittle fractures and fatigue fractures, ductile fractures have not been studied in the same depth. However, some excellent studies have been conducted. Overload is the cause of ductile failures and the stresses causing such fractures can be predicted mathematically unlike brittle failures which are generally unpredictable. From a practical standpoint, ductile fractures are important in certain metal-forming operations.

From a *macroscopic* standpoint, a ductile fracture exhibits the following characteristics:

- A large amount of plastic deformation precedes the fracture
- Shear lips may be present
- The fracture may appear to be fibrous, or have a matte or silky texture
- The cross section at the fracture may be reduced by necking
- Crack growth will be slow.

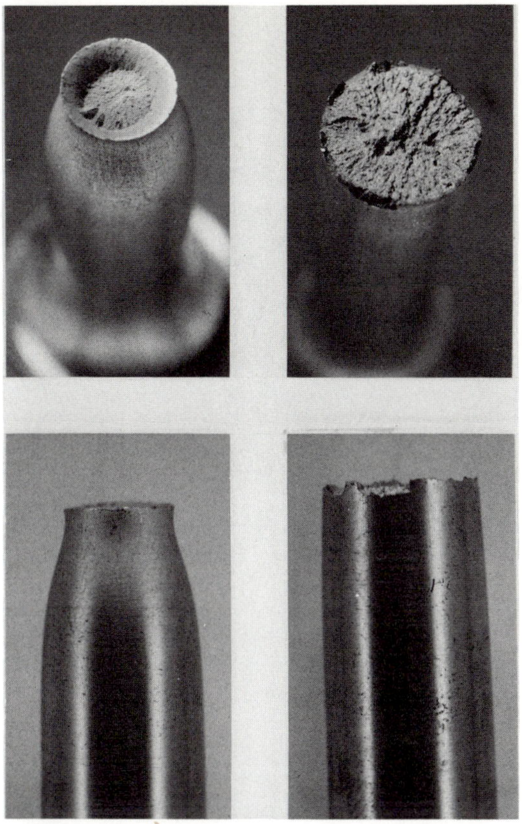

Fig. 22. Appearance of ductile (left) and brittle (right) tensile fractures.

Figure 22 shows ductile and brittle fractures in tensile specimens. Note that the ductile tensile fracture has the classic cup-and-cone failure, with considerable necking (plastic deformation) prior to fracture. On the other hand, the brittle tensile fracture is flat, with little or no necking before fracture.

Figures 23, 24, 25 and 26, photographs of controlled laboratory fractures [4] in full-size line pipe, illustrate several important features of ductile and brittle fractures. When line pipe fails catastrophically, failure occurs because of rapid lengthening of an initially small crack in these large, continuously welded structures. When a crack makes its appearance in a line pipe, the possibility that it will not grow, grow slowly, or propagate catastrophically (with or without arrest in some reasonable pipe length) depends on (1) the toughness of the pipe material and (2) the magnitude of the potential driving force present in the pipe as a function of operating stress, pipe dimensions and the nature of the fluid within the pipe.

Full-scale testing of pipe revealed good correlation between the crack speed and the 50 percent shear-area transition temperature, as determined by the Battelle Drop Weight Tear Test (DWTT). The fracture appearance of the pipe gives a good indication

TEST TEMP. +56°F | DWTT +48°F | CRACK SPEED 279 fps

Fig. 23. Fracture of X–60 grade line pipe tested 8° F above the 50% shear-area DWTT transition temperature.

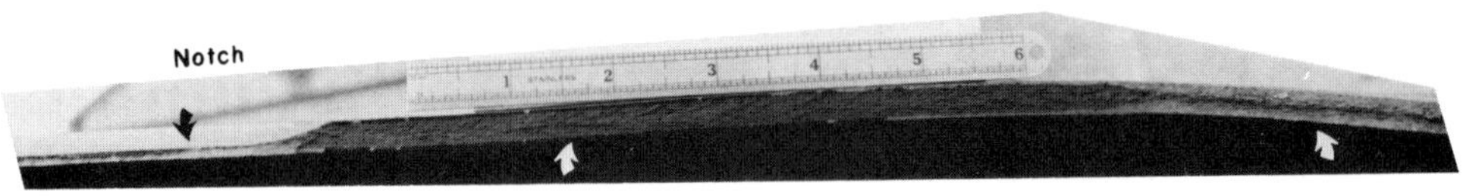

TEST TEMP. 0°F | DWTT + 2°F | AVG. CRACK SPEED 566 fps

Fig. 24. Fracture of X–60 grade line pipe tested 2° F below the 50% shear-area DWTT transition temperature.

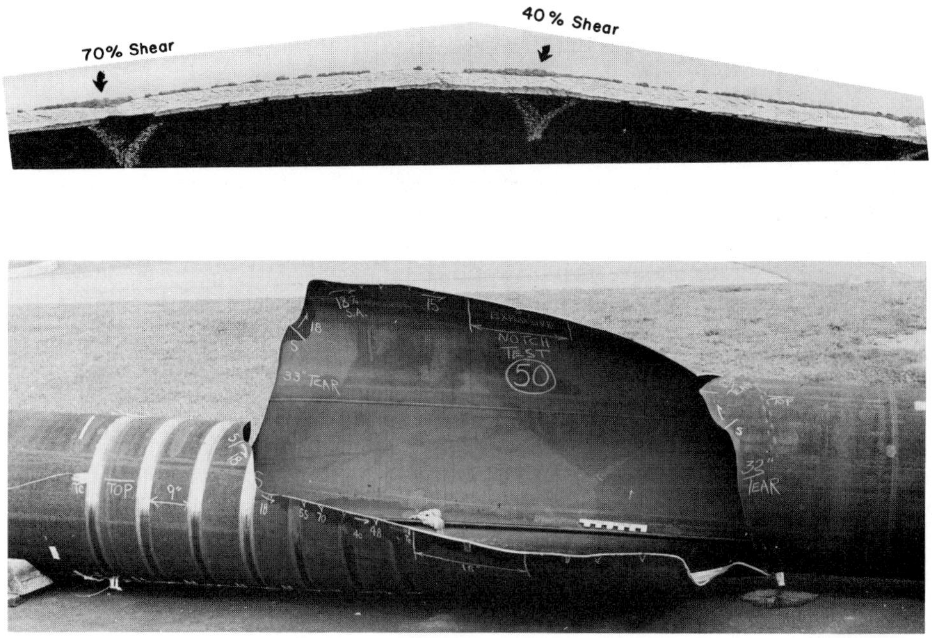

TEST TEMP. + 3°F | DWTT +13°F | CRACK SPEED 1550 fps

Fig. 25. Fracture of X–60 grade line pipe tested 10°F below the 50% shear-area DWTT transition temperature.

TEST TEMP. −15° | DWTT +25° | CRACK SPEED 2215 fps

Fig. 26. Fracture of X–60 grade line pipe tested 40°F below the 50% shear-area DWTT transition temperature.

of the crack speed, as follows:

- Less than 10 percent shear area corresponds to brittle fracture with a fast-running crack.
- Greater than 40 percent shear area corresponds to ductile fracture with a crack speed below 900 fps.

Figures 23 through 26 show full-scale tests of X—60 grade line pipe where the test temperature was varied with respect to the 50 pct shear-area DWTT transition temperature. In each instance, the pipe was loaded to 40 percent of the yield strength. A 30-grain charge was detonated beneath an 18-in. long notch cut in the pipe.

Figure 23 shows a ductile fracture in a pipe tested at 8°F above the 50 percent shear-area DWTT transition temperature; resultant crack speed was 279 fps. From the 18-in. notch, the crack propagated 33 in. in full shear, and then 18 in. in tearing shear before stopping.

Figure 24 shows the result of testing 2°F below the 50 percent shear area DWTT transition temperature; average crack speed was 566 fps. The crack began by brittle cleavage with 15 percent shear area, and progressed 8 in. in this manner before changing to 100 percent shear (ductile) for 37 in. before stopping. The macrograph shows the initial fracture area.

Figure 25 shows the result of testing a pipe 10°F below the 50 percent shear area DWTT transition temperature. The fracture propagated by cleavage with a 15 to 18 per cent shear area fracture, with small patches having shear areas as high as 70 percent (area shown in macrograph). The crack propagated straight along the top of the pipe for 27 in., and then developed a wave pattern. Only a half-wave was completed before the crack changed to 100 percent shear, and then tore circumferentially for 33 in. before stopping. Actual crack speed was 1550 fps; straight line crack speed was 1250 fps.

Figure 26 shows the result of testing 40°F below the 50 percent shear area DWTT transition temperature. The fracture traveled in a wave pattern for a full-wave by cleavage with less than 10 percent shear area present, and then changed to full shear and tore circumferentially 19 in. before stopping. Maximum straight line speed was 1760 fps, and actual crack speed was 2215 fps.

If a component is subjected to cyclic loading involving tensile stresses below the statically determined yield strength of the material, the component may eventually fail by a process known as "fatigue". While fatigue fractures in metals are most commonly known, nonmetallic materials, such as polymers, also can fail by fatigue. Understanding the fatigue process is of great technological importance since fatigue accounts for a large proportion of all service failures.

The tensile stresses required for fatigue failures will be much lower than would be required for static fractures. After some critical number of stress cycles, a crack will form in the region of highest stress concentration. With continued cycling, the crack will grow in length. Growth will occur in a direction perpendicular to the applied tensile stress. A fatigue failure occurs in four steps:

- microcrack nucleation
- microcrack propagation (Stage I)
- macrocrack propagation (Stage II)
- final rupture

Macroscopically, the fracture is brittle in nature and a fatigue fracture can be considered as a special type of brittle fracture. After some amount of crack growth, the cross-sectional thickness will be reduced until it can no longer support the load and the fracture will be completed. Hence, fatigue failures will contain a portion of cyclically-grown fracture and a portion of overload fracture. The apparent ductility or brittleness of the overload fracture will vary depending on the strength, ductility and toughness of the material, the temperature, etc. Macroscopically, there will be no evidence of necking associated with the fracture. Also, fatigue fractures cannot be successfully remated, a fact which never seems to inhibit such attempts!

Some examples of fatigue failures are shown in Figs. 27—30. Note that the fatigue portion of the fracture can be quite small or can extend over most of the fracture. This depends on the load involved, the brittleness of the material, etc. *Macroscopically,* fatigue fractures are generally characterized by "beach", or "clam shell" markings. It is possible that other types of fractures can have similar markings; also, not all materials will exhibit such marks when fatigue occurs. For example, such marks may not be visible on cast iron fatigue fractures. Laboratory fatigue failures also do not show such marks even in materials which ordinarily exhibit such marks on service fractures. These markings represent an outline of the fatigue crack front at various arrest points during its growth and can represent changes in loading which retard or accentuate crack growth. Laboratory test samples generally are run continuously until fracture in a rather sterile environment compared to that of field failures.

The presence of these fracture surface markings enable the investigator to easily determine the origin(s) of the failure. Some fatigue fractures exhibit a pattern of fan-like marks similar to the radial fracture markings present on chevron fractures. In this case, many small fatigue cracks have joined up at shear steps [5]. This is a case of multiple fatigue cracks initiating from a common location.

Examination of the fatigue beachmarks will enable the investigator to determine the fracture origin and make estimates of the nominal stress level, the type of loading, and the importance of stress concentrators. Guides to interpreting fatigue fracture markings have been shown schematically by Lipson [6] and by Jacoby [7] and are given in References 8 and 9. Figures 31 and 32 show schematically the fatigue fracture surface markings produced in smooth or notched components with circular cross sections or for square or rectangular cross sections under various loading conditions.

Many investigators have observed a surface roughening which is peculiar to fatigue loading. Thin slivers of metal were observed protruding from the surface at some of the dense slip bands. Further study of this phenomenon showed that there were intrusions as well as extrusions and that fatigue cracks started from the intrusions and propagated into the slip band. This effect has been observed in a great many materials and appears to be an essential step in creating fatigue fractures. Indeed, studies [10] have shown that fatigue life can be vastly improved if the metal surface is periodically removed (only a minor amount of metal need be removed); however, such a procedure is not easy to apply commercially.

Fig. 27. Failure of this ASTM A36 member began by fatigue (arrows) but progressed only a short distance before brittle fracture occurred.

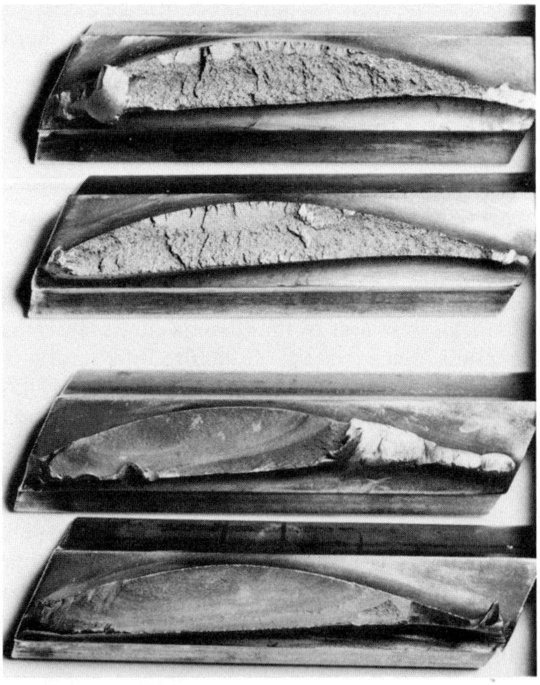

Fig. 28. Fatigue fractures of turbine blades.

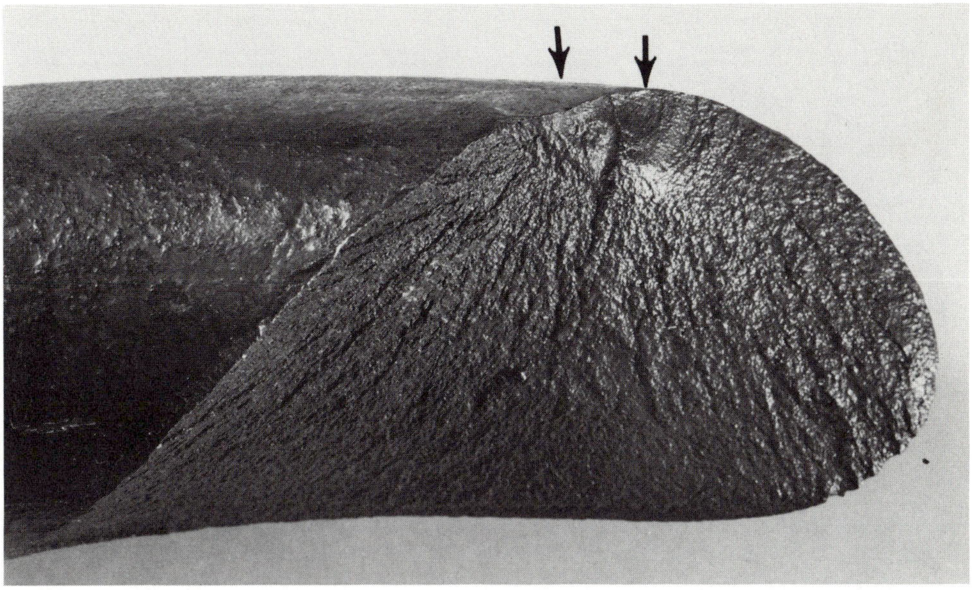

Fig. 29. Torsion failure in AISI 51B60. Railroad Spring (1 5/8 in. diameter) began by fatigue at abraded area on top (arrow).

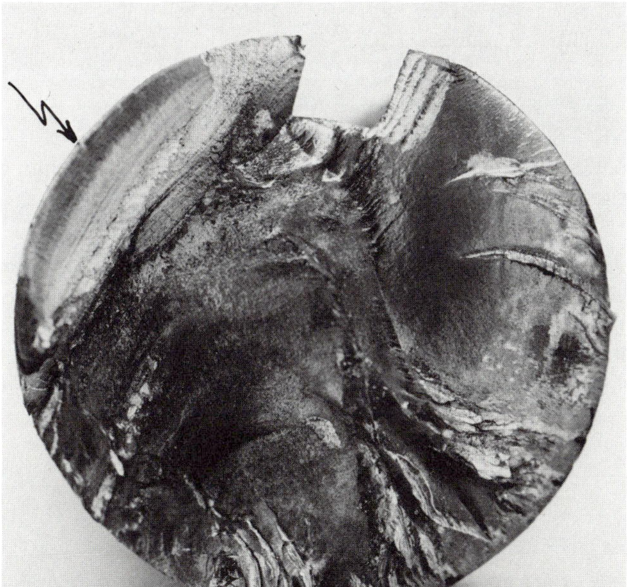

Fig. 30. Failure of this AISI 8620 armature shaft (pinion-end) began by fatigue (arrow). Note fatigue marks around sharp keyway corners. The fatigue failure progressed only a short distance before torsional stresses took over causing final rupture.

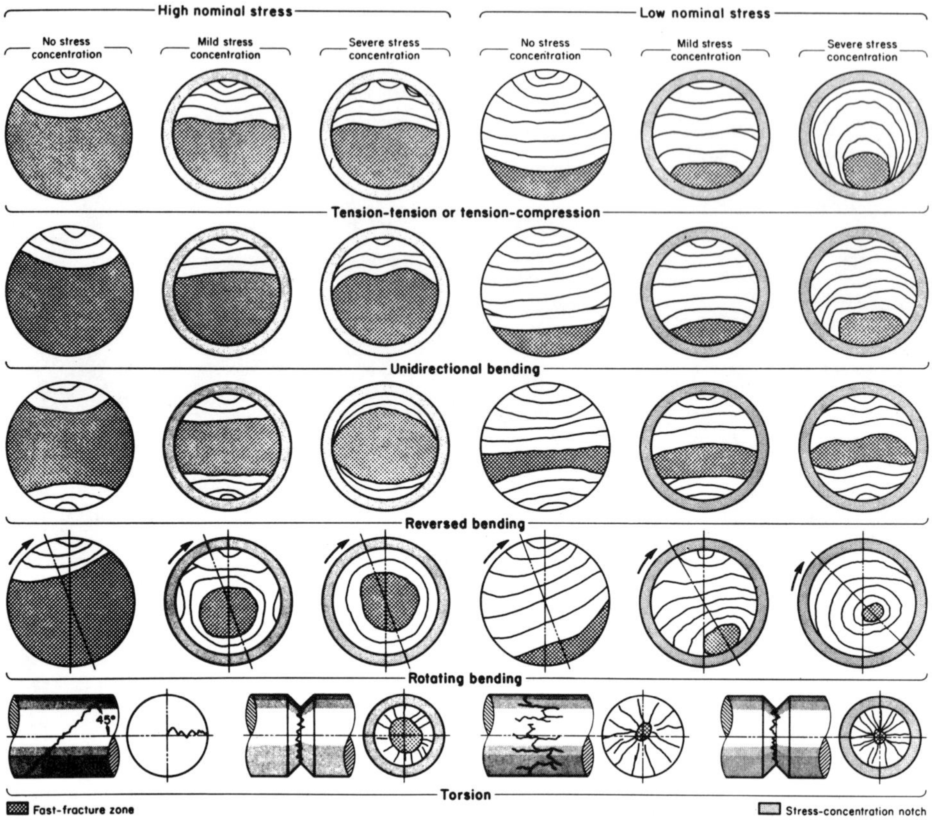

Fig. 31. Schematic of fatigue—fracture markings in smooth or notched round components under different loading conditions. (Reference 8)

In some cases, the fracture will be destroyed and the investigator must rely on other information to discern the responsible mechanism. As an example, Fig. 33 shows one half of a broken axle. After rupture, the fracture was smeared and burnished by rubbing. Examination of the surface of the axle showed well-defined torsion-twist marks. Microstructural examination showed that the axle had not been heat treated but was used as-annealed.

SUMMARY

As the state of art of failure analysis has developed over many decades, standard practice has come to recognize that macroscopic examination must in many cases be combined with other procedures, such as microfractography and microstructural examination, in order to confirm a suspected mechanism or to detect environmentally assisted fracture mechanisms. For example, fractures that are macroscopically brittle may not propagate by cleavage, but may show characteristics of ductile fracture when examined microscopically. Fatigue failures do not always exhibit clam shells and such marks may be created

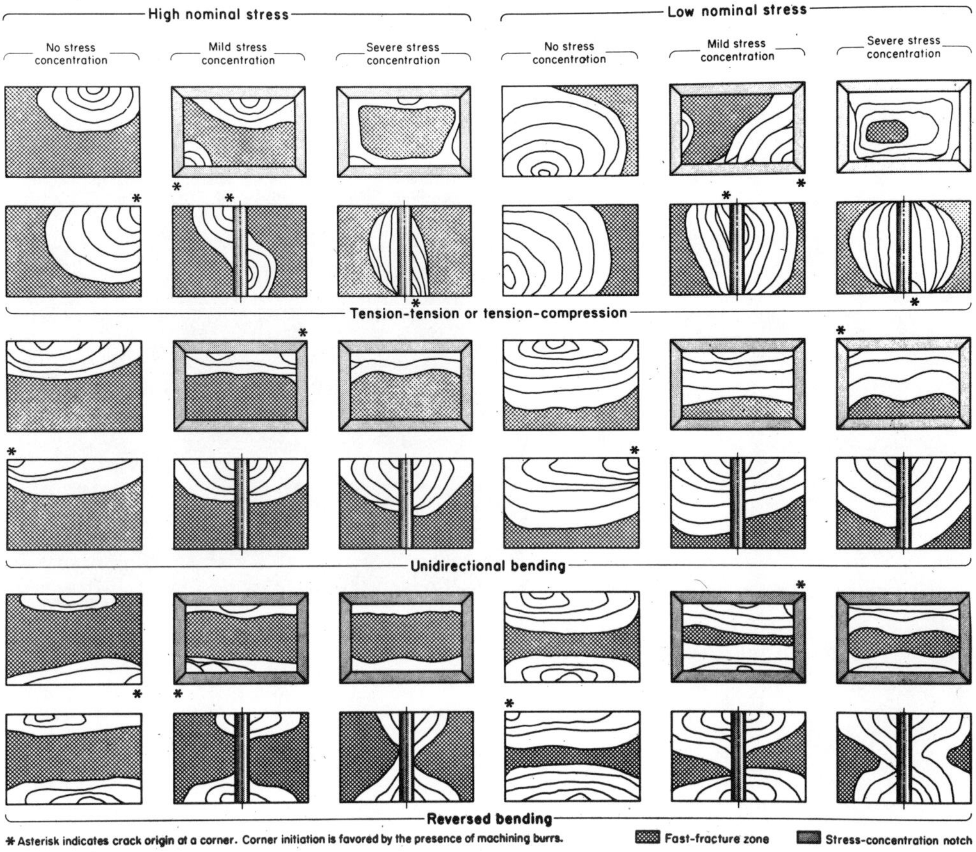

Fig. 32. Schematic of fatigue—fracture markings in square, rectangular or plate components (smooth or notched) under different loading conditions. (Reference 8)

by other mechanisms. Therefore, microfractography is required to obtain conclusive evidence of fatigue by observation of striations; however, not all fatigue failures, for example, those in ultra-high-strength steels, will exhibit striations. Hence, the investigator must approach each problem with all available tools and with experience and knowledge gained through careful observation and experimentation. Thus, implicit in the modern application of macroscopic examination techniques is the awareness of just how and when these techniques may need to be combined with other procedures. By itself, the macroscopic examination will usually not give conclusive identification of many environmentally assisted fracture mechanisms but may indicate the possibility of certain types. Other tests must be brought to bear until the responsible mechanism can be clearly identified.

The macroscopic examination sets the tone of any failure analysis. If the techniques are properly applied and the "signs" are correctly interpreted, chances are excellent that the analysis will be successful. Although visual, or macroscopic, examination may be sufficient to identify the fracture propagation mechanism and to identify initiators, the analyst must be prepared to use this type of analysis as a lead-in to the whole range of techniques available in the modern practice of failure analysis.

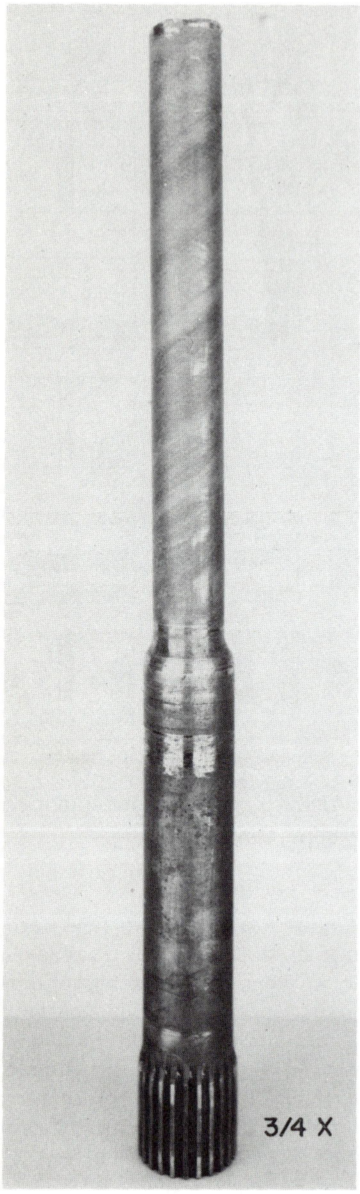

Fig. 33. This "super-duty" axle shaft, made from AISI S7 tool steel, failed in torsion (note torsion marks on axle shaft). The axle was not properly hardened. The hardness, 22/27 HRC, and the microstructure of spherodized carbides and ferrite indicate that it was underaustenitized.

REFERENCES

1. J.C. Russ and G.A. Miller, "Effect of Oxidation on the Electron Fractographic Interpretation of Fractures in Steel", *JISI*, p. 1635 (December, 1969).
2. C.F. Tipper, "The Study of Fracture Surface Markings", *JISI*, V. 185, p. 4 (1957).
3. G.M. Boyd, "The Propagation of Fractures in Mild Steel Plates", *Engineering*, V. 175, p. 65–69 (January 16, 1953); p. 100–102 (January 23, 1953).
4. J.B. Cornish and J.E. Scott, "Fracture Study of Gas Transmission Line Pipe", *Mech. Working & Steel Processing,* V.VII, AIME, p. 222 (1969).
5. D. McIntyre, "Fractographic Analysis of Fatigue Failures", *Trans. ASME, J. of Eng. Materials & Technology,* p. 194 (July 1975).
6. C. Lipson, *Basic Course in Failure Analysis,* Penton Publishing Co., Cleveland, Ohio, (1969).
7. G. Jacoby, "Fractographic Methods in Fatigue Research", *Experimental Mechanics,* p. 65 (March 1965).
8. *Fractography and Atlas of Fractographs,* Metals Handbook, V. 9, 8th ed., ASM, Metals Park, Ohio, (1974).
9. *Failure Analysis and Prevention*, Metals Handbook, V. 10, 8th ed., ASM, Metals Park, Ohio,(1975).
10. N. Thompson, N.J. Wadsworth, and N. Louat, "The Origin of Fatigue Fracture in Copper", *Phil. Mag.,* V. 1, p. 113 (1956).

ANALYSIS OF FRACTURES UTILIZING THE SEM

Marion Russo*

INTRODUCTION

The Scanning Electron Microscope (SEM), a versatile and useful tool for failure analysis investigations, has since the early 1960's steadily replaced the transmission electron microscope (TEM) for fracture surface analysis. This instrument is particularly useful in fracture analysis because (1) it extends in useful magnification from the light microscopescope to that of the TEM, and (2) fractures are viewed directly, eliminating the tedious and time-consuming replication process required for TEM examination. With this instrument, an experienced engineer can determine the mode of failure of a broken part in a relatively short period of time. By itself, however, the SEM cannot uncover all the facts leading to the identification of the cause of a component failure. Most often, it offers enough information about a failure so that together with other test results, a more complete understanding of the cause of failure is determined. This paper illustrates the usefulness of SEM in the analysis of fractures and the importance of correlating other test results, such as a light microscope fracture surface examination and metallographic examination, with SEM results to determine a cause of failure.

SEM Features

Several features of the SEM provide capabilities which make it particularly useful for fracture surface examinations of failed parts. Listed below are some of the primary features.

1) *Resolution* — The SEM has the necessary resolution (100—200 Å) to resolve fracture surface features that identify the fracture mode of a failed part. With the recent developments, some SEM manufacturers are now claiming 70 to 80 Å resolutions. Although this low resolving power is not essential in fracture surface analysis to identify the mode of failure, the higher image brightness and lower accelerating voltage that can be used for examination reduces over-voltage and specimen charging effects that tend to hide surface features. This effect can be seen on the images shown in Fig. 1 where additional surface features are revealed with decreasing voltage. Features hidden by bright regions that occur from over-voltage at 25 kV are revealed when examined at 5kV accelerating voltage. This effect should be taken into account when examining fracture surfaces of light element materials such as aluminum or magnesium alloys that have a rough, irregular appearance.
2) *Magnification Range* — The magnification range of the SEM extends from the low magnification capability of the light microscope to high magnification of several thousand diameters. This feature of the SEM is used in fracture surface

*McDonnell Aircraft Company, McDonnell Douglas Corporation, St. Louis, Missouri, USA.

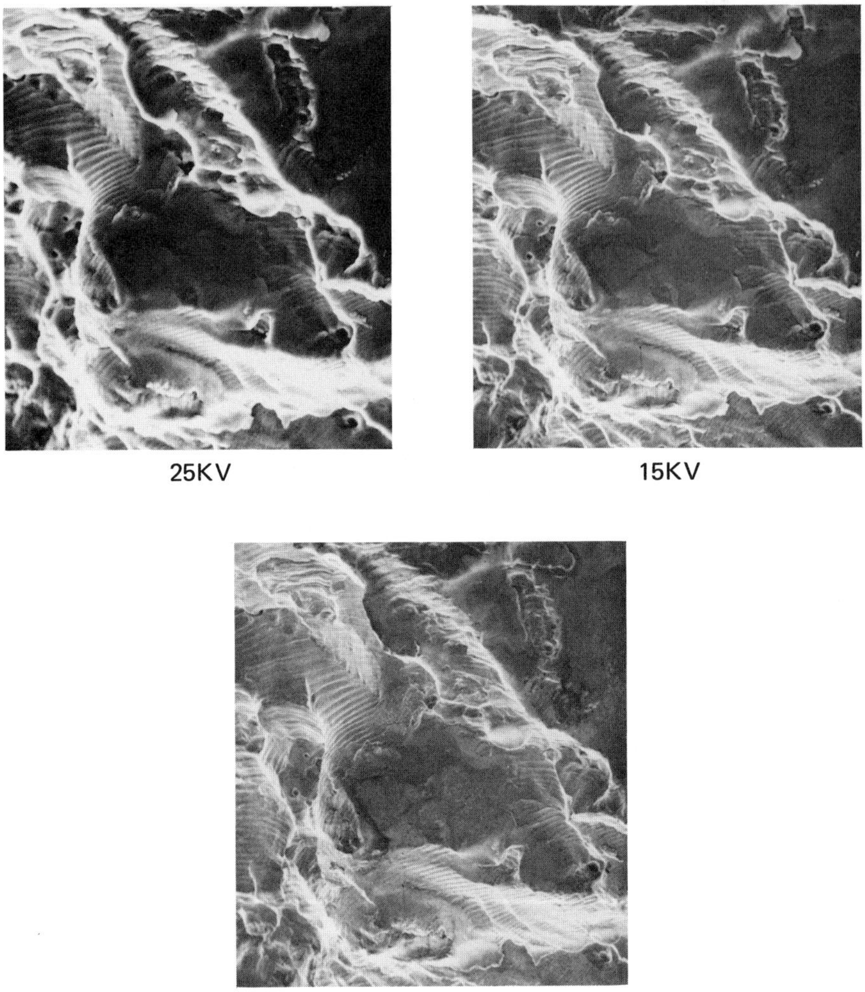

Fig. 1. SEM photographs of the same area on a fatigue fracture at decreasing accelerating voltages. Original magnification 300X. (Reduced approximately 38 percent for reproduction).

analysis for direct correlation of macroscopic features, which are important in maintaining orientation of the fracture during examination, to microscopic features for identification of the fracture mode. Figure 2 shows photographs of the same fracture surface taken with the SEM and a light microscope at 25X magnification that illustrate the low magnification capability of the SEM as well as the similarity of surface features.

3) *Photographic Documentation* — With the use of Polaroid-type film, rapid documentation can be obtained eliminating the need for darkroom photographic facilities.

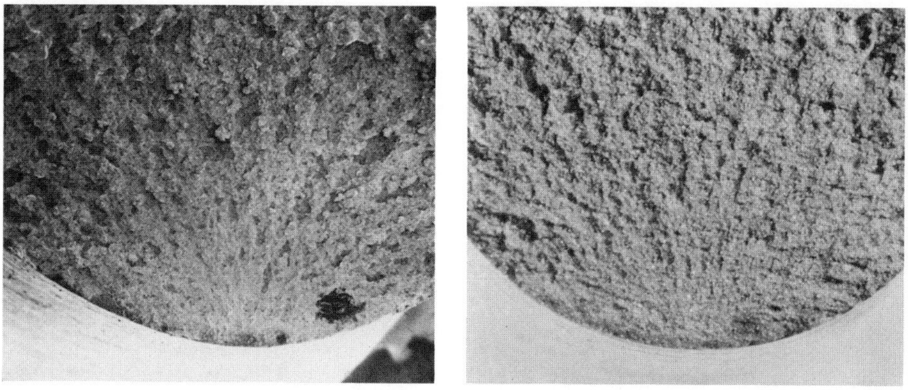

SEM Light microscope

Fig. 2. Photographs of the same fracture surface taken with the SEM and light microscope. Original magnification 25X. (Reduced approximately 40 percent for reproduction).

4) Specimen Preparation — Little preparation is needed prior to viewing with the SEM. In most cases, only cutting the specimen to accommodate the specimen holder and cleaning the fracture surface with replication tape and a solvent in an ultrasonic cleaner are necessary. When corrosion products or other nonconductive contaminants cannot be extracted by cleaning, a thin coating of gold, 200 to 400 Å thick, must be applied to the fracture to prevent electrical charging effects. Two common methods of applying the gold coating are vacuum vapor deposition and ion sputtering.

An important consideration that should be taken into account prior to cleaning a fracture surface or depositing a gold coating is the identification of the foreign material on the fracture. Valuable information pertaining to the failure of a part is often obtained by identifying elements or compounds present in a contaminant on or adjacent to a fracture surface.

5) Specimen Size Limitation — Many new-model SEM's can accept specimens as large as a 3—in. cube, which often enables the investigator to examine the complete fracture surface of the failed part. Many small parts with fractured surfaces too small to analyze with a light microscope can be examined in total with the SEM. In Fig. 3, for example, the fracture surface of a 0.001—in. diameter aluminum wire from an electrical component is examined in the SEM. It was determined by SEM examination that the wire failed in a brittle manner as evidenced by the absence of plastic deformation adjacent to the fracture (necking) and the flat brittle appearance of the fracture surface. Corrosion was an obvious suspect because of the presence of foreign material on and adjacent to the fracture surface.

6) X—Ray Spectrometer Attachments — Characteristic x-ray analysis, either by energy dispersive (ED) or wavelength dispersive (WD), is a valuable SEM attachment for failure analysis investigations. The ED x-ray analyzer, which is probably more often used with the SEM, is capable of providing an x-ray energy spectrum for identification of elements present in micro- or macroscopic areas in a matter of minutes. With this attachment, contaminants, inclusions or corrosion products that have contributed to the failure of a part can be easily identified. Computer attachments to both types of x-ray analyzers for quantitative analysis

(A)

(B)

Fig. 3. A fracture of a 0.001—in. diameter aluminum wire. (A) Original Mag. 800X and (B) Original Mag. 6000X. (Reduced approximately 35 percent for reproduction).

and recent advances in detectors for light element detection have greatly extended their usefulness in failure analysis. An example of ED x-ray analysis used in an in-service failure investigation is shown in Fig. 4. A failed cadmium-plated steel fitting was examined with the SEM, and the fracture surface revealed brittle intergranular features typically observed for delayed failure. However, ED x-ray analysis of small foreign particles present on the fracture surface revealed cadmium. From these results it was concluded that the fitting had cracked during or prior to cadmium plating and failure was not the result of service conditions.

7) *Literature Reference* — Many publications document fracture surface characteristics of most engineering alloys. Two recent publications that offer well-documented fracture modes are the *ASM Handbook, Volume 9* [1], and the *SEM—TEM Handbook* published by Battelle [2]. These reference books, along with many other periodically published technical journals offer important assistance to the investigator in the interpretation of fracture features observed with the SEM. This is especially true in those laboratories where failure analyses are performed on many different alloys. An investigator, in most cases, is not totally familiar with all the fracture surface features that characterize the various fracture modes of all alloys.

BASIC FRACTURE MODES

The SEM can be used to identify the following basic features that is helpful in analyzing a failed part.

1) *Fatigue* — Since fatigue in general accounts for most failures, it is a well-researched fracture mode. Forsythe [3], an early researcher of fatigue crack propagation, determined that a fatigue crack propagates in two separate stages of growth. The initial stage of crack growth, State I, follows a definite crystallographic plane, while Stage II, the stress-dependent stage, follows a path perpendicular to the stress axis and produces parallel striations, each a result of a single cycle of stress. It was later shown [4] that Stage II cracks can also follow crystallographic paths when a low stress amplitude condition prevails. Figure 5 illustrates

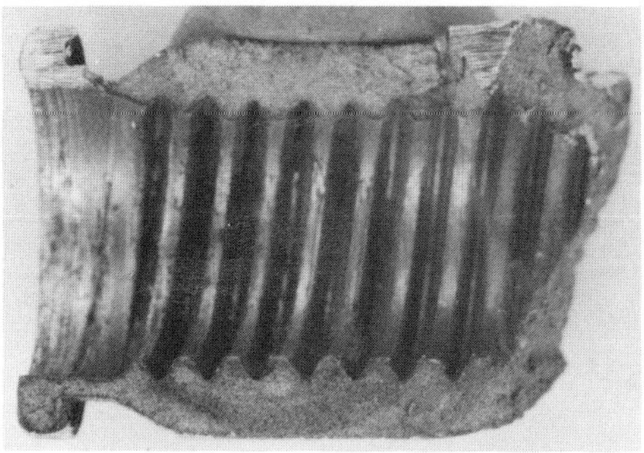

(A)

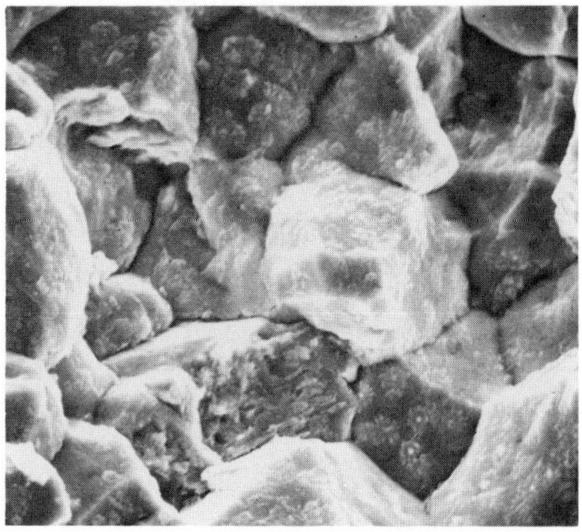

(B)

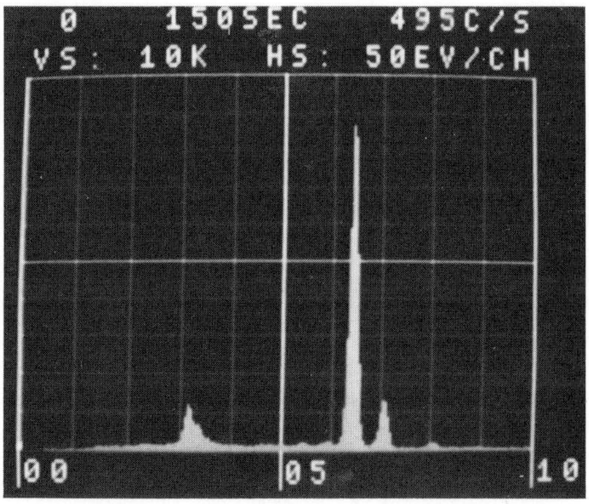

(C)

Fig. 4. SEM and ED x-ray analysis of a failed steel fitting. (A) Light microscope fracture surface appearance, (B) SEM, and (C) X-ray energy spectrum. (A) original Mag. 4X and (B) original Mag. 3000X. (Approximate reductions for reproduction are (A) 20 percent and (B) 32 percent).

the stages of fatigue crack propagation. The significance of the stages of fatigue crack growth to failure analysis is that when features on the fracture surface of a fatigue failure are present that characterize a crystallographic fracture mode, then failure can be attributed to high cycle fatigue.

In examining the fracture surface of a fatigue crack with the SEM, additional information can be obtained from the spacing and definition of the fatigue striations. As shown in Fig. 6, when the cyclic loads producing the fatigue crack are constant amplitude, then the striation spacing is uniform (A). A random-loading cyclic condition produces striations that have varied spacings (B). If

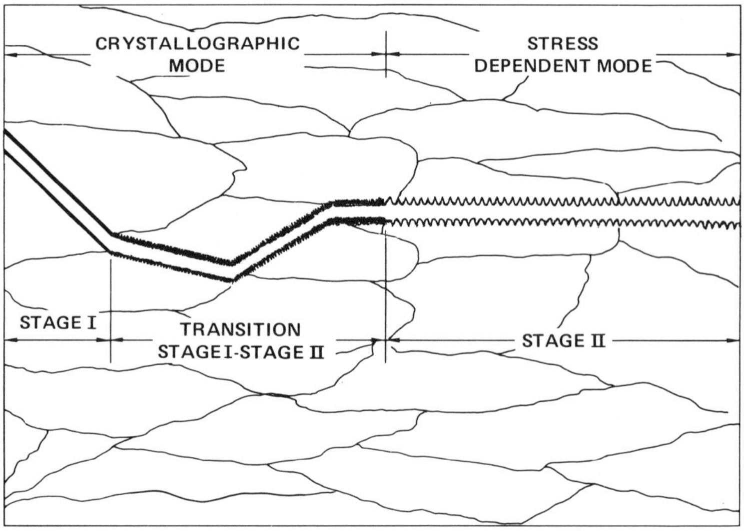

Fig. 5. An illustration showing the stages of fatigue crack growth.

the striation spacing is very fine throughout most of the fatigue region, then a low-load, high-cycle condition prevails (C). A wide spacing from beginning to end of the fatigue region indicates a high-load, low-cycle condition (D). The clarity or definition of the striation also yields important information. It may be important to know whether pure tension loads were responsible for a fatigue failure, which would produce well-defined striations (E), or whether compression loads were encountered which would reveal a rubbed appearance to the striations with small cracks (F).

Since each striation is a result of a single cycle of load, by determining the number of striations present within a fatigue region, a close approximation can be made of the number of load cycles applied to a part after a crack initiates. Jacoby [5] has described a technique whereby the total number of striations (cycles) can be determined from a crack growth plot based on striation spacing measurements. The number of striations per inch is measured at intervals from the origin to the final fatigue depth from SEM fractographs, Fig. 7. The equation of the line resulting from a plot of the number of striations per inch versus crack length is integrated giving the number of striations (cycles) versus crack length. The information that can be obtained from a crack growth plot, such as this, may be helpful in determining inspection intervals for service components as well as replacement times. Obviously, a steep crack growth curve showing that once a crack initiates, only a short time is required to propagate the crack to final size, or one having an inflection point indicating a sudden increase in crack growth rate would be of concern, Fig. 8.

2) Overload — A fracture surface produced by overload also offers a distinct feature which is referred to as dimples. Dimples are related to ductile fracture and are a concave structure resulting from the fracturing of many small voids that initiate and grow in the process of plastic deformation. As with striations in fatigue, there is also much information about an overload failure that can be gained by proper interpretation of the dimple features. Figure 9 shows the characteristic shape of dimples that are associated with types of overload fracture. When a fracture occurs by tensile overload, the fracture plane is perpendicular to the loading axis which produces a dimple shape that is predominantly equiaxed (A). When the fracture plane of an overload fracture is other than perpendicular

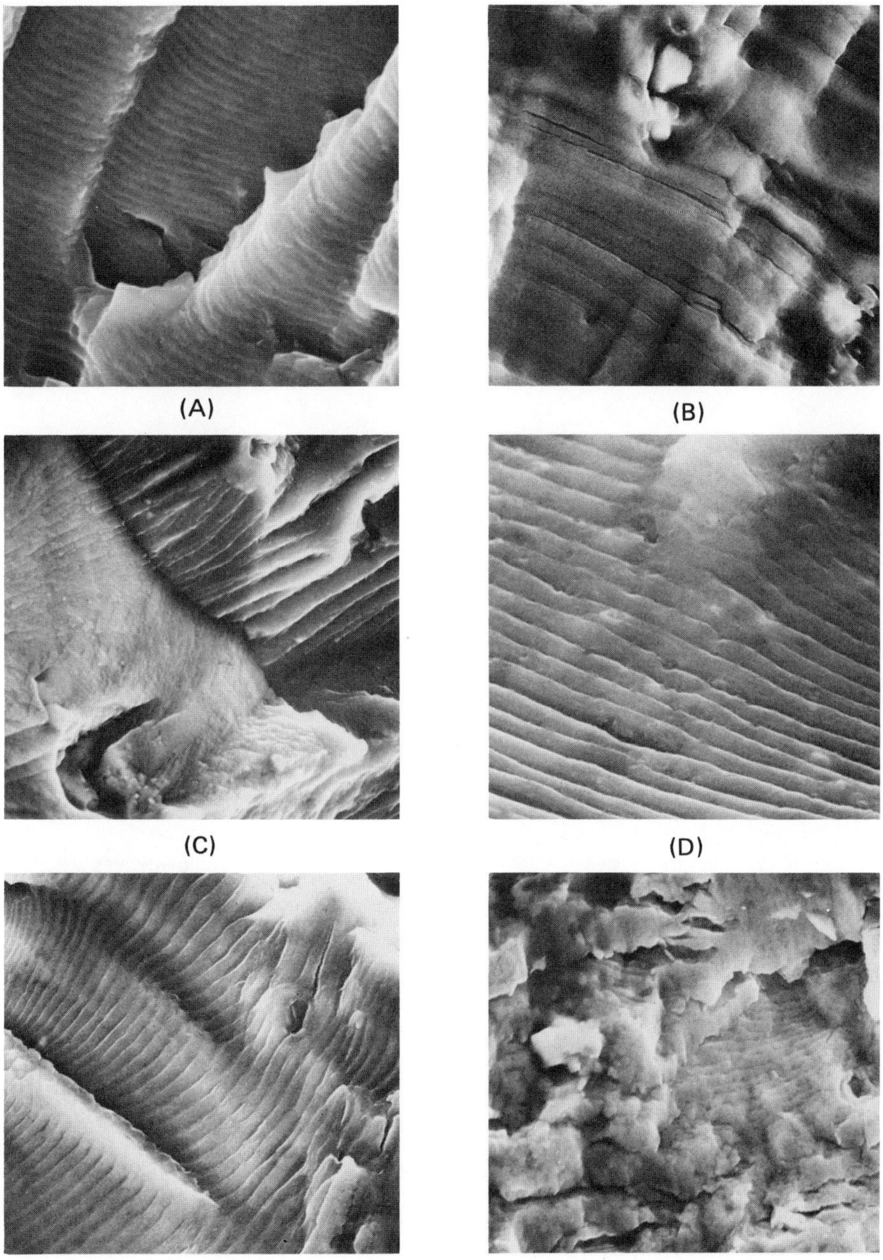

Fig. 6. Fatigue striations resulting from different loading conditions. (A) Constant amplitude, Original Mag. 3000X, (B) Random loading, Original Mag. 5000X, (C) High cycle, Original Mag. 2000X, (D) Low cycle, Original Mag. 2000X, (E) Tension-tension loading, Original Mag. 3000X and (F) Compression loading, Original Mag. 5000X. (Reduced approximately 40% for reproduction).

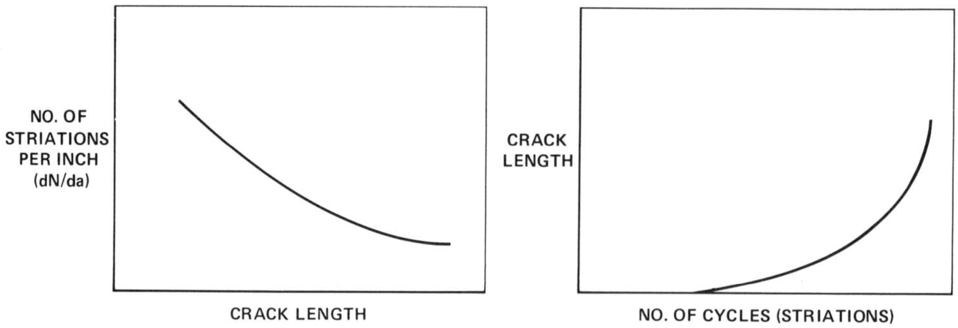

Fig. 7. Crack growth curves determined from striation spacing.

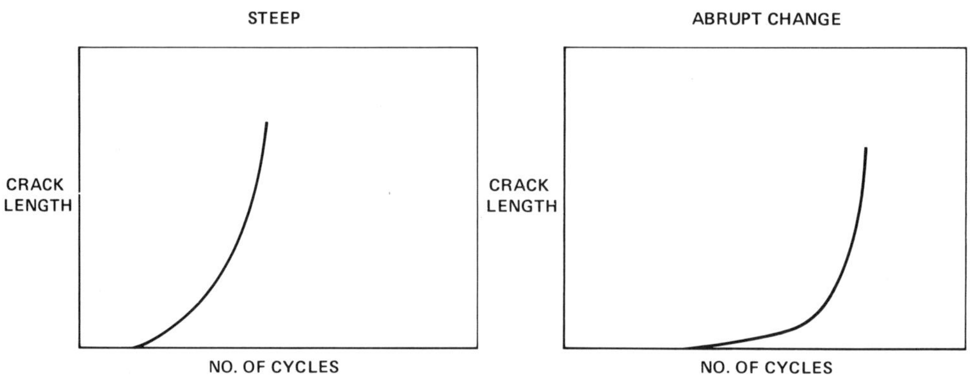

Fig. 8. Crack growth curves showing a steep growth rate and an abrupt increase in crack growth rate.

to the stress axis, such as with shear, torsion and tear, the dimple features are elongated in a parabolic shape oriented with the axis of the resolved stress. In fracture by shear, the dimples are aligned along the fracture plane pointing in opposite directions (B) on each fracture surface. The dimple pattern of a torsion feature also points in opposite directions on each fracture, but follows the curved fracture plane (C). When fracture occurs by tear overload, the dimples on both fracture halves point in the same direction, opposite to the fracture direction (D).

Dimples are a common feature found on most fracture surfaces of failed parts, even though the primary fracture mode was other than overload. They are the primary feature of the rapid fracture region that extends from the slow growth portion of the fracture resulting from some other fracture mode. In cases where the fracture origin of a failed part is not well distinguished from the rapid fracture region on a macro or light microscopic scale, dimple features can

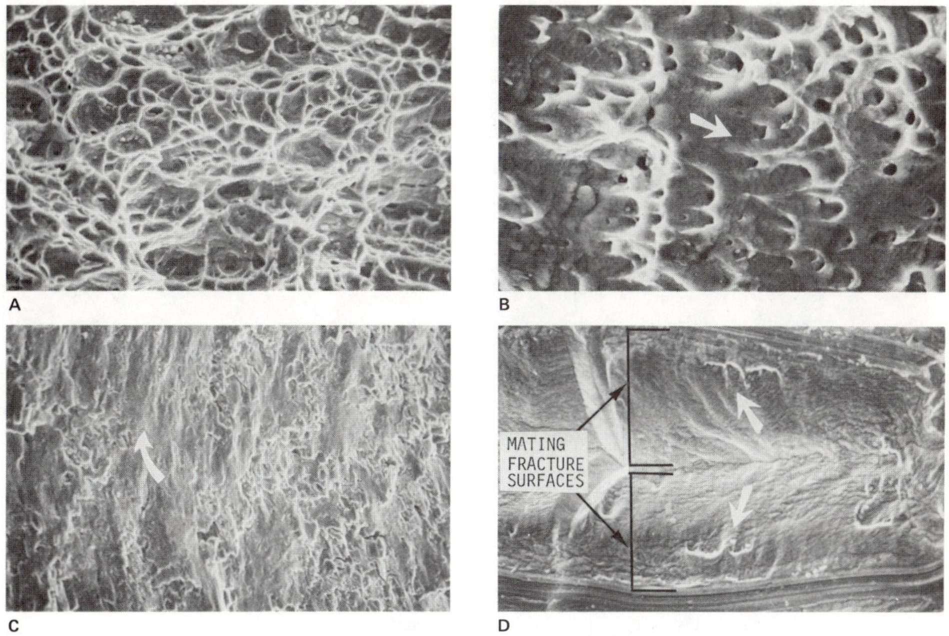

Fig. 9. Characteristic shapes of dimples that are associated with various types of overload fracture. (A) Tensile, Original Mag. 1000X, (B) Shear, Original Mag. 1800X, (C) Torsion, Original Mag. 500X, (D) Tear, Original Mag. 500X. (Reduced approximately 40 percent for reproduction).

be helpful in determining the location and extent of the slow growth region. Figure 10 shows an example of such a condition where an ill-defined fatigue region in 300M steel was revealed with the SEM by observing the change from dimple features resulting from rapid fracture to somewhat smoother transgranular texture characteristic of fatigue in high strength steel.

The dimple features of a rapid fracture region can also be of help in determining the direction of crack propagation. The elongated dimples that occur in the rapid fracture region of thin sheet materials are oriented with the apex of the parabola pointing in the opposite direction of crack propagation [6]. This technique can be of particular benefit in determining the origin of failure of thin-wall components that do not reveal macro features to determine fracture direction. Figure 11 shows an example of a failed thin-wall aluminum alloy hat section in which the fracture origin was determined from the dimple orientation.

Although dimples are the one feature commonly associated with overload failure and the rapid fracture region, they are not the only features present on these types of fractures. In addition to large variations in dimple size, other prominent fracture surface features, such as cleavage and intergranular fracture, may also be evident and even more extensive than dimples. An example of variation in overload features is shown in Fig. 12 where a cold-worked 70-copper, 30-zinc brass that failed in overload showed a distinct difference in fracture surface appearance when compared to an overload fracture of the same material in the

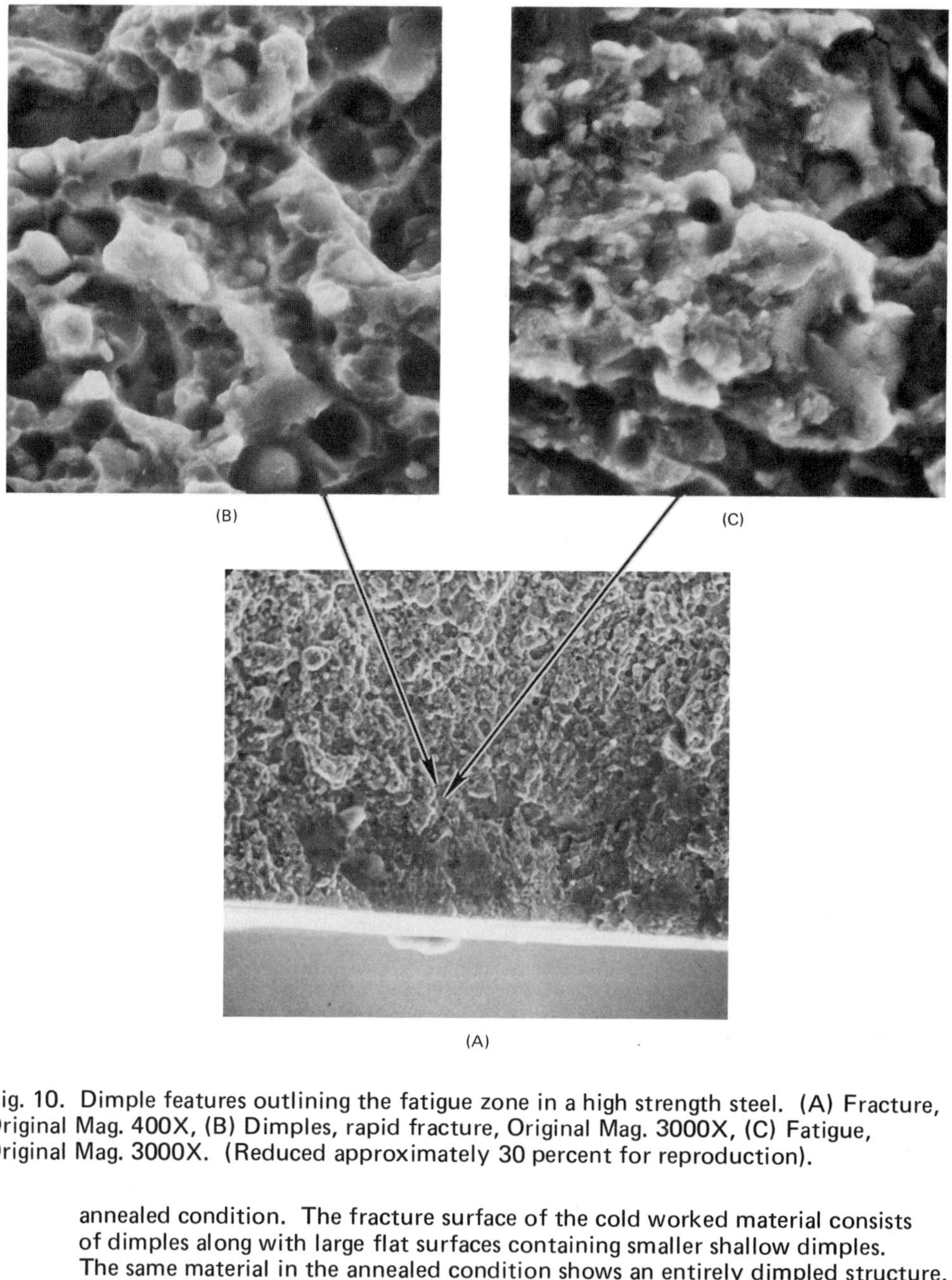

Fig. 10. Dimple features outlining the fatigue zone in a high strength steel. (A) Fracture, Original Mag. 400X, (B) Dimples, rapid fracture, Original Mag. 3000X, (C) Fatigue, Original Mag. 3000X. (Reduced approximately 30 percent for reproduction).

annealed condition. The fracture surface of the cold worked material consists of dimples along with large flat surfaces containing smaller shallow dimples. The same material in the annealed condition shows an entirely dimpled structure. Another factor that can affect the extent of a dimple formation is the material form, that is, a rolled sheet, forging or a casting. Figure 13 shows the fracture surfaces of three different aluminum alloy tensile specimens. The sheet specimen having a small elongated grain structure is entirely dimpled (A); the specimen from a thick forging with a large grain size shows a mixture of different size dimples (B); and the cast specimen having a dendritic microstructure reveals a

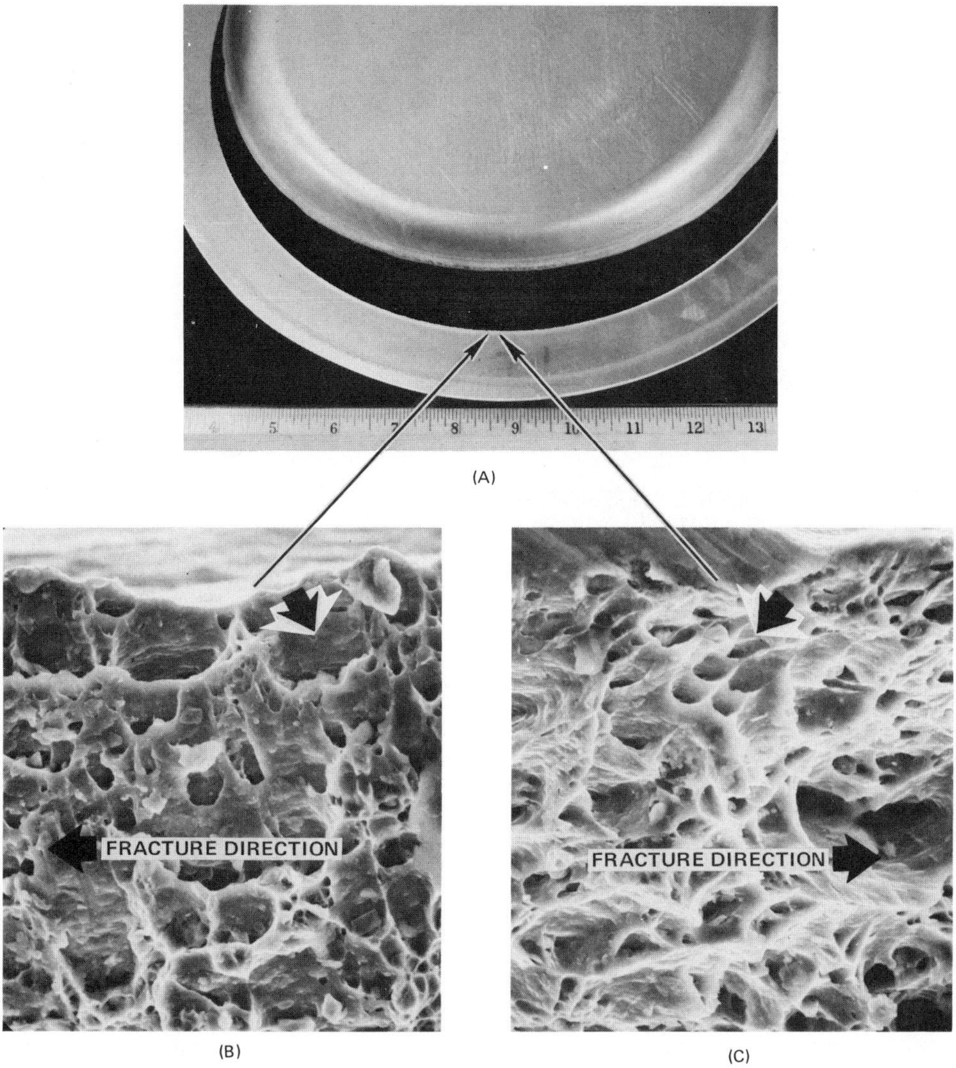

Fig. 11. Dimple features revealing the fracture direction and fracture origin on a failed thin wall hat section. (B) and (C) Original Mag. 600X. (Reduced approximately 30 percent for reproduction).

brittle appearance in which failure occurred along interdendritic paths (C). The heat-treat condition of some alloys can also result in different fractographic features of overload fracture. Figure 14 shows the dimple structure of a titanium specimen annealed below the beta transformation temperature (A) and above the beta transformation temperature (B). The fracture surfaces showed distinct differences. Uniform equiaxed dimples were predominant over the entire surface of the specimen annealed below the beta temperature, while large flat surfaces, containing small shallow dimples, were mostly evident on the fracture surface of the specimen annealed above the beta temperature.

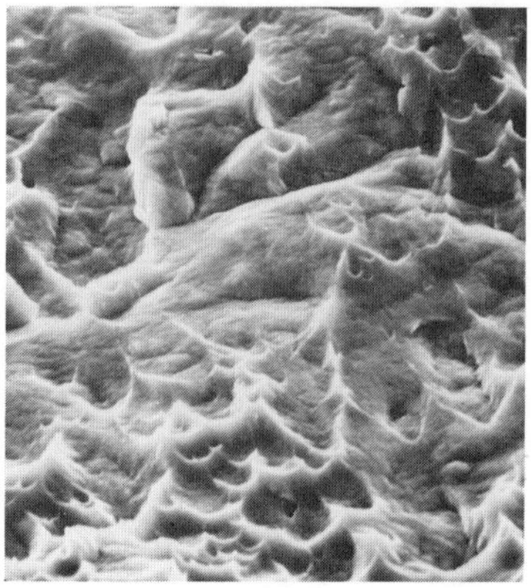

(A)

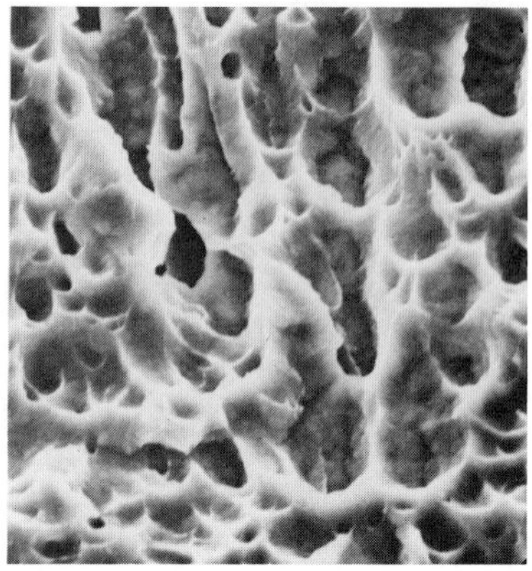

(B)

Fig. 12. Fracture surface appearance of overload failure in a cold-worked and annealed 70 copper—30 zinc brass. (A) Cold-worked, Original Mag. 3000X and (B) Annealed, Original Mag. 3000X. (Reduced approximately 45 percent for reproduction).

From this discussion of fractographic features related to overload and rapid fracture, it should be evident that knowledge of the material form and heat treatment can be beneficial in making proper fractographic interpretation. This information can usually be obtained through metallographic examination and hardness measurements. Another technique that is also helpful in making a fractographic interpretation is to produce a known overload fracture in the failed part. The fracture surface features of the known overload failure observed macroscopically and with the SEM can then be compared with the primary fracture surface of the service failure.

3) *Delayed Failure* — This type of failure is a result of one or more crack fronts that develop and propagate over a period of time under static load condition. Two common fracture modes in this category are stress corrosion and cracking (SCC) and hydrogen embrittlement (HE). Both are considered brittle fracture and evidence plastic deformation only in the rapid fracture region of the failure. Unlike fatigue and overload, not all materials are susceptible to delayed failures. Some materials are more susceptible to stress corrosion cracking or hydrogen embrittlement than others. Some alloys, however, are susceptible to both. For example, aluminum alloys are susceptible to SCC but not to HE; titanium alloys, except at elevated temperature, are more prone to fail by HE than SCC; steels, on the other hand, are susceptible to both HE and SCC.

The brittle fractographic features of SCC and HE fracture surfaces, as observed with the SEM, display either an intergranular or cleavage appearance as shown in Fig. 15. This type of feature, however, is not unique to SCC or HE as are striations with fatigue or dimples with overload. There can be other reasons why a material fails by an intergranular or cleavage fracture mode. In steel, for example, in which SCC and HE are intergranular, a case-hardened layer, grinding cracks and quench cracks will also reveal an intergranular fracture surface appearance. Aluminum alloys exhibit an intergranular appearance in SCC and also show a similar fracture appearance when failure is due to eutectic melting. Titanium, unalloyed or alloyed to form an all alpha or alpha-beta microstructure, is probably one of very few materials that reveals a characteristic feature that can distinguish HE from SCC. Titanium hydrides, small rod-shaped

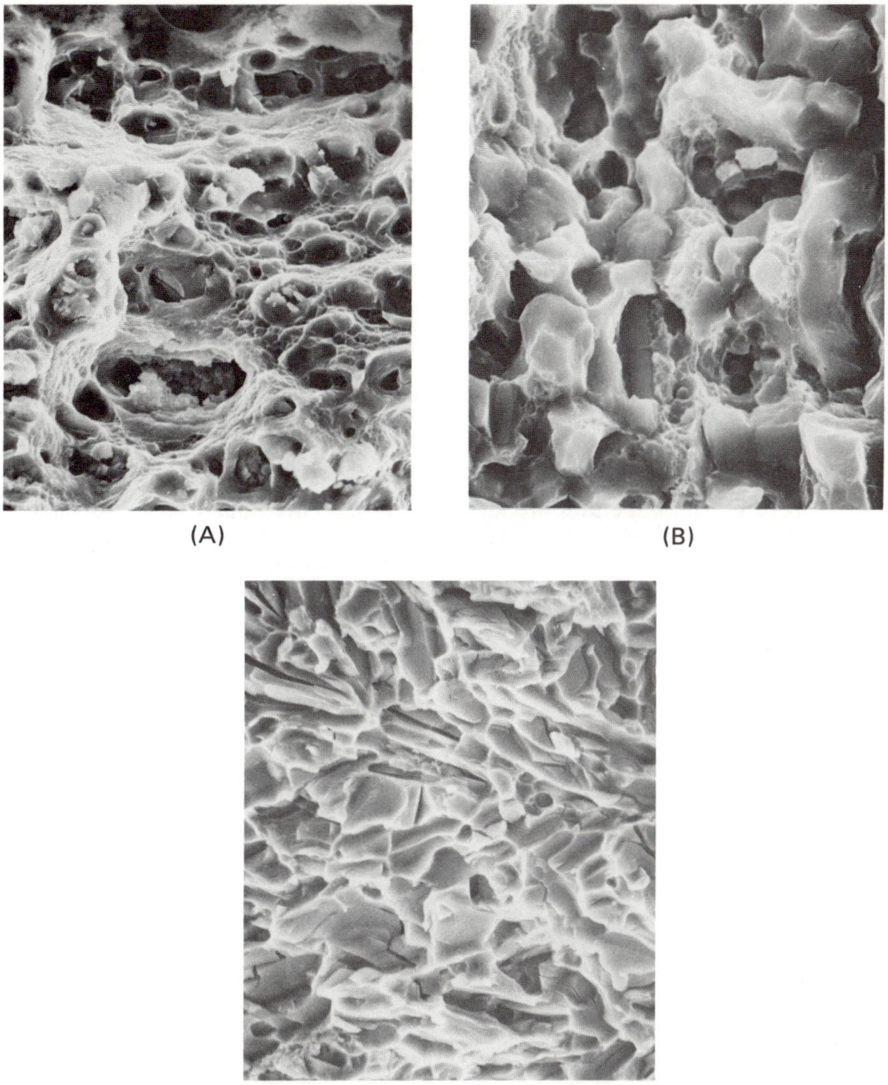

(A) (B) (C)

Fig. 13. The fracture surface appearance of three aluminum alloy tensile specimens. (A) Sheet Material, Original Mag. 1200X, (B) Forging, Original Mag. 2000X and (C) Casting, Original Mag. 500X. (Reduced approximately 30 percent for reproduction).

particles, precipitate in a preferred orientation within the titanium grains, resulting in a drastic loss in ductility and notched tensile strength. Although the particles are not easily detected in an SEM fractograph, when their concentration increases sufficiently, fracture occurs and the fracture surface has a brittle transgranular appearance consisting of a step-like structure, Fig. 16.

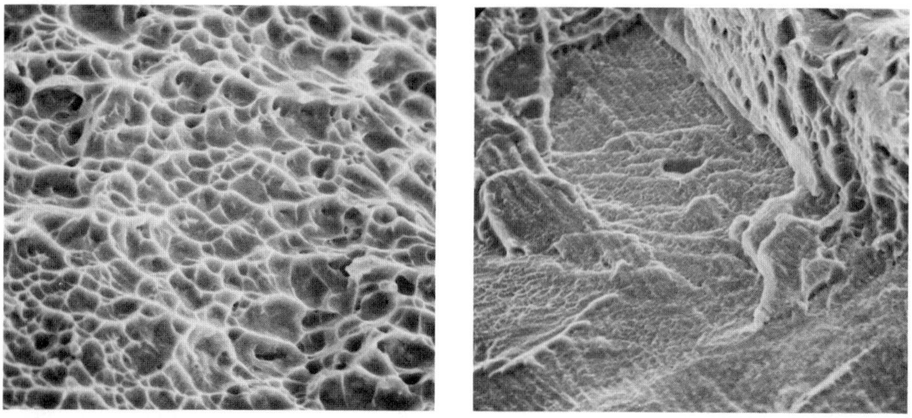

(A) (B)

Fig. 14. Fracture surface features produced by overload in titanium 6Al–4V alloy annealed (A) above and (B) below the beta transformation temperature. (A) and (B) Original Mag. 1000X. (Reduced approximately 37 percent for reproduction).

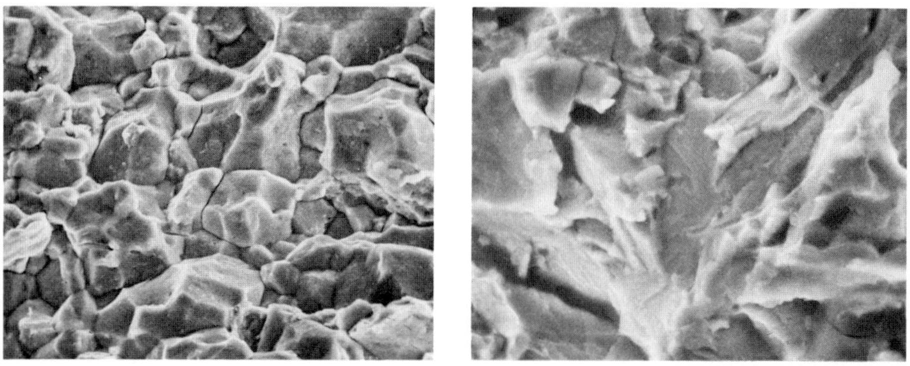

(A) (B)

Fig. 15. Brittle fractographic features of SCC and HE. (A) Intergranular, Original Mag. 1800X and (B) Transgranular cleavage, Original Mag. 1800X. (Reduced approximately 35 percent for reproduction).

SCC in titanium resulting from hot-salt exposure also produces a brittle fracture appearance. The appearance of the fracture surface produced by this fracture mode is dependent upon the microstructure resulting from the applied heat treatment. As shown in Fig. 17, fracture surfaces of hot-salt SCC produced

(A) (B)

Fig. 16. Brittle transgranular appearance of an unalloyed titanium fracture resulting from hydrogen embrittlement. (A) Original Mag. 500X and (B) Original Mag. 2000X. (Reduced approximately 22 percent for reproduction).

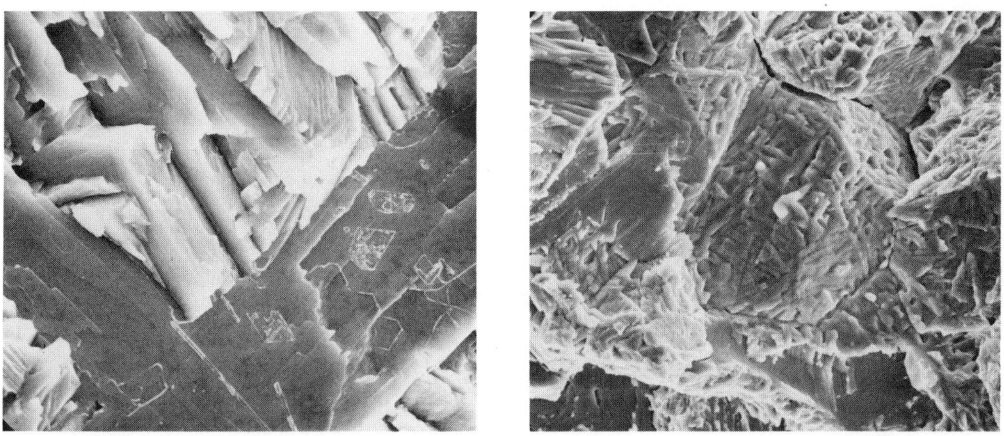

(A) (B)

Fig. 17. Fracture surfaces of hot salt SCC failures in titanium 6Al–4V annealed (A) above and (B) below the beta transformation temperature. (A) Original Mag. 1800X and (B) Original Mag. 1000X. (Reduced approximately 35 percent for reproduction).

in titanium 6Al–4V annealed above the beta transformation temperature reveal a brittle-plate-like structure, whereas the same material annealed below the beta transformation temperature results in a fracture having an intergranular appearance. In neither case will titanium hydrides be evident.

Because of the similarity in fracture surface features of delayed fracture and other modes of failure, extreme care must be used in SEM fractographic interpretation. In order to identify the type of delayed failure service history, fabrication processes and the combined test results of different analyses, such as light microscope fracture surface examination and metallographic examination, may have to be utilized. For example, time-in-service failures resulting from HE usually occur in a shorter period of time than SCC failures. This is because hydrogen absorption is not usually related to service conditions (exclusive, of course, to pure hydrogen environment) but to a previous processing of a material; therefore, upon loading the embrittling process begins immediately. Common processes that can result in hydrogen absorption are welding, acid pickling, and electrolytic plating. SCC, unless in a highly aggressive environment, will take time to initiate because before exposure of the material to a corrosive environment can occur, a breakdown of a protective coating or component must take place. Corrosion products on the fracture surface are another important factor; the fracture surface of steel parts failed due to SCC will usually evidence a corroded appearance, the amount depending on the severity of the environment and the time in service. Pitting, a common form of corrosive attack, is often evident and can be responsible for initiation of SCC as shown in Fig. 18. Some caution, however, should be exerted when associating the amount of corrosion products with SCC. As shown in Fig. 19, an SCC failure in a 4340 steel heat-treated to a 260/280 ksi ultimate strength shows a clean fracture surface appearance, while an HE failure in a steel fastener reveals a corroded appearance. Corrosion occurred on the HE failure as a result of handling of the fastener after and prior to examination.

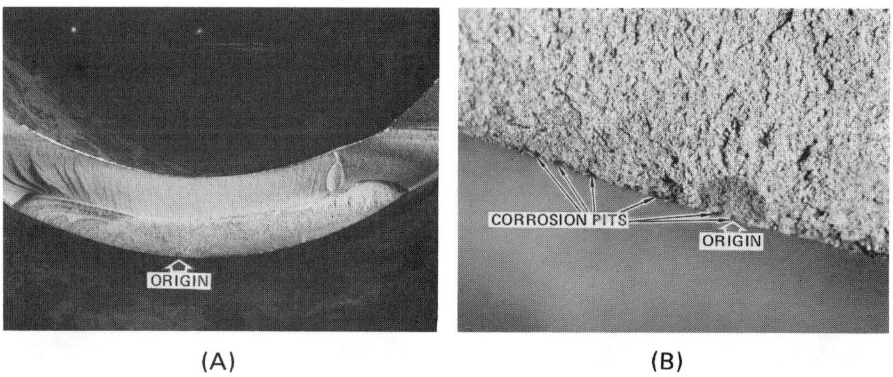

Fig. 18. SCC in a high strength steel initiating from corrosion pits. (A) Original Mag. 1.3X and (B) Original Mag. 12X. (Reduced approximately 45 percent for reproduction).

The macroscopic and microscopic fracture surface appearance of a delayed failure, as observed through a light microscope, also reveals features that allow identification of this mode of failure. A relatively clean fracture surface of a delayed failure of a steel alloy failed by either SCC or HE will reveal shiny facets within the slow growth region that result in a somewhat speckled appearance. The shiny facets are apparently reflections of grains exposed by the fracture. Figure 20 shows this speckled appearance in the slow growth region of a HE failure of high strength steel.

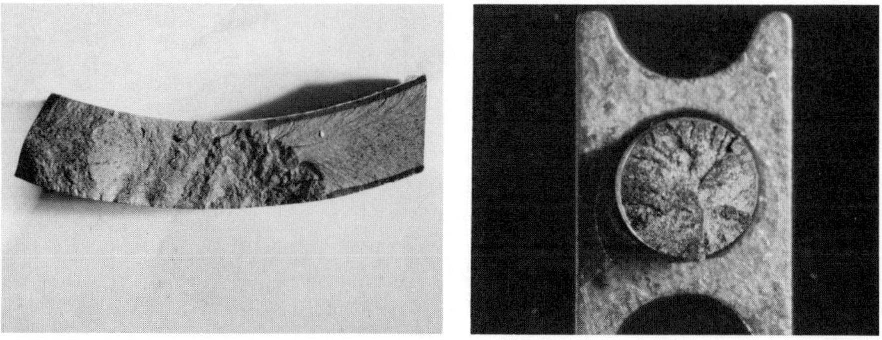

(A) (B)

Fig. 19. Photographs showing (A) the clean fracture surface appearance of a SCC failure and (B) a discolored corroded appearance of a HE failure in steel. (A) Original Mag. 2X and (B) Original Mag. 5X. (Reduced approximately 45 percent for reproduction).

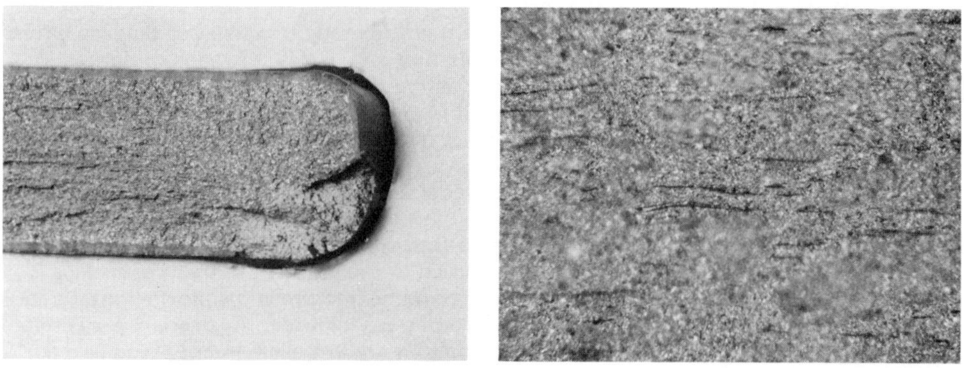

(A) (B)

Fig. 20. A clean fracture surface within the slow growth region of a delayed failure showing a shiny speckled appearance. (A) Original Mag. 6X and (B) Original Mag. 50X. (Reduced approximately 45 percent for reproduction).

The shape of the slow growth region is another important feature of a delayed failure. When the slow growth region of a delayed failure is bordered by a shear lip, a subsurface fracture origin is indicated which is a characteristic of a failure due to HE. The slow growth portion of a fracture surface resulting from SCC will reveal that the failure initiated from the surface of a part by showing no evidence of a shear lip along that surface. Figure 21 presents photographs of fracture surfaces showing the slow growth fracture region of an HE failure having a subsurface origin bordered by a shear lip and of an SCC failure showing that the fracture initiated at the surface of the part.

(A) (B)

Fig. 21. Photographs showing the slow growth fracture surface of a (A) HE and (B) SCC failure. (A) Original Mag. 16X and (B) Original Mag. 12X. (Reduced approximately 45 percent for reproduction).

A metallographic examination of a failed part can reveal important microstructural information that can also aid in interpreting electron fractographic features, as well as help identify the fracture mode of a delayed failure. Often the fracture surface of a service failed part is so badly corroded or contaminated that even after extensive cleaning, characteristic features related to a fracture mode cannot be resolved with the SEM to determine whether the part failed in a brittle intergranular or transgranular manner. With failures such as these, the only method of determining the fracture path is through metallographic examination of a cross section that intersects the fracture surface at the failure origin.

Aluminum alloys which fail as a result of SCC are known to reveal branching secondary intergranular cracks as shown in Fig. 22. If the fracture surface is not too corroded, the profile of the primary fracture will follow an intergranular path. In aluminum alloys this can be of particular help since grain size is relatively large and intergranular fracture features are not always easily recognized during SEM examination.

Metallographic examination of high-strength steel parts that have failed by SCC do not always reveal evidence of the characteristic secondary branching crack. This, of course, makes it difficult to separate SCC from HE by metallographic examination. Figure 23 shows the profile of the resulting fracture surfaces of both modes of delayed failure. In each case, the fracture profile is intergranular with little evidence of secondary cracks.

An examination of the microstructure of titanium alloys, on the other hand, can benefit interpreting fractographic features, as well as distinguishing HE from SCC. As previously discussed, the characteristic feature of HE in titanium alloys is the precipitation of titanium hydrides which are not easily identified on a fracture surface when examined with SEM. They are, however, easily observed in the microstructure of polished and etched sections adjacent to the fracture surface. Figure 24 shows a typical microstructure of an unalloyed titanium failure resulting from HE. The rod-shaped titanium hydrides show a general alignment with the fracture path.

Fig. 22. A photomicrograph of SCC in an aluminum alloy showing an intergranular fracture profile and secondary intergranular branching cracks. Original Magnification 100X. (Reduced approximately 30 percent for reproduction).

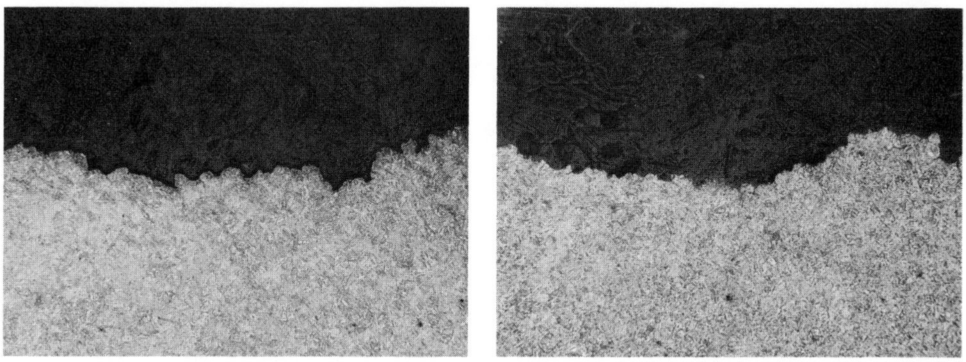

(A) (B)

Fig. 23. Photomicrographs showing the intergranular fracture profile of high strength steel failed by (A) HE and (B) SCC. (A) and (B) Original Magnification 200X. (Reduced approximately 45 percent for reproduction).

The differences in fracture surface appearance that result from hot-salt SCC in titanium solution treated above and below the beta transformation temperature, can also be explained by a microstructural analysis. Figure 25 shows the microstructures of the two specimens that failed due to hot-salt SCC shown in Fig. 16. The microstructures are as distinctly different as the SEM fractographic features. The microstructure of the specimen annealed above the beta transformation temperature shows plate-like alpha grains, while the specimen annealed below the beta transformation temperature reveals an equiaxed alpha grain structure.

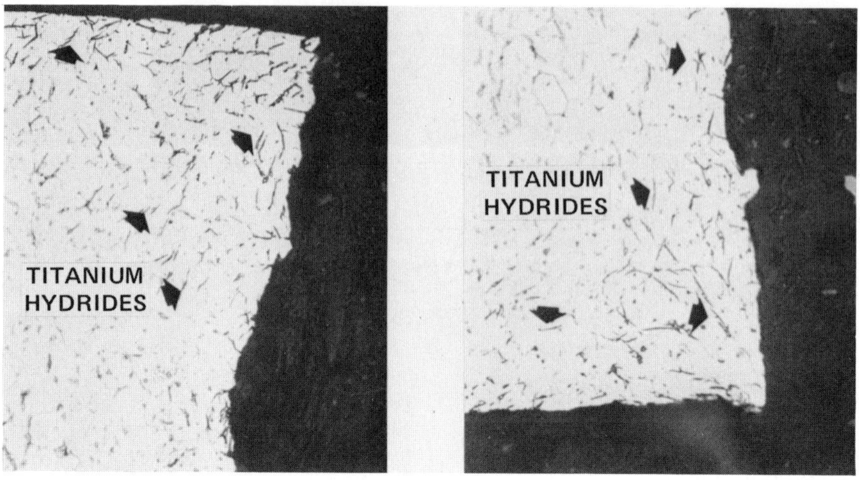

Fig. 24. Microstructure of unalloyed titanium showing evidence of titanium hydrides. Original Magnification 400X. (Reduced approximately 20 percent for reproduction).

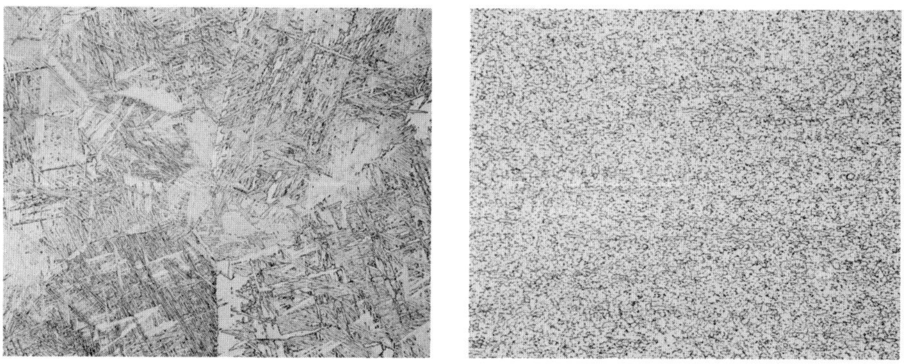

(A) (B)

Fig. 25. Microstructure of hot salt SCC specimens annealed (A) above and (B) below the Beta transformation temperature. Original Magnification 200X. (Reduced approximately 45 percent for reproduction).

SUMMARY

In summary, the following section presents two failure analysis investigations that illustrate the value of the SEM for fractographic examination and show the importance of correlating results of other tests with SEM results to ultimately determine the cause of a failure.

EXAMPLE 1 – FAILURE ANALYSIS OF A HYDRAULIC TUBE

The objective of this examination was to determine the mode and initiating cause of failure of a coiled hydraulic tube. The tube, fabricated from titanium 3 aluminum 2.5 vanadium alloy, contained a crack on the inside diameter of the coil that extended circumferentially approximately 160 degrees (see arrow in Fig. 26). A dye penetrant inspection was initially performed to determine if any secondary cracks were present. Crack indications were detected on both sides of the primary crack. The indications were small and were located within the encircled areas of the tube as shown in Fig. 27. An examination of the crack indications with a light microscope showed evidence of smeared metal, typically observed for a scratch or gouge, Fig. 28. The tube was also inspected optically, and small scratches were evident randomly dispersed over the entire surface. Figure 3A shows a photograph of a typical area containing scratches away from the primary crack location. The scratches were all oriented in a circumferential direction, suggesting that they may have resulted from a belt polishing or buffing operation.

The fracture surface of the primary crack was exposed and is shown in Fig. 29. A relatively smooth textured fracture surface with well-defined crack arrest lines (beach marks) was evident, characteristic of a failure resulting from fatigue. The concentric beach marks that resulted from variations in crack growth rate indicated a single point of origin.

SEM Examination

A section of the tube containing the fracture surface and the dye penetrant indications was removed for examination with the SEM. SEM photographs that document the fracture surface features observed are shown in Figs. 30 and 31. The fracture origin, shown in Fig. 30, shows a deformed region at the o.d. surface, indicating that fatigue initiated from a scratch mark. Striations evident on the fracture surface confirmed fatigue as the mode of failure, Fig. 31.

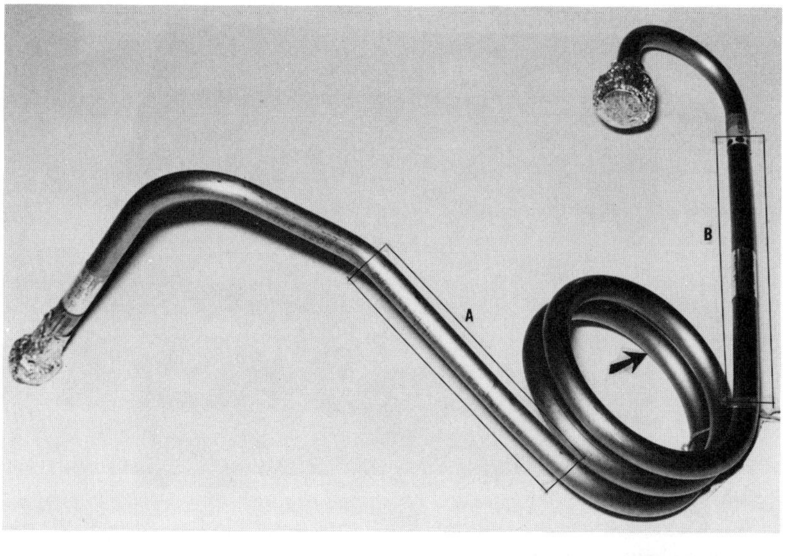

Fig. 26. Failed hydraulic tube as received for examination.

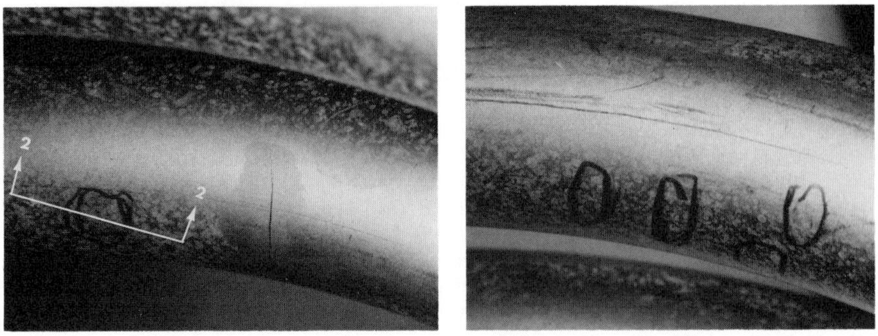

(A) (B)

Fig. 27. Closeup view of hydraulic tube showing the primary crack and the location of dye penetrant crack indications (circled). (A) Original Mag. 4.25X and (B) Original Mag. 5X. (Reduced approximately 50 percent for reproduction).

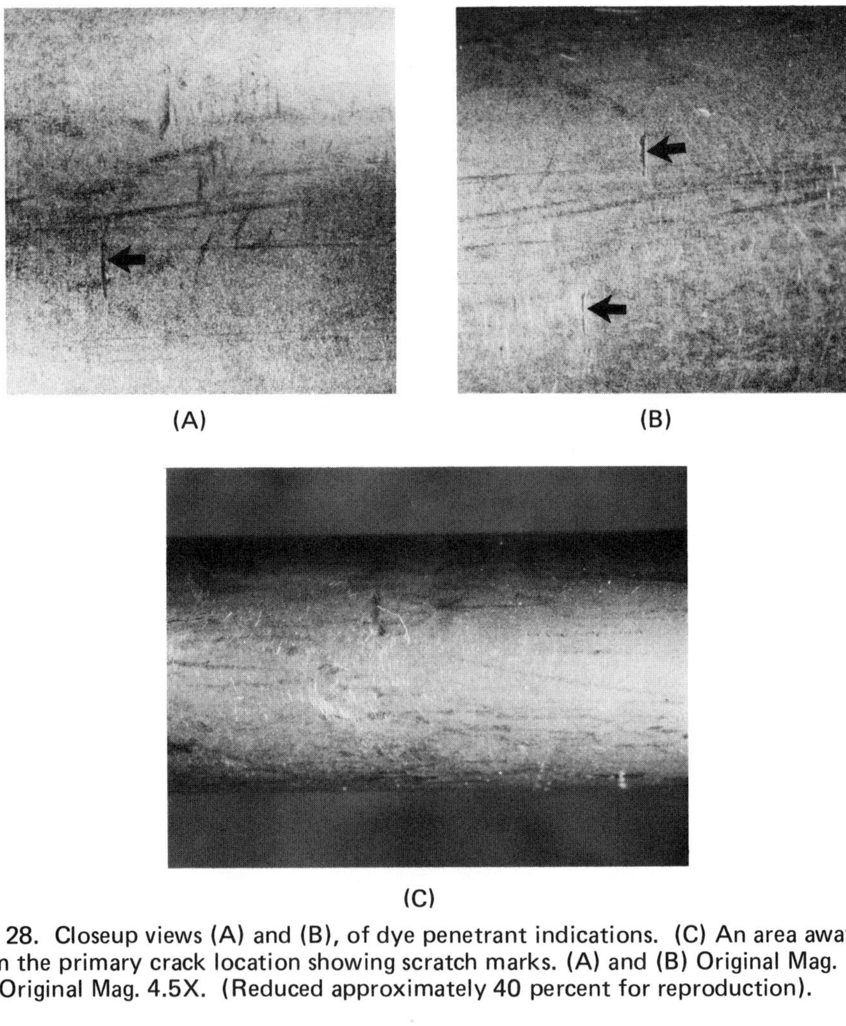

(A) (B)

(C)

Fig. 28. Closeup views (A) and (B), of dye penetrant indications. (C) An area away from the primary crack location showing scratch marks. (A) and (B) Original Mag. 12.5X, (C) Original Mag. 4.5X. (Reduced approximately 40 percent for reproduction).

(A) (B)

Fig. 29. Exposed fracture surface of primary crack. Arrow points to fatigue origin. (A) Original Mag. 5X and (B) Original Mag. 10X. (Reduced approximately 45 percent for reproduction).

(A) (B)

Fig. 30. SEM photographs of the fracture origin. (A) Arrow points to fracture origin. Original Mag. 45X. (B) Arrow points to scratch mark. Original Mag. 300X. (Reduced approximately 35 percent for reproduction).

The surface of the tube adjacent to the fracture surface was also examined in detail with the SEM. Several scratch marks were observed, similar to those shown in the light microscope examination, Fig. 32. The surface appearance of the marks appeared somewhat irregular, possibly resulting from pickling of the tube, the final step in fabrication.

Metallographic Examination

A metallographic examination was performed at the fracture origin (Section 1–1, Fig. 29) and also at one of the dye penetrant indication sites (Section 2–2, Fig. 27). The microstructure at the fracture origin, shown in Fig. 33 appeared normal for Ti–3Al–2.5V in the cold-worked, stress-relieved condition. In Section 2–2 Fig. 33, a secondary crack was present that initiated at a scratch mark on the surface of the tube. The crack, having propagated in a transgranular manner and perpendicular to the surface of the tube, is believed to be a secondary fatigue crack.

(A) (B)

Fig. 31. SEM photographs of the fracture surface showing fatigue striations, (A) near origin and (B) away from origin. Original Magnification 3000X. (Reduced approximately 35 percent for reproduction).

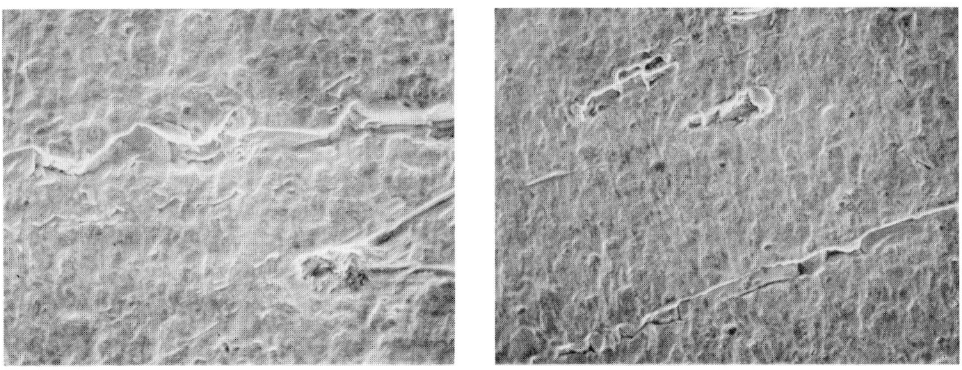

(A) (B)

Fig. 32. SEM photographs showing scratch marks on the tube surface near the fracture origin. (A) Original Mag. 300X and (B) Original Mag. 240X. (Reduced approximately 45 percent for reproduction).

Mechanical Properties

Two straight sections of the tube (A and B) were removed from the locations shown in Fig. 26 for determination of mechanical properties. The results of this test, presented in the table below, show that the mechanical properties of neither specimen met the requirements. Both ultimate (Ftu) and yield (Fty) strength were below the minimum requirements.

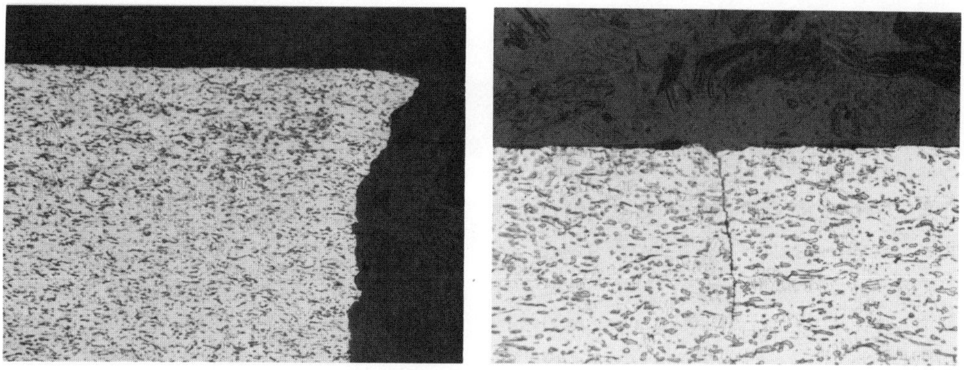

(A) (B)

Fig. 33. Photomicrographs showing the microstructure at the primary fracture location and at secondary crack. (A) Section 1—1 and (B) Section 2—2. (A) Original Mag. 100X and (B) Original Mag. 400X. (Reduced approximately 35 percent for reproduction).

Specimen	Ftu (ksi)	Fty (ksi)	% Elongation
A	111.5	91.7	13.0
B	112.5	92.3	11.0
Material Specification	125.0 (min)	105.0 (min)	10.0 (min)

Conclusion

Based on the results of this examination, it was concluded that failure of the coiled hydraulic tube was due to fatigue. The initiating cause of fatigue was a scratch mark which was believed to have occurred during fabrication of the tube; it was also shown that the tube material did not meet the specified mechanical property requirements.

EXAMPLE 2 — FAILURE ANALYSIS OF A HYDRAULIC CYLINDER

The objective of this examination was to determine the mode and initiating cause of failure of a hydraulic cylinder. The cylinder was die forged from 7075—T73 and sulfuric acid anodized on the bore surface.

A crack resulting in an oil leak was located near one end of the cylinder, Fig. 34, and was marked by cracked and lifted paint along one edge of the parting plane of the forging. A borescope inspection revealed a large crack at the inside anodized surface opposite the lifted paint, and oriented parallel to the long axis of the cylinder.

After opening the crack, an examination of the fracture revealed a classic smooth-textured, parabolic-shaped region, typical of fatigue, Fig. 35. The radiating fracture ridges indicated that the crack initiated at the inside surface of the cylinder at a flat, featureless area.

SEM Fractography and X-Ray Analysis

The fracture and the adjacent inside surface of the cylinder was then examined

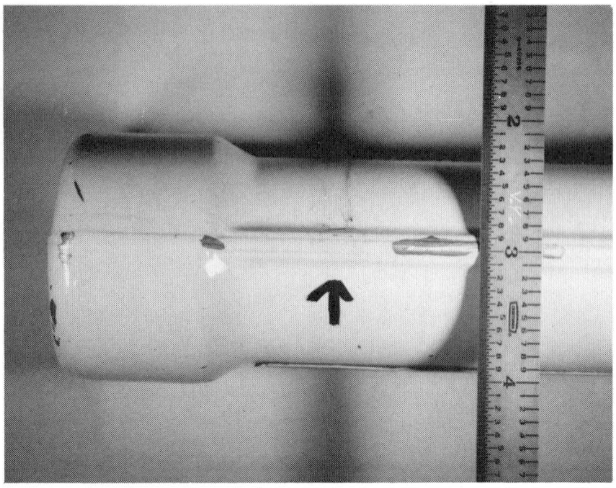

Fig. 34. The closed end of the leaking cylinder. The arrows indicate the crack in the paint where the leak occurred.

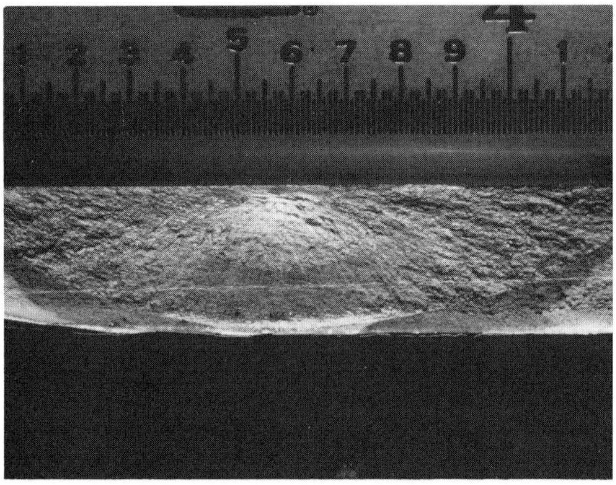

Fig. 35. The fracture appearance of the crack.

in the SEM. The general area of the fracture origin is shown in Fig. 36. The fracture initiated from a region at the inside surface which was covered with a non-conducting, amorphous, striated and occasionally mud-cracked deposit, Fig. 37. The crack propagated from the origin by a combination of fatigue (Figs. 38 and 39) and cleavage/decohesive rupture through stringers of non-metallic inclusions, Fig. 40.

Energy dispersive x-ray analyses of the deposits at the origin, as well as the sulfuric acid anodized inside surface of the cylinder, presented a strong sulfur signal. Sulfur was not detected elsewhere on the fracture. An x-ray map of the general sulfur distribution in the origin region is shown in Fig. 41. The presence of sulfur at the origin indicated that those areas were exposed to the sulfuric acid during the anodizing process. Point scans of the large number of particles exposed on the fractured surface indicated

Fig. 36. Appearance of the fracture origin area. Original Magnification 25X. (Reduced approximately 35 percent for reproduction).

Fig. 37. Striated and mud-cracked deposits at the fracture origin. The lower edge of the fracture is along the inside surface of cylinder. Original Magnification 1000X. (Reduced approximately 35 percent for reproduction).

that they were rich in a combination of Mg, Si, Cu or Fe. All of these elements are characteristic of the various non-metallic inclusions or second phases normally present in this aluminum alloy. Some similar x-ray signatures within the origin suggested that a few of these non-metallic particles were present under the amorphous deposit.

SEM examination of the inside surface of the cylinder adjacent to the fracture origin revealed rows of pits as shown in Fig. 42. The pits appear to be remnants of inclusion stringers, exposed by the machining of the cylinder. There were only a few particles remaining in the pits. Perhaps they were pulled out during the machining or etched out by the subsequent anodizing process.

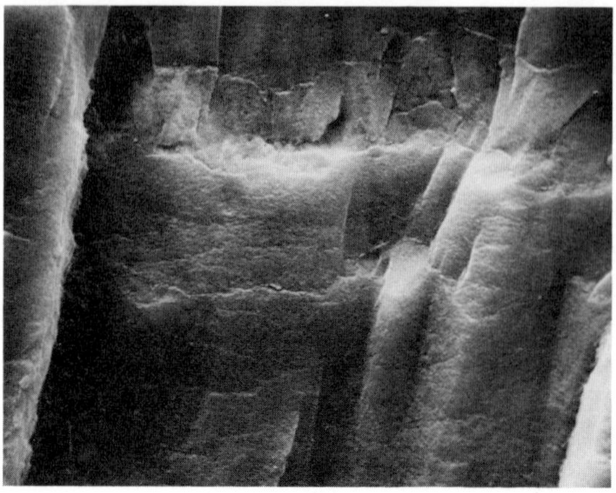

Fig. 38. Fatigue fracture adjacent to origin region. The fatigue striations are very fine and difficult to resolve. Original Magnification 2000X. (Reduced approximately 35 per cent for reproduction).

Fig. 39. Fatigue fracture away from origin region. Fatigue striations are distinct and easily identified. Original Magnification 2000X. (Reduced approximately 35 percent for reproduction).

Metallographic Examination

A metallographic examination was performed on a polished and etched cross-section at the fatigue origin normal to the fracture surface. Photomicrographs representative of the microstructure observed are shown in Figs. 43 and 44. As shown in Fig. 43, there was no abnormal structure or disturbance of the normal grain flow pattern at the fracture origin that may be associated with the initiation of the fatigue crack. The typical microstructure near the fracture origin, shown in Fig. 44, appears normal for this forging.

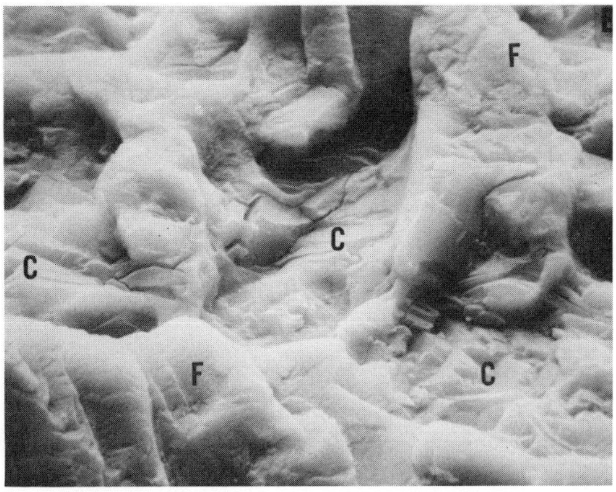

Fig. 40. An area of fragmented particles within the fatigue region. Fatigue fracture (F) and cleaved particles (C). Original Magnification 2000X. (Reduced approximately 35 percent for reproduction).

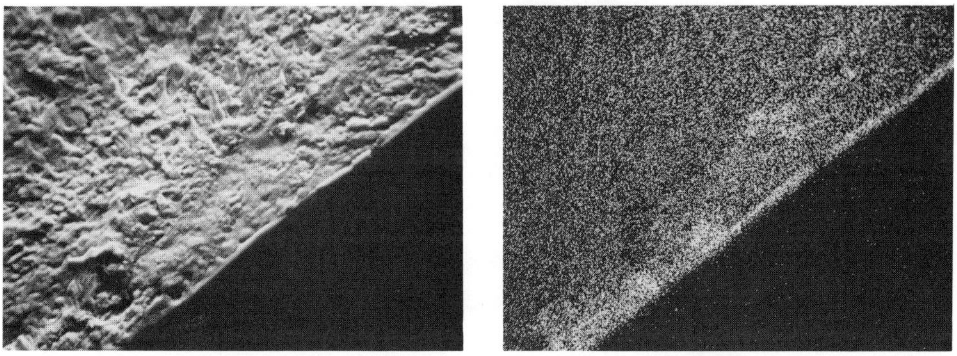

(A) (B)

Fig. 41. Fracture origin region (A) and x-ray map (B) showing sulfur distribution. Original Magnification 350X. (Reduced approximately 45 percent for reproduction).

Conclusion

From the results of this examination, it was concluded that the forged aluminum hydraulic cylinder failed as a result of fatigue and that the fatigue crack initiated from a row of pits that were open to the inside surface of the cylinder during the anodizing process.

ACKNOWLEDGMENT

Acknowledgment is expressed to J.L. Evans for the metallography and R.R. Wilcox and V. Kerlins for scanning electron microscope fractographs.

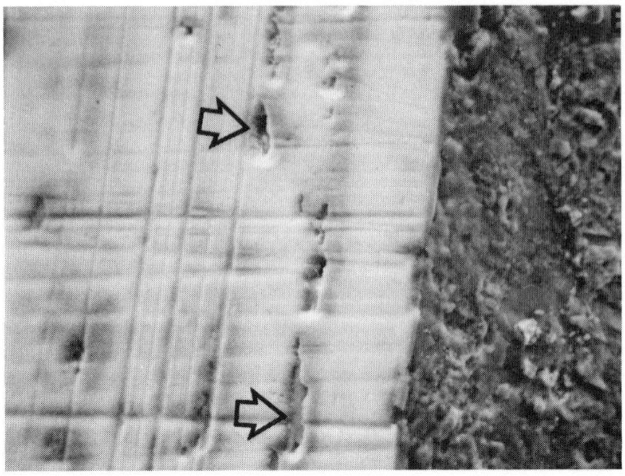

Fig. 42. Appearance of the anodized surface adjacent to the fracture origin. Dark region at right is fracture origin area. Open arrows point to rows of pits at inside surface of cylinder. Original Magnification 600X. (Reduced approximately 35 percent for reproduction).

Fig. 43. Fracture profile at origin. Original Magnification 500X. (Reduced approximately 35 percent for reproduction).

REFERENCES

1. *Metals Handbook, 8th Edition, Volume 9, Fractography and Atlas of Fractographs,* (August 1974).
2. Pittinato, G.F., Kerlins, V., Phillips, A., Russo, M.A., *SEM/TEM Fractography Handbook,* McDonnell–Douglas Astronautics Company, Huntington Beach, California; Published by Metals and Ceramics Information Center, Battelle Columbus Laboratories, Columbus, Ohio, (December 1975).
3. Forsyth, P.J.E., *A Two Stage Process of Fatigue Crack Growth,* Proceedings, Crack Propagation Symposium, Cranfield, England, Volume 1, (1961).

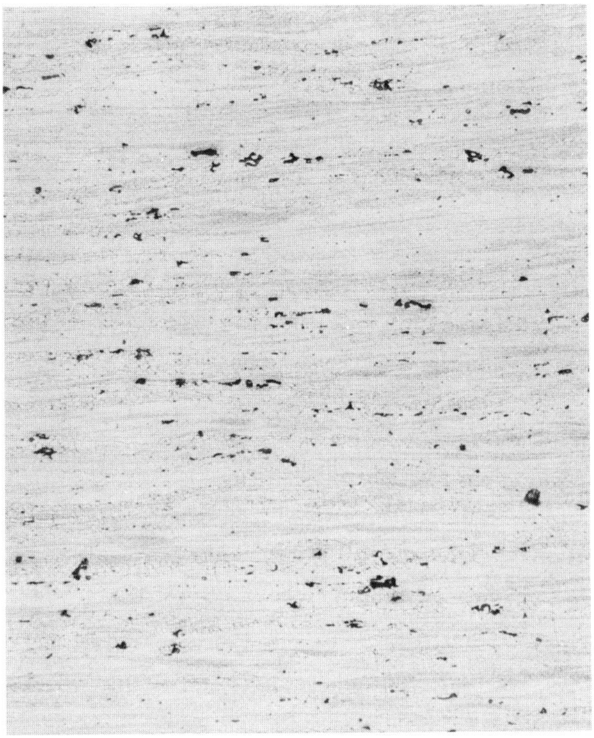

Fig. 44. Typical microstructure near fracture origin. Original Magnification 200X. (Reduced approximately 10 percent for reproduction).

4. Stubbington, C.A. and Forsythe, P.J.E., *On the Metallography and Crystallography of Shear Mode Fatigue Fracture in Aluminum Alloys, Metallurgia,* Volume 74, (July 1966).
5. Jacoby, G., *Fractographic Methods in Fatigue Research, Experimental Mechanics,* (March 1965).
6. Phillips, A., Kerlins, V., Rawe, R.A., Whiteson, B.V., *Electron Fractography Handbook, Specific Applications of Electron Fractography,* AFML–TR–64–416 Supplement 1, (December 1966).

METALLOGRAPHIC CHARACTERIZATION OF FATIGUE FAILURE ORIGIN AREAS

J. Mogul*

INTRODUCTION

Improving product performance, reliability, and cost effectiveness is the surest way of insuring customer satisfaction. High performance machines, however, inevitably experience failures during engineering test or in service unless they are so overdesigned as to be non-competitive. During the last decade industry has realized that failures provide some of the most valuable, albeit costly, sources of engineering information. Materials engineers performing failure analyses have come *out of the closet* and are publishing their findings for the education of the whole engineering community. Publication of the failure analysis experience of many investigators can only help to improve designs, materials selections, quality control, and service engineering. Most important, it will enable us to prevent, as well as analyze, failures.

Metal components can fail by a wide variety of mechanisms. Identifying the primary fracture mode and locating the fracture origin fall within the scope of *fractography,* the first and most powerful diagnostic technique in failure analysis. At this point in an investigation the materials engineer is ready to employ *metallography,* the second most effective weapon in the failure analysis arsenal.

The fracture mechanism most likely to be encountered in a failure analysis is metal fatigue. In fatigue failure analysis the center of attention is the fracture origin, and it is here that the metallographic analysis is focused.

Understanding the basic causes of fatigue in metals and the importance of the fatigue fracture origin is essential to fatigue failure analysis.

FATIGUE

Practically all mechanical systems are subjected to repeated load applications in operation. The loads may be in the form of bending, pulling, pushing, or vibration. The stresses set up in the structure by such loads are usually far below the static fracture strength of the metal. Nevertheless, when such stresses are cyclic in nature, and exceed a critical threshold value, the structure will have a finite cyclic life. That is, after a number of load cycles, varying from the tens to the millions, a crack will develop and propagate. Catastrophic failure can occur if this *fatigue crack* reduces the cross-sectional area to the extent that the structure can no longer support the static load.

---

*Curtiss-Wright Corporation, Wood—Ridge, New Jersey, USA

Except at high temperatures, where other mechanisms interact, fatigue is cyclic rather than time related. Failure can occur after only minutes, or after years of operation. The time to failure is dependent on both the magnitude and rate of cyclic load application.

We can define metal fatigue as the progressive fracture of a structure under the application of repeated or cyclic loads. The stress level necessary to initiate and propagate a fatigue fracture is much lower than that required to cause yielding or fracture under a single load application.

OCCURRENCE OF FATIGUE

Fatigue was recognized as a distinct type of metal failure in the early 1800's in the examination of broken carriage and locomotive axles. Laboratory fatigue tests were being run as early as the 1860's. Many thousands of technical papers have been published on fatigue and hundreds of handbooks contain fatigue properties on almost every type of metal alloy. A reasonable understanding even exists on the micromechanisms of fatigue crack initiation and growth.

Despite this mountain of information, fatigue is by far the most frequent cause of failure in machinery and structures. Figure 1 is a dramatic reminder of this fact. The headline *METAL FATIGUE BLAMED IN CRASH* appeared in a New York newspaper a few days after a helicopter accident at a roof-top heliport.

A reasonable question is why fatigue failures still occur with such regularity, and in so many varied industrial products. One reason is that there are always uncertainties about the loads and environment a part will experience in service. Another is that the scatter in fatigue properties is probably greater than that of most other metal properties. The predominant reason, however, is that complex metal structures may contain hidden or unrecognized areas of stress-concentration or "stress-raisers". The majority of component fatigue failures originate at some form of stress-raiser.

Fig. 1. Dramatic reminder that fatigue is the most frequent cause of failure in machinery and structures.

THE ROLE OF STRESS—RAISERS IN FATIGUE

As the name implies, a stress-raiser is an irregularity in shape, section, or surface which causes the peak stress at or near the irregularity to be higher than the average stress in the surrounding area.

Stress-raisers can be either mechanical or metallurgical. Mechanical stress-raisers involve actual discontinuities or separation in the metal, such as notches and cracks. Metallurgical stress-raisers result from localized differences in properties, such as hard and soft spots.

Most stress-raisers can also be classified by their origin:

- Those resulting from design requirements, such as: holes, corners, keyways, threads
- Those caused inadvertently during fabrication, such as: laps, seams, tears, cracks
- Those carried over from the metal refining process, such as: pipe, inclusions, segregation
- Those incurred as a result of the operating or maintenance environment, such as: corrosion, stress-corrosion, stress-rupture, foreign object damage

A large part of fatigue failure analysis involves the detection and identification of stress-raisers at the fracture origin.

FATIGUE FRACTURE IDENTIFICATION

Conveniently for materials engineers, fatigue fractures generally have a distinctive appearance which, in most metals, permits not only identification of the fracture mechanism, but location of the fracture origin. The characteristic features of the fatigue zone are usually:

- Smooth, polished fracture surfaces
- The absence of significant plastic deformation
- Concentric rings or stop marks that focus back to the origin
- Lines or shallow ridges that appear to radiate outward from the origin

The fracture appearance will be greatly influenced by the load spectrum and the metal properties. *Textbook* examples of classic fatigue fractures are illustrated in Figs. 2 and 3.

THE ROLE OF METALLOGRAPHY IN THE ANALYSIS

Metallographic analysis is concentrated at the fatigue origin because it is the area where the metallurgical characteristics of the material are most important. This most often involves detecting and identifying a stress-raiser, which may be a *defect* in design, material, or workmanship. In metal fatigue, however, even *perfect* parts can fail. The role of metallography lies in assessing the degree of perfection. Not only is the presence or absence of a stress-raiser important, but whether the macrostructure and microstructure of the metal are consistent with the specified processing requirements of the part. Information derived from a failure analysis, even in a *perfect part*, can be an important part of a redesign effort. On the other hand, failure analysis can prevent unnecessary and costly redesign if a material or processing defect is responsible for the failure.

The microscopic characterization of fatigue fracture origins is the most challenging task of the metallographer. Sections must often be made through pinpoint origins. The metallographic preparation must retain edges and be free of artifact and staining, and

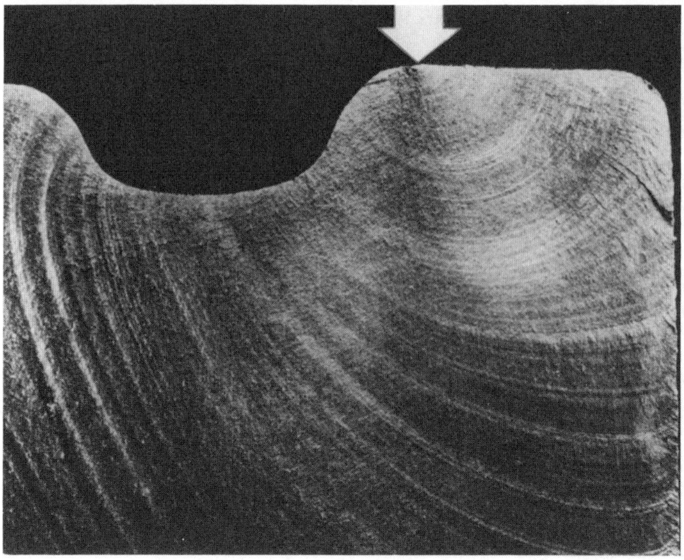

Fig. 2. Classic fatigue fracture in forged and hardened steel shaft. The origin is marked with the arrow.

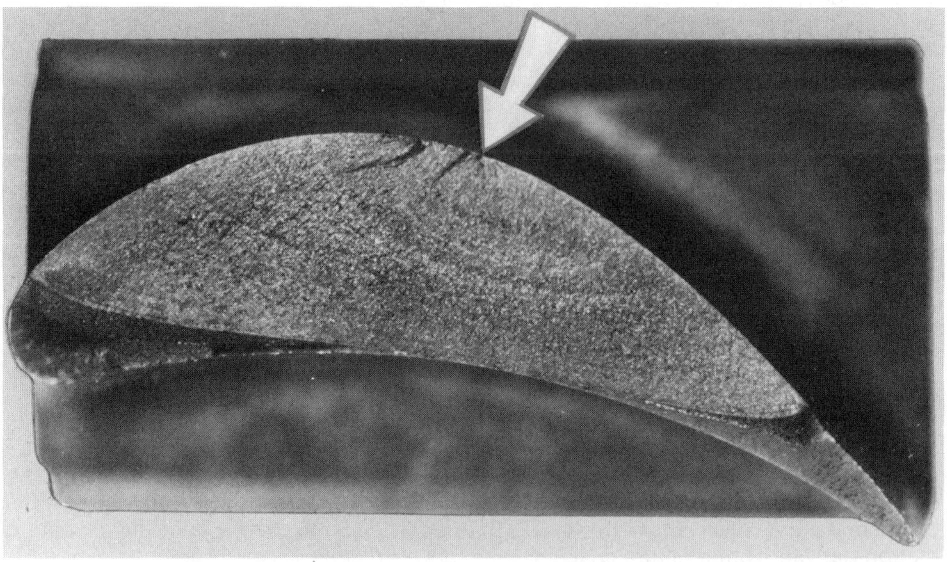

Fig. 3. Classic fatigue fracture in forged nickel-base alloy turbine blade. The origin is marked with the arrow.

the photomicrographic evidence must be clear and free from misinterpretation. All of this is generally accomplished under the pressure of short cycle times. The quality of the failure analysis lies very much in the talented hands of the metallographer.

CASE STUDIES

Every discourse on failure analysis must eventually present case studies. This is particularly true of fatigue failure and the metallographic characterization of the fatigue origin. There is a universal need to observe the metallurgical conditions which have resulted in fatigue failure in actual hardware. Metallic components cannot be made that are completely free of impurities, inhomogeneties, and imperfections, because the complex processes of metal refining, metal treating, and metal fabrication are not themselves perfect. Terms such as purity and accuracy, however, are relative and depend on the end use of the component, the expected loads in service, and the life expectancy of the product. In addition, the best designs often can not foresee all the environmental hazards a component may be exposed to during service and maintenance.

Assembled in the following case studies are a ***rogues gallery*** of stress-raisers that resulted in fatigue failure, either during testing or in service. They were detected or confirmed by metallographic analysis at the fatigue origin. Though hardly all inclusive, they are broad enough so that engineers engaged in fatigue failure analysis can expect to meet most of them face-to-face during their career.

Machining Defects

Most metal components are placed in service only after having been sized and shaped by one or more of the many machining processes. Mechanical and metallurgical damage to the surface can drastically increase the possibility of fatigue failure without affecting static properties.

Figures 4 and 5 illustrate how a sharp, unradiused corner or break-edge caused fatigue fracture of the blade attaching tenons in a forged steel compressor disc from an experimental jet engine. This is one of the few illustrations where metallography played only a minor role in the analysis since the stress-raiser was clearly visible on fracture examination.

In Fig. 6 fatigue fractures can be seen originating in the wall of an oil hole drilled in a hardened steel shaft. The metallographic examination, (Fig. 7) revealed many small cracks and a re-hardened zone of untempered martensite at the origin. This damage occurred due to severe overheating of the metal during drilling, probably due to a dull tool or inadequate coolant.

Abusive grinding of hardened steel parts is one of the most common causes for fatigue failure. Figure 8 illustrates a steel shaft with fatigue originating from a chromium plated thrust shoulder. The shoulder had been ground at overhaul to remove a chafed surface. Following grinding the shoulder was rebuilt with chromium plate. It appeared at first that the fatigue strength reducing effect of the chrome plate was responsible for the failure. Metallographic analysis, (Fig. 9) however, showed a rehardened zone containing many small cracks under the plate. Chrome plate evident in some of the cracks proved that they were present prior to plating (Fig. 10). Grinding cracks and burns such as these can be readily detected by magnetic particle inspection and etching after grinding and prior to plating.

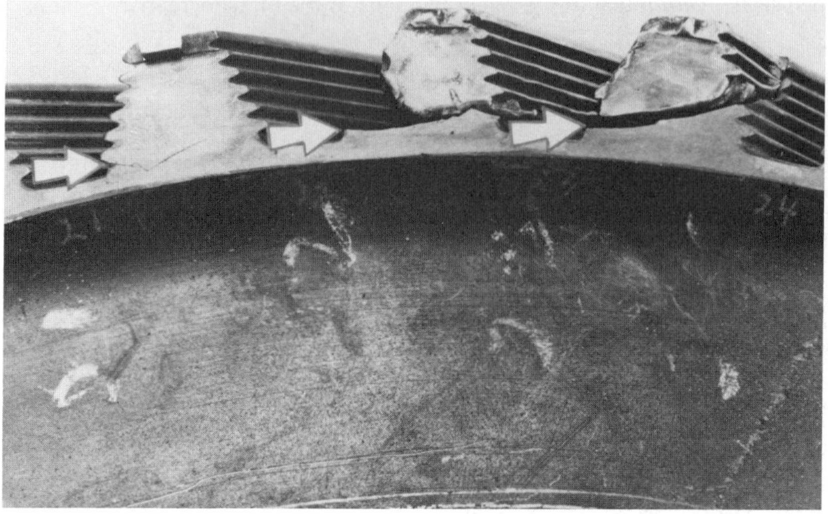

Fig. 4. Failure of forged steel compressor disc through blade attachment tenons. The origins are marked with arrows.

Fig. 5. Fatigue of disc originating at sharp unradiused corner. The origin is marked with the arrow.

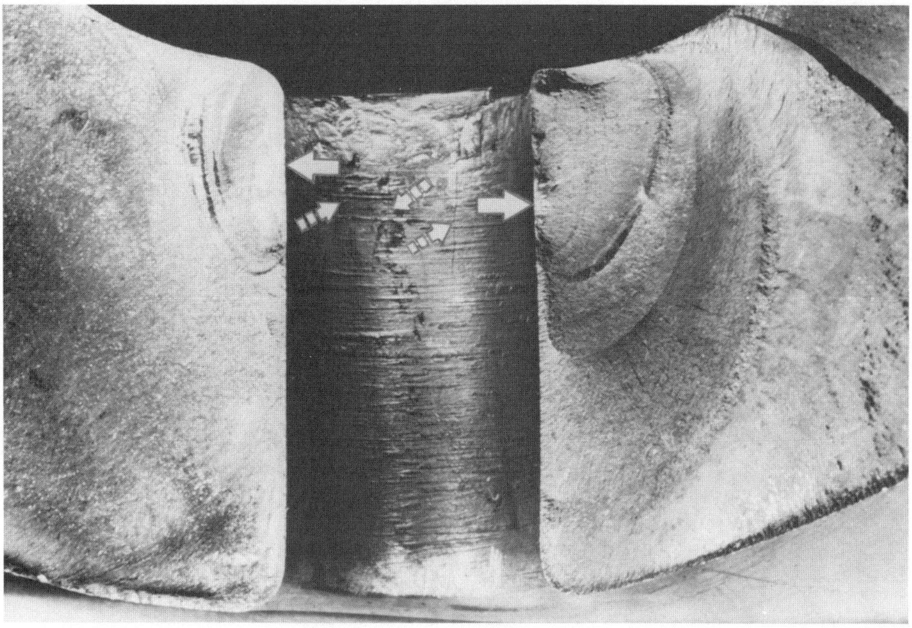

Fig. 6. Fatigue of hardened steel shaft through drilled oil passage hole. Origins are marked with arrows.

Fig. 7. Metallographic section through damaged areas of steel shaft shown in Fig. 6. Arrows mark small cracks in re-hardened zone of untempered martensite.

Fig. 8. Fatigue fracture of steel crankshaft originating at chrome-plated shoulder. Arrow marks origin.

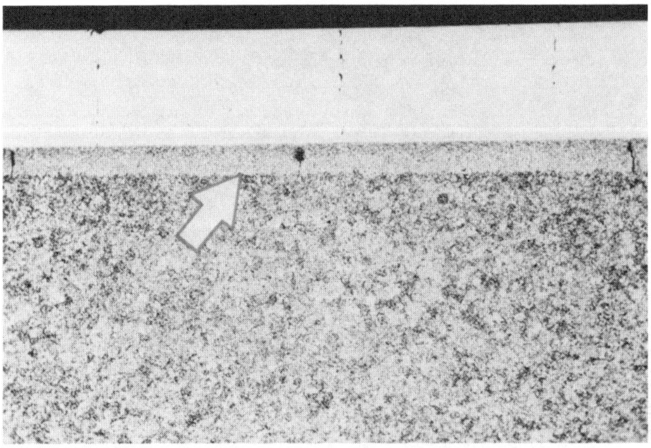

Fig. 9. Grinding cracks and martensitic burn in steel under chrome plate. Arrow marks extent of martensitic burn.

Defects From Thermal–Mechanical Working

All wrought metal components are subjected to thermal-mechanical processing which shapes the metal by plastic deformation, either by hammering, squeezing, or drawing. The movement of metal during these processes make them a common source of defects.

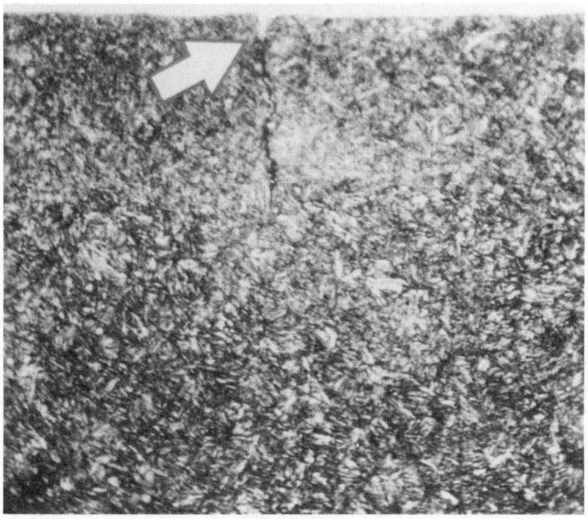

Fig. 10. Chrome plate in crack in martensite burn zone shown in Fig. 9.

Figure 11 illustrates a fatigue failure of a forged turbine blade made from a nickel-base superalloy. The failure occurred in a very short time during experimental testing in a gas turbine engine. The fatigue originated at an apparent defect at the convex airfoil surface (Fig. 12). The metallographic evidence, (Fig. 13), shows the angular orientation of the defect and the surface filled with high temperature oxide scale. This is characteristic of a forging lap, a surface defect caused by folding over of one surface onto another without fusion. The cause of the lap was traced to imperfection in the forging die. Etching followed by penetrant inspection can usually detect such laps.

A forged metal component is generally exposed to its highest temperature during the heating operation prior to forging. Exceeding the incipient melting point can cause grain boundary fusion, oxide films and voids.

Figure 14 illustrates the fatigue failure of a forged steel valve rocker arm shaft. The fatigue originated at a dark, coarse grained surface (Fig. 15). The metallographic analysis showed the characteristic grain boundary damage associated with burned steel, i.e., large grains, incipient melting, and intergranular oxidation (Fig. 16).

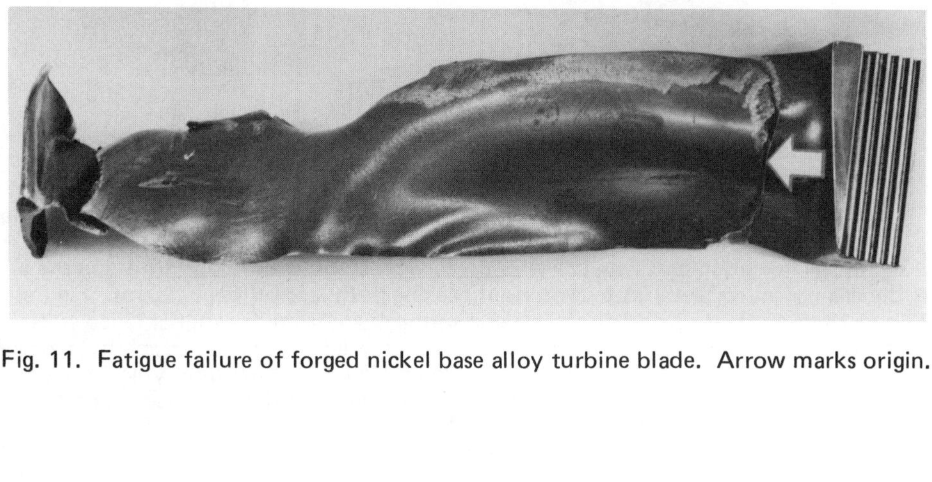

Fig. 11. Fatigue failure of forged nickel base alloy turbine blade. Arrow marks origin.

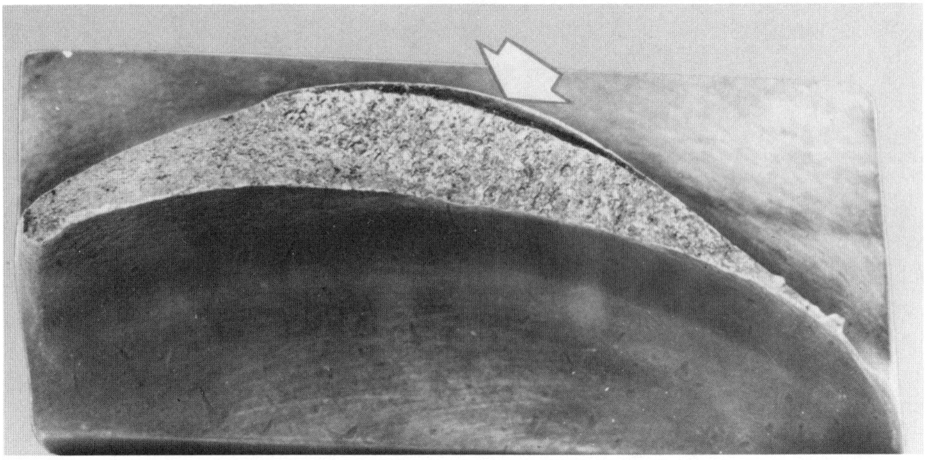

Fig. 12. Fracture of failed turbine blade shown in Fig. 11. Arrow marks origin.

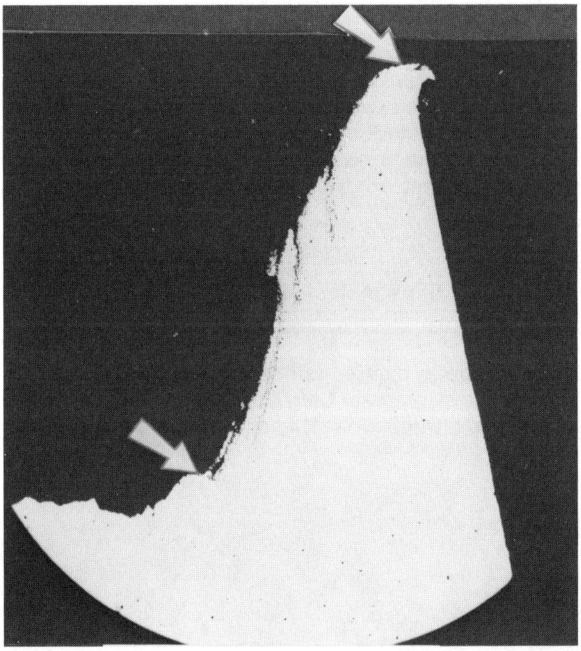

Fig. 13. Forging lap at turbine blade fatigue origin. Note oxide scale on fracture surface.

In the hot or cold drawing of wire through dies, surface discontinuities in the wire or die can cause the metal to fold on itself causing both laps or longer seams. Figures 17 and 18 show the fatigue failure of a coil spring made from drawn, hardened steel wire. The fatigue originates at a line defect, accentuated in the photograph by a magnetic particle inspection. The metallographic section (Fig. 19) shows the typical folded over appearance of a seam. The gray scale in the seam formed during the over heat-set after spring coiling.

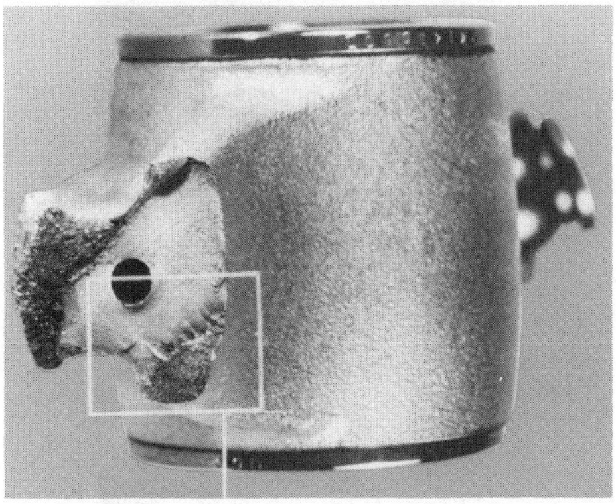

Fig. 14. Fatigue failure in forged steel rocker arm.

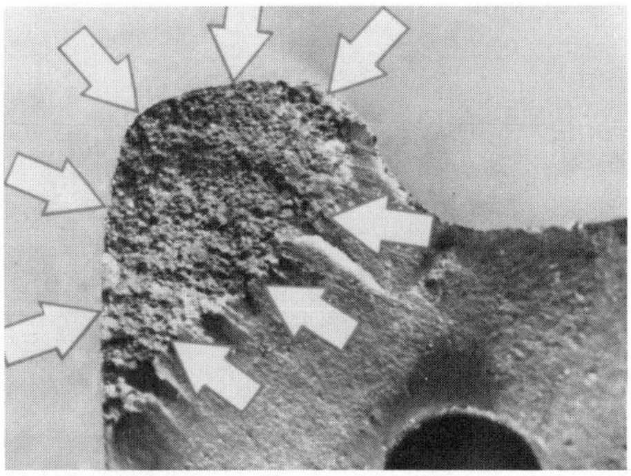

Fig. 15. Fracture of fatigue failure shown in Fig. 14.

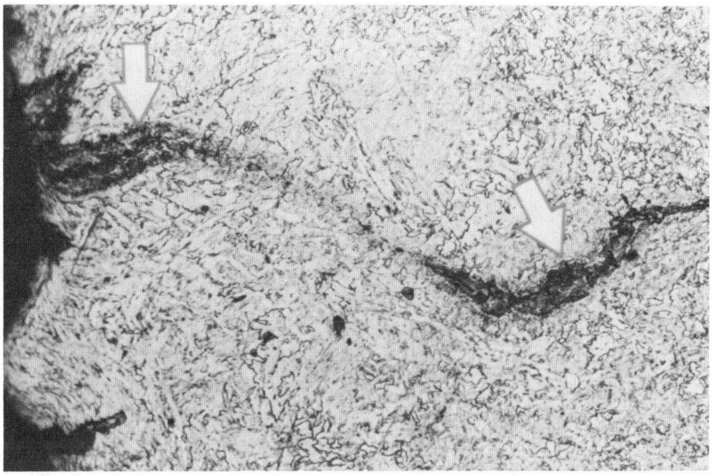

Fig. 16. "Burned" steel at origin of failure shown in Fig. 14.

Casting and Welding Defects

Since both casting and welding involve solidification of molten metal, many of the defects that plague one process also occur in the other. Stress-raisers such as porosity, shrinkage, slag inclusions, hot tears, and cold shuts or lack of fusion are possible, and all can contribute to fatigue failure.

Figure 20 shows a fatigue fracture through the butt section of a cast, cobalt-base alloy turbine bucket. While no defect is discernible on the fracture, a microstudy (Fig. 21) reveals heavy casting shrinkage at the fatigue origin.

In Fig. 22, the fatigue fracture of a cast, magnesium alloy housing is illustrated. The fatigue clearly originates at some defect that extends well into the casting. Metallographic examination (Fig. 23) showed it to be a large, entrapped dross inclusion.

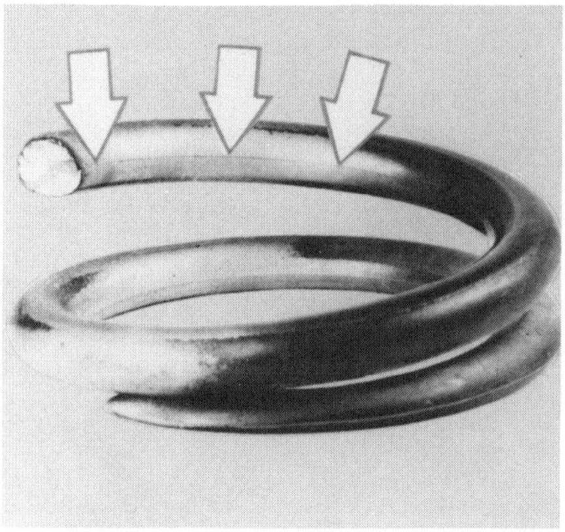

Fig. 17. Hardened coil spring fatigue failure originating at drawing seam.

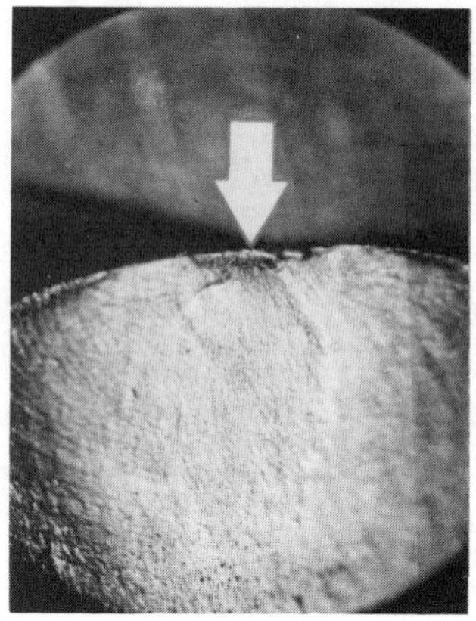

Fig. 18. Fracture of coil spring shown in Fig. 17. Arrow marks the origin.

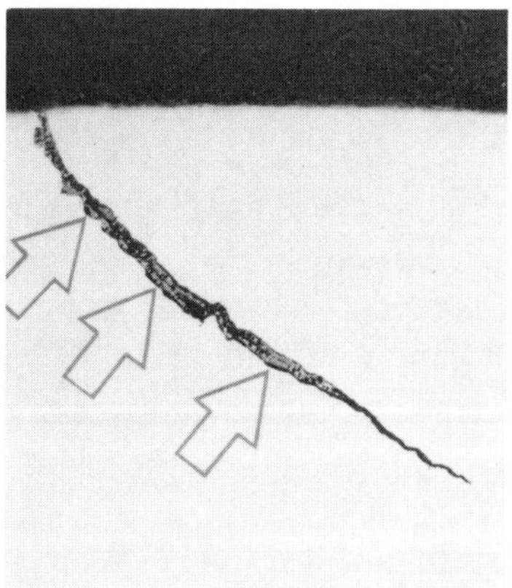

Fig. 19. Cross section through seam shown in Fig. 17. Note the gray scale in the seam.

Dross is the scum that forms on the surface of molten metals largely because of oxidation, but sometimes due to the rising of impurities in the melt to the surface. In this case the casting design acted as a natural trap for any dross in the molten metal.

Another stress–raiser often encountered in castings and welds is hot-tears. They are cracks that occur due to external or internal stresses acting on the metal after solidification has taken place, but while the metal is still at a high temperature and has low strength. Figure 24 shows the fatigue fracture of a cast, cobalt-base alloy turbine bucket. The fracture originates at the airfoil leading edge where metallographic examination (Fig. 25) revealed a repair weld containing many hot cracks. The high restraint associated with a small weld at the airfoil base set up high stresses during and after solidification.

Fig. 20. Fatigue failure of cast cobalt-base superalloy turbine blade. Arrow marks origin.

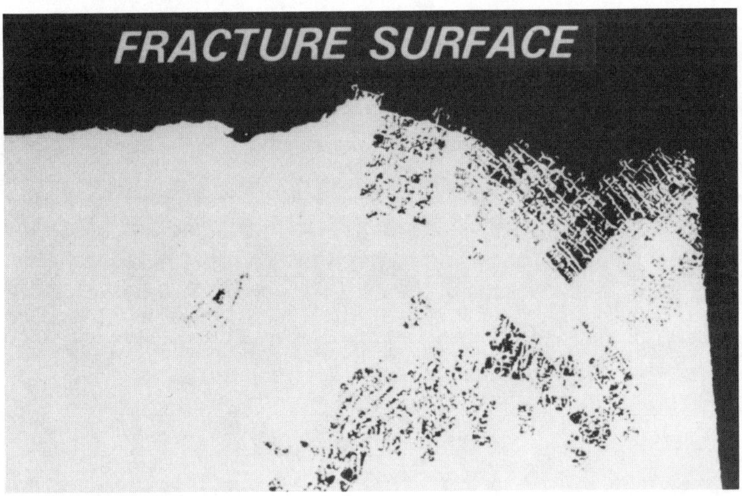

Fig. 21. Cross section of Fig. 20 showing casting shrinkage at the fatigue origin.

Defects Associated with Heat Treatment

Many of the alloys used in highly loaded components undergo some form of heat-treating operation to improve properties. Heat treatment always involves controlled heating and cooling operations. Fatigue strength may be drastically affected by defects arising from improper temperature, uneven rates of heat application or heat removal, improper preparation of surfaces or shapes prior to heat-treatment, and improper furnace atmospheres. Every heat-treater is familiar with quenching cracks in steel caused by the sudden volume changes that occur in hardening. Though most are easily detected, occasionally they are missed, and result in fatigue failure in operation.

Fig. 22. Fatigue failure of cast magnesium alloy housing. Arrow marks origin.

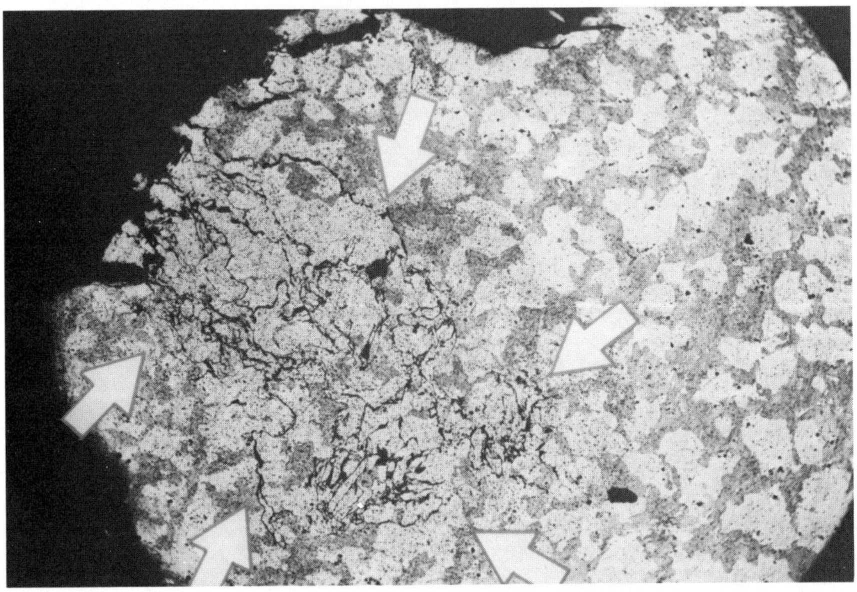

Fig. 23. Cross section of Fig. 22 showing entrapped dross at origin area.

Figure 26 illustrates a more subtle type of heat-treating problem that caused a fatigue failure. Shown is the fatigue fracture of a steel coil spring washer made from a hardened, high carbon steel. The fatigue originated at several points in the shank to flange fillet. The photomicrograph (Fig. 27) shows a shallow decarburized surface layer of ferrite. The fatigue strength reduction due to a decarburized layer can be over 50 per cent. The copper plate used in hardening to prevent decarburization was probably inadequate only in the fillet area since the rest of the part was satisfactory. A macro-etch of after hardening can detect such *spotty* decarburization.

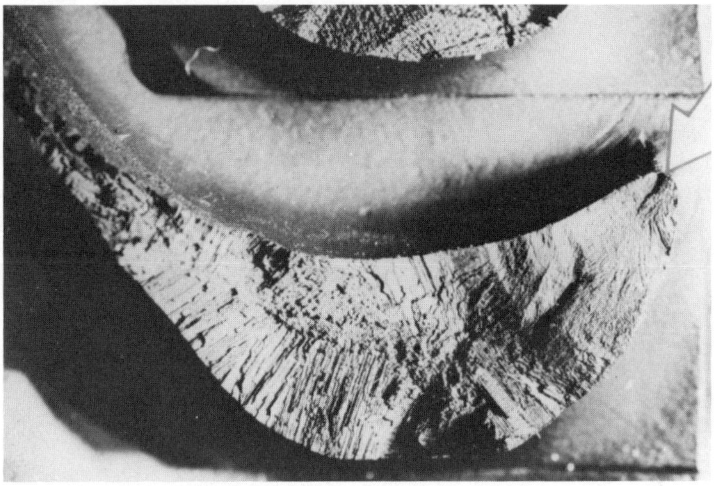

Fig. 24. Fatigue failure of cast cobalt alloy turbine blade. Arrow marks origin.

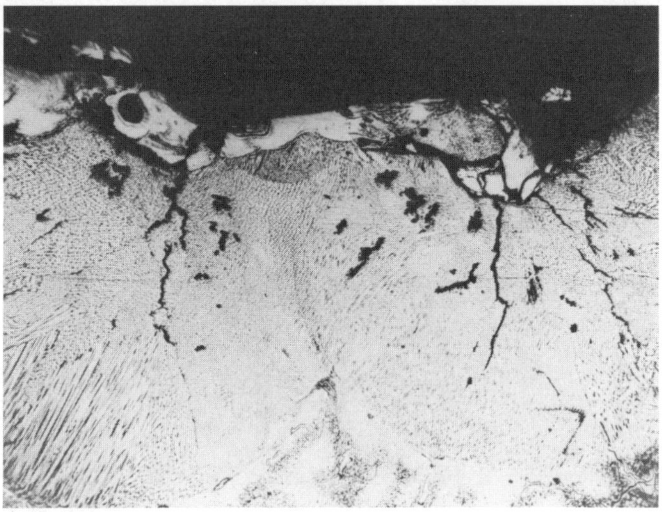

Fig. 25. Cross section through origin of failure shown in Fig. 24. A "poor" weld repair is located at the origin.

Heat-treated aluminum alloys are solution-heat treated at temperatures higher than the forging temperature. In fact, the most favorable temperature for such heat-treating is very close to the eutectic melting temperature. Unless very close temperature control is exercised, incipient melting can occur in an intergranular network and within the grains in areas called rosettes. Figure 28 shows a forged aluminum alloy compressor blade that failed by fatigue in a test engine. The metallographic analysis (Fig. 29) shows characteristic ***eutectic melting***. The brittle fused eutectic drastically reduces the fatigue strength of the alloy.

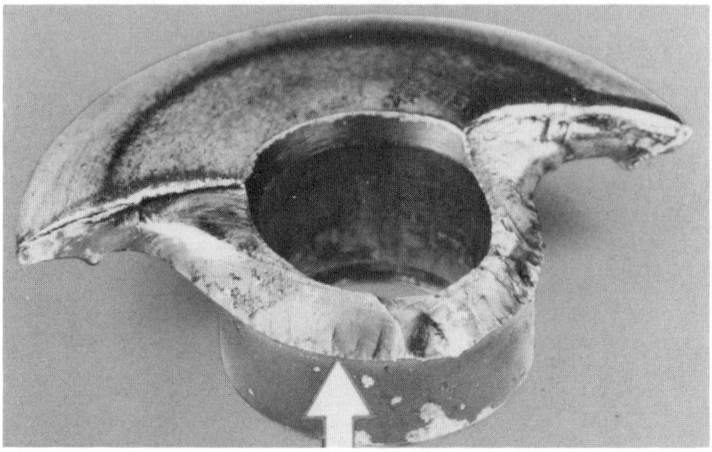

Fig. 26. Fatigue failure of hardened steel spring washer.

Fig. 27. Cross section through origin of steel washer shown in Fig. 26. Note the decarburized surface.

Non–Metallic Inclusions

All industrial metals and alloys contain non-metallic impurities or contaminants from the metal refining process. Large inclusions if located in critically stressed areas can cause fatigue failure. Each metal has its own characteristic inclusions. In ferrous alloys they are generally oxides, sulfides, silicates or complex combinations of each. In wrought metals the inclusions are deformed or strung out in the direction of the metal flow. The fatigue strength of high strength metals is more seriously degraded by inclusions than low strength metals.

Fig. 28. Fatigue failure of forged aluminum alloy compressor blade.

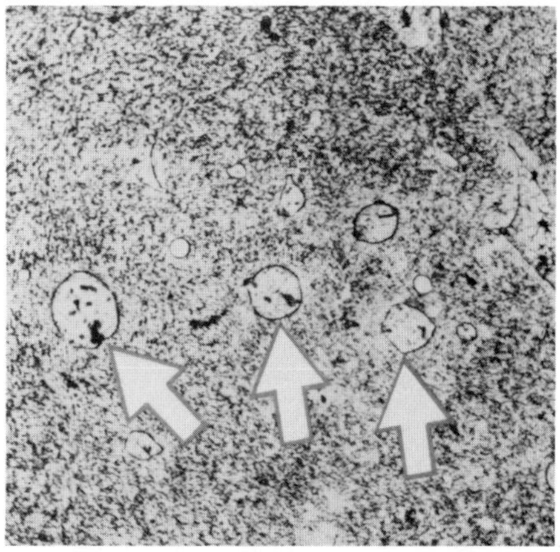

Fig. 29. Cross section of compressor blade shown in Fig. 28. Shown are areas (arrows) of eutectic melting.

Figure 30 shows the fracture surface of a hardened steel coil spring with the fatigue originating from a sub-surface nucleus. Very fine metallographic preparation through this pinpoint origin revealed a large non-metallic inclusion stringer (Fig. 31). The inclusion was too far beneath the surface to be detected by magnetic particle inspection.

Figure 32 illustrates fatigue flaking on the inner race of high hardness steel roller bearing. Although the flaked area is relatively small, in terms of a failure analysis it has already progressed to an advanced state. A section of the race adjacent to the flaking luckily intersected an incipient fatigue crack emanating from a large inclusion (Fig. 33). Vacuum melting and even double vacuum melting have found wide application in highly loaded springs and bearings in efforts to achieve very low inclusion sizes and frequency.

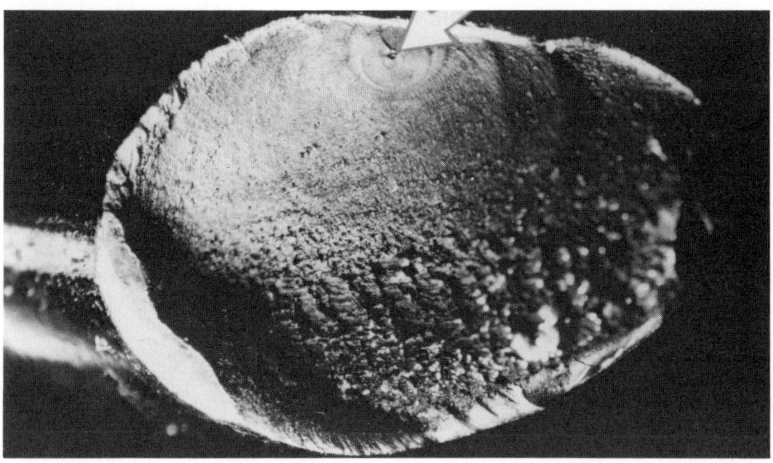

Fig. 30. Fatigue failure of hardened steel coil spring originating at a sub-surface inclusion (arrow).

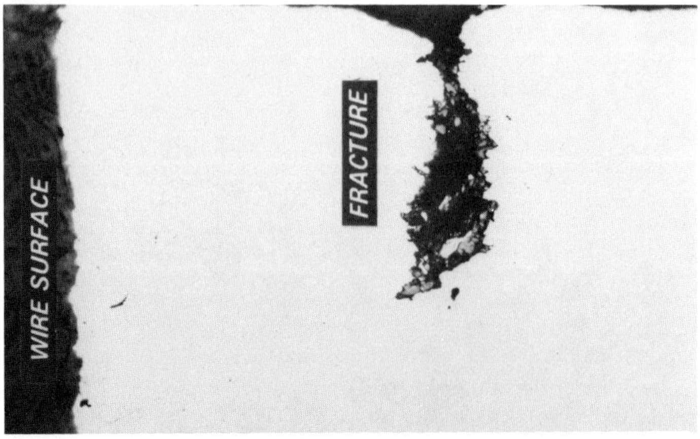

Fig. 31. Cross section through origin of coil spring failure shown in Fig. 30. The inclusion at the origin is revealed.

Fig. 32. Fatigue spalling in roller bearing race. Incipient fatigue spall crack originating at non-metallic inclusion (arrow).

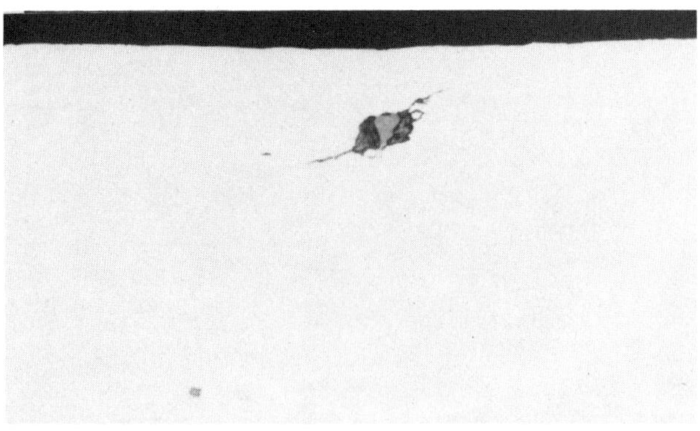

Fig. 33. Cross section through inclusion located at origin of failure shown in Fig. 32.

Chemical Segregation

The distribution of chemical elements in wrought metal alloys is not always uniform. Some regions of the metal may be enriched in certain elements, while some regions are impoverished. When such segregation is elongated due to mechanical working, the material is said to be *banded.* Often such banding is not critical unless mechanical properties in the transverse direction are important to the application. In some cases, however, segregation can lead to fatigue failure.

Figure 34 shows the fatigue fracture of a carburized and hardened steel cam tappet roller. The fatigue originated beneath the roller surface, near the carburized case-core interface (Fig. 35), then progressed circumferentially, stripping the case from the roller. Metallographic studies showed that the microstructure contained heavy retained austenite banding (Fig. 36). This resulted from chemical segregation in the extruded bar stock from which the roller was fabricated. The austenite retention was particularly heavy in the carburized case because of the high-carbon. It was deduced that transformation of some retained austenite to martensite occurred under service loads. The resultant volume expansion caused micro-cracks to form in the low ductility case-core interfacial zone. The micro-cracks then progressed by fatigue. This type failure is a reminder that chemical segregation must be evaluated in each individual component before it is considered as acceptable.

Fig. 34. Sub-surface fatigue failure of carburized steel roller caused by retained austenite banding.

Fig. 35. Cross section of tappet roller of Fig. 34 showing case-core interface.

Fig. 36. Retained austenite banding in microstructure of tappet roller.

Arc Burns

Arc burns are caused by electric current arcing between a metal component and some current carrying electrode. In steel parts, these burns in addition to causing localized areas of fusion and pitting, will always result in a brittle re-hardened zone of untempered martensite, sometimes containing cracks.

Figure 37 illustrates the fatigue fracture of a hardened steel connecting rod. It is interesting because it illustrates some of the unforeseen hazards of a system overhaul. The rod had been marked as to its position in the engine by an electric etch pencil in an overhaul facility. A fatigue failure occurred in the engine a very short time after reinstallation. It originated in the zone of cracking, melting, and martensitic burning made by the electric etch arc burn (Fig. 38).

Stress—Raisers from Operation

Up to this point we have concentrated on fatigue failures arising from man-made stress-raisers associated with metal refining, fabrication, and overhaul. Stress-raisers can also form during operation that result in fatigue failure. Figure 39 shows a series of fatigue cracks in the plated surface of a steel-backed, silver-plated sleeve bearing. The cracks radiate outward from a chip embedded in the silver. The metallographic section (Fig. 40) graphically shows a steel chip in the surface and the fatigue crack progression from it. The chip probably entered the bearing cavity with contaminated lube oil where it became embedded in the soft silver.

Figures 41 and 42 illustrate the fatigue failure through the airfoil of a forged, martensitic stainless steel compressor blade. Metallographic examination (Fig. 43) revealed intergranular stress-corrosion cracks at the fatigue origin and in adjacent areas. To achieve the required strength level this alloy was tempered in an *embrittlement* range, which also seriously impairs the stress-corrosion resistance of the material. A change to a different alloy was required.

Fig. 37. Fatigue failure of hardened steel connecting rod originating in melted and burned area caused by electro-etched number.

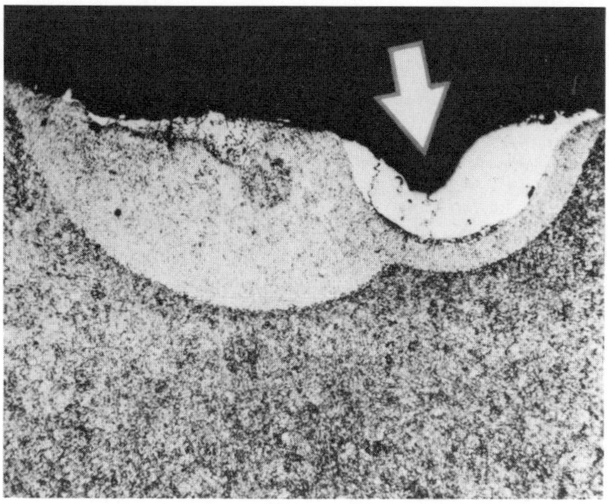

Fig. 38. Zone of cracking, melting (arrow) and martensitic burning in connecting rod shown in Fig. 37 caused by electro-etching.

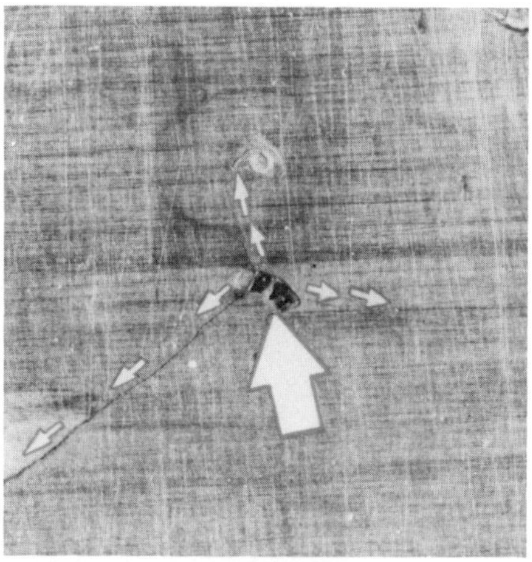

Fig. 39. Fatigue cracks in silver plated sleeve bearing originating at embedded steel chip (arrow)

Fig. 40. Cross section through embedded chip in silver-plated sleeve bearing shown in Fig. 39.

Figure 44 shows the bending fatigue failure of a gear tooth on a carburized and hardened pinion. Following the tooth failure the gear also fractured radially by fatigue (Fig. 45). Corrosion pits at the fatigue origin, in the gear tooth root, were detected by the metallographic examination (Fig. 46). In this instance the corrosion was traced to contaminated lubricating oil.

In Fig. 47, the high temperature fatigue fractures of several of the blade retaining tenons on a steel turbine disc are shown. Microstudy showed that the fatigue was initiated by stress-rupture cracks in the tenon *fir-tree* roots (Fig. 48). The disc rim operated at high temperatures and was life limited due to stress-rupture. For extended life the material was changed to a more costly, but unlimited life superalloy.

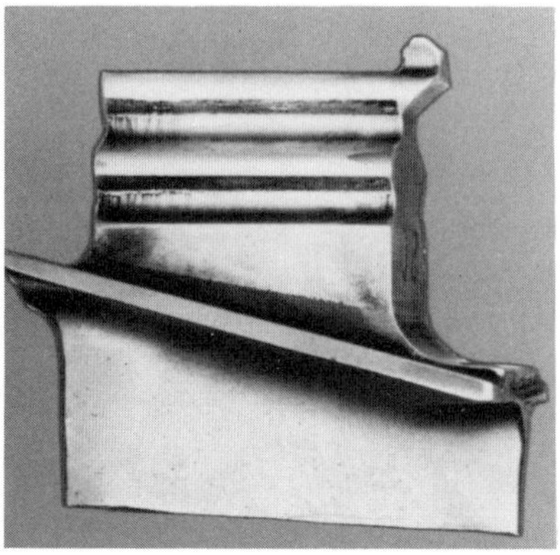

Fig. 41. Fatigue failure of forged and hardened martensitic stainless steel compressor blade originating at stress-corrosion cracks.

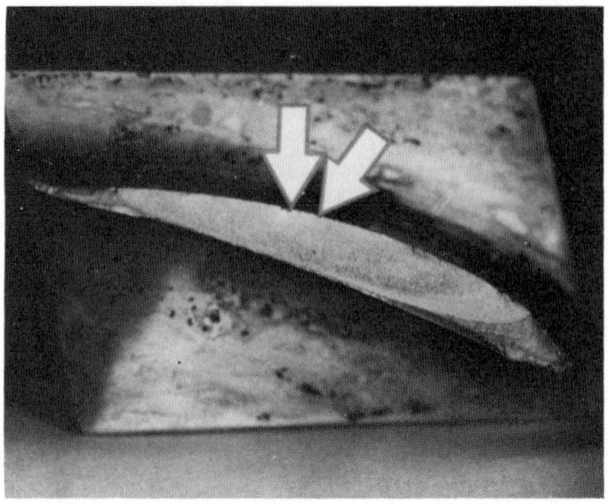

Fig. 42. Fracture of failure shown in Fig. 41. Arrows show area of stress-corrosion cracking.

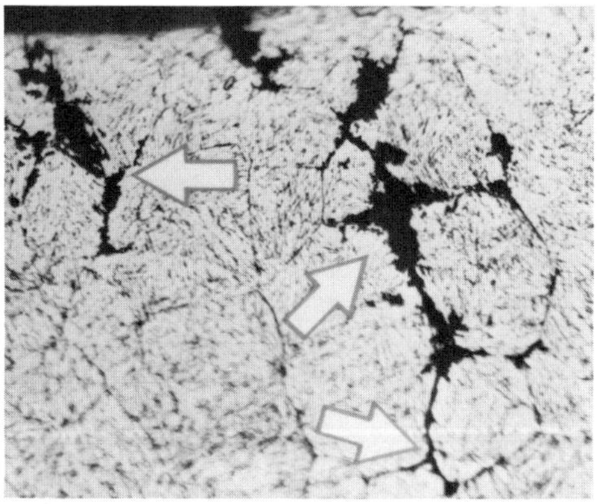

Fig. 43. Intergranular cracking in area of stress corrosion cracking of Fig. 42.

Fig. 44. Fatigue failure of carburized and hardened steel gear originating at corrosion pits (arrow).

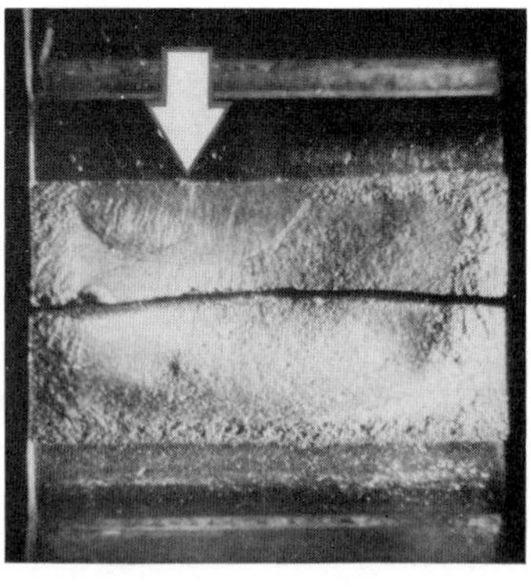

Fig. 45. Fracture surface of failure shown in Fig. 44. Arrow marks origin.

Fig. 46. Cross section through failure shown in Fig. 44 revealing corrosion pits (arrows).

Fig. 47. Fatigue failure of steel turbine disc tenons originated by stress-rupture cracks in blade attachment area. Arrows mark origins.

SUMMARY

The roles of fractography, and particularly of metallography, in establishing the cause of metal fatigue failures have been reviewed and should be clear. There is, however, a third powerful failure analysis tool that cannot be overlooked. That *tool* is deductive reasoning. The ability to carefully assemble all the available evidence, separate facts from opinion, and arrive at a definitive conclusion and corrective action through step-by-step, unbiased deduction is the mark of the mark of the professional failure analysis engineer. It is a discipline not easily acquired.

Many fatigue failure investigations are quite complex and solutions require the combined skills of materials, design, stress, quality, and service engineers. In addition, fatigue failure analysis is often complicated by extensive secondary damage and too much, often

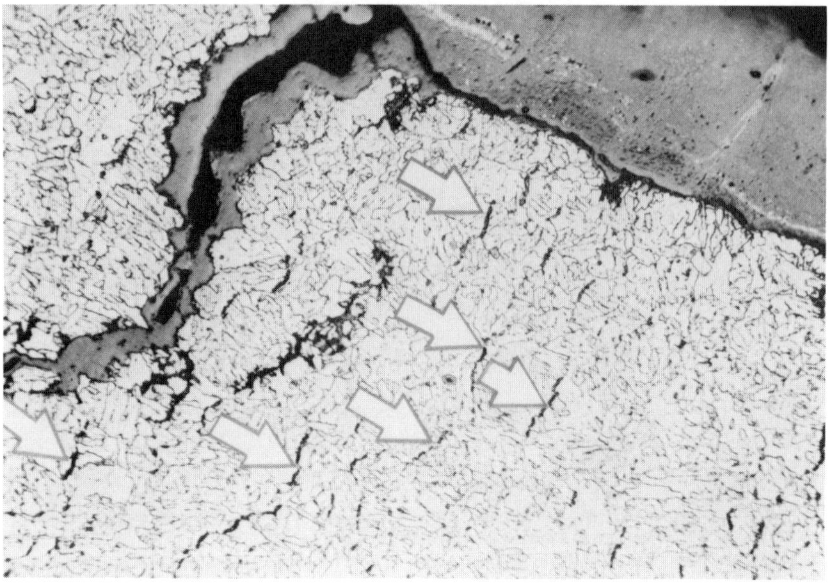

Fig. 48. Cross section through origin area of failure shown in Fig. 47. Stress rupture cracks (arrows) are revealed.

conflicting operating and historical information. In such cases it is often difficult to determine both the sequence and cause of failure. An approach for handling these situations was set down by a 14th century philosopher theologian. It is the principle of *Occam's Razor*. Simply it says:

- Shave away all extraneous details.
- When several explanations are possible, the simplest is probably the correct one.

SELECTED REFERENCES

1. Barer, R.D. and Peters, B.F., *Why Metals Fail,* Gordon and Breach Science Publishers, New York, (1970).
2. Parker, R.F., *Elements of Metallurgical Failure Analysis,* Curtiss–Wright Corporation, Wood–Ridge, N.J., (1956).
3. Wulpi, D.J., *How Components Fail,* American Society for Metals, Metals Park, Ohio, (1966).
4. Bennett, J.A. and Quick, J.A., *Mechanical Failures of Metals in Service,* National Bureau of Standards Circular 550, (Sept. 27, 1954).
5. *Failure Analysis and Prevention, Metals Handbook,* Volume 10, American Society for Metals, Metals Park, Ohio, (1975).

METALLOGRAPHIC ANALYSIS OF CORROSION FAILURES

V. Colangelo *

INTRODUCTION

The proper analysis of a component failure can provide valuable assistance in determining the validity of a product design. Errors in selection and design as well as materials defects and shortcomings in processing can frequently be revealed. These problems can often be detected in prototype testing or early in production, which results in substantial savings in time and money. However, such analysis is often neglected. In many small organizations sophisticated materials and process engineering groups are not available. Too often, the responsibility falls to an engineer who lacks the background to interpret the available information.

Nevertheless, from an engineering standpoint, the proper application of failure analysis techniques can provide a valuable feedback to design problems and material limitations. The optimum design is one in which the requirements are slightly exceeded by the capabilities in all circumstances. This aim is seldom realized because of the obvious difficulty in recognizing or defining precisely the various demands that the system must be called upon to meet. This latter aspect of the design requirements is generally met by a sound engineering device, the application of safety factors. However, how much of a safety factor is appropriate? To grossly overdesign the component is economically extravagant and can inadvertently overload other parts of the structure. Underdesigning of the component leads to its premature failure, is economically wasteful, and, most important, could jeopardize life. The role failure analysis plays in the overall design and production of a component is shown in Fig. 1. It is in this role, as a design adjunct, that failure analysis can play a maximum part, since the most sophisticated simulation testing can never duplicate the varied and interacting conditions found in actual service.

Over the years the fundamental factors causally related to failure or shortening of service life have been identified:

- Design
- Improper selection of material
- Heat treatment
- Fabrication
- Improper machining and assembly
- Improper service conditions or usage
- Defects in material

*Dept. of the Army, Watervliet Arsenal, Watervliet, New York 12189 USA.

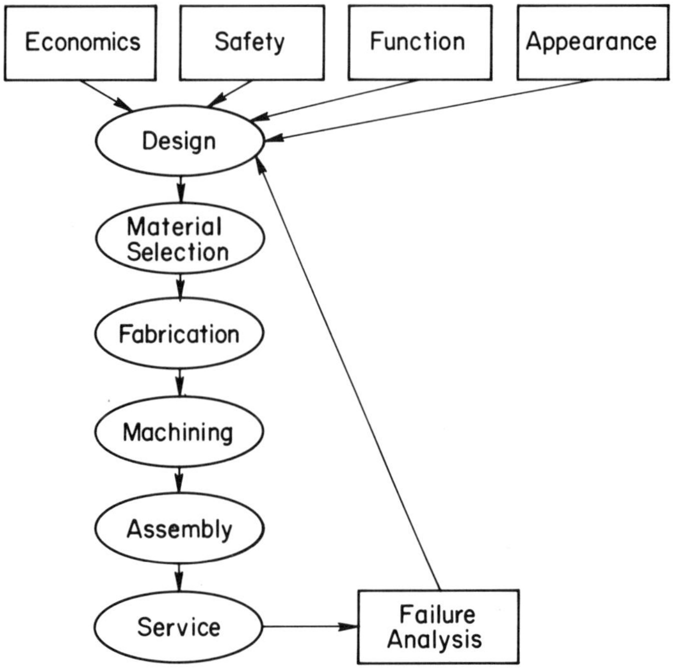

Fig. 1. Relationship of failure analysis to design.

The actual failure may be due to any one of these factors acting independently or to the interaction of several of them. The exact cause is often not easy to ascertain and can only be resolved after an intensive investigation.

The unsuccessful performance of a structure or component can, in general, be traced to the following modes of failure: ductile or brittle fracture, fatigue, creep, corrosion, or wear. Failure may result via the independent action of any of these modes; however, the final failure is often caused by the simultaneous or sequential activity of several mechanisms. One mechanism may create stress raisers while another may promote the initiation of a crack and its subsequent growth. Thus although there may be one primary mechanism, contributory mechanisms also exist.

Our concern here is with the corrosion mechanisms and with the recognition of those mechanisms through metallographic evidence. General corrosion has not been discussed since by definition the attack appears as a relatively uniform dissolution of the surface.

PITTING

Pitting is a form of localized corrosion in which the attack is confined to numerous small cavities on the metal surface. The cavities created may vary in size and form, however, it is commonly held that a true pit has a length/depth ratio equal or greater than unity. Figure 2 shows the results of pitting attack on a 416 stainless steel probe in an aqueous, mildly acidic environment as seen on the surface. Figure 3 shows the cross section and the profile of the pits occurring.

Fig. 2. Pitting attack on a 416 grade stainless steel probe. (Courtesy of P.A. Thornton).

Fig. 3. Metallographic section of pit, etched with mixed acid. Original Mag. 250X. (Reduced approximately 10 percent for reproduction). (Courtesy of P.A. Thornton).

The fact that only very slight metal losses may result in perforation means that failure can be quite rapid and occur without warning, making pitting an extremely dangerous form of attack. Pits also can contribute to general failure in another way. In highly stressed components, these pits can act as stress raisers, resulting in the creation of fatigue cracks.

Pitting can occur with a number of metals, but stainless steels and aluminum alloys are the commonest structural materials which are susceptible to this type of degradation. Pitting occurs most frequently in solutions of near neutral pH, containing chloride bromide or hypochlorite ions. Fontana and Greene [1] have pointed out that pitting is intermediate between the condition of general corrosion and complete immunity. Consequently, those factors favoring general corrosion, e.g., low pH and increased temperatures, do not, in general, favor pitting attack. Greene and Fontana [2] have tabulated the effect of a number of these environmental variables on the rate of pitting attack.

Visual examination of the surface, even if possible, does not reveal the extent of pitting damage since the pit orifice can be quite small, bordering on the microscopic, while penetration may be severe. Pitting can be best evaluated by making metallographic sections of representative components to determine maximum pit depth or by the use of ultrasonic inspection. Since the former method is destructive, it is best adapted to laboratory evaluations. However, as Fontana and Greene [3] point out, the probability of finding deeper pits increases with increasing surface area and laboratory results on small specimens must be interpreted advisedly.

CREVICE CORROSION

Crevice corrosion is a type of corrosive attack which is, in particular, associated with the confined spaces or crevices formed by certain mechanical configurations such as threaded joints, gasket interfaces, tubular sleeves, etc.

Crevices can exist in any assembly but there appears to be geometrical requirement. In order for crevice corrosion to occur, the crevice must be close fitting having dimensions of only a few thousandths or less. Although the limits of the gap have not been defined, it is known that crevice corrosion does not occur in larger spaces.

It is not necessary for both approximating surfaces to be metal in order for crevice corrosion to occur. Crevice corrosion has been reported in crevices formed by a number of nonmetallic materials, e.g., polymers, glasses, rubber in contact with metal surfaces. The fact that this can occur is of particular importance in the application and selection of gasketing materials. The corroded area on the Hastelloy B cylinder shown in Fig. 4 occurred in the region of contact between it and a glass basket in less than 600 hours at 30°C in 0.9% NaCl solution. Figure 5 shows a surgical hip pin which has suffered severe crevice corrosion. Figure 6 depicts crevice corrosion on a corresponding plate together with a fatigue crack emanating from the corroded region.

SELECTIVE DISSOLUTION

Selective dissolution is the term applied when a phase is selectively attacked in an alloy or when one element is preferentially dissolved from a solid solution. This type of corrosion occurs in several common alloy systems. Among them are dezincification of

Fig. 4. Crevice corrosion in Hastelloy B.

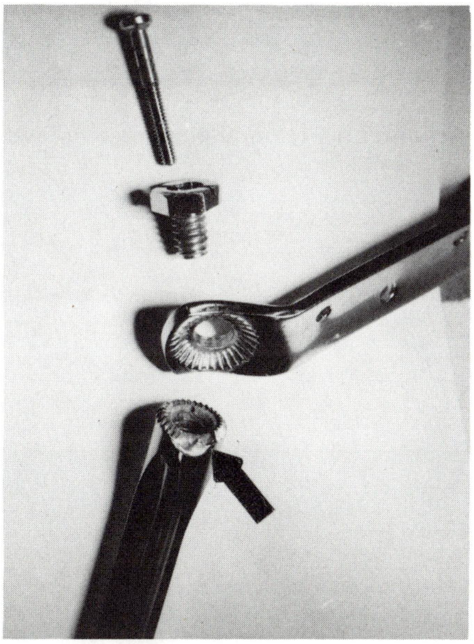

Fig. 5. Surgical hip pin suffering from crevice corrosion.

brass alloys, preferential attack of ferrite in austenitic [4] and martensitic stainless steels [5], and of ferrite in grey cast irons (graphitization). While these are the commonest examples, selective dissolution also has been reported in a number of other systems (under highly specific conditions), e.g., the loss of nickel from Monel [6], of aluminum from aluminum bronze [7], of copper from 70:30 cupro-nickel alloy [8].

Figure 7 shows a macroscopic view of a dezincified 70/30 brass while Fig. 8 illustrates the microstructure obtained when ferrite has been selectively dissolved from a grey cast iron. The microstructural appearance of a dezincified structure would be similar ex-

Fig. 6. Crevice corrosion on a femoral plate together with a fatigue crack.

Fig. 7. Macroscopic view of a dezincified structure.

cept that the remaining phase would be network of copper and copper oxide. Figure 9 depicts the dissolution of a discontinuous phase, ferrite in an austenitic matrix.

Modern technology has created another group of materials where selective dissolution can produce serious consequences; that is, in the area of metallic composite structures. These are usually constructed with the fibers and matrices possessing widely differing chemical and mechanical characteristics. They may be severely degraded and show sharp losses in strength if the reinforcing fibers are selectively attacked.

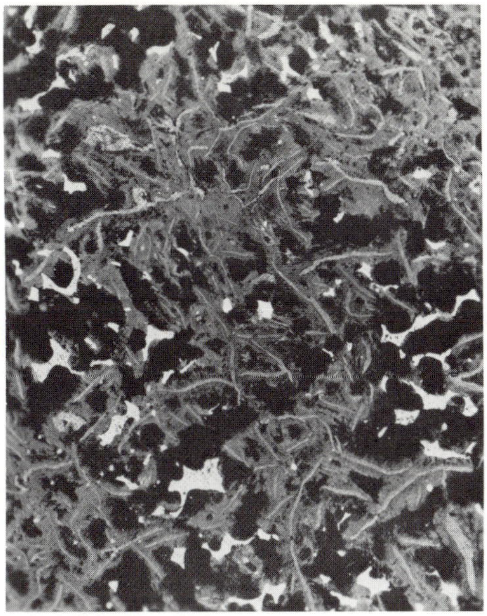

Fig. 8. Microstructure illustrating the removal of ferrite from a grey cast iron (graphitization).

The form of the attack may vary from system to system but, in general, the severest degradation results when the attacked species or phase is present in a continuous network. There may be little change in the overall configuration or geometry of the component but the mechanical properties are adversely affected in a most serious manner. On the other hand, if the resistant phase is continuous, it is left behind together with the corrosion products as a porous network.

INTERGRANULAR CORROSION

Metallic components are usually polycrystalline. In certain cases, the grain boundaries existent in such aggregates are more susceptible to corrosive attack than the grain interior. The preferential dissolution suffered by these areas may be related to several factors depending on the particular circumstances.

The primary cause of intergranular attack (IGA) is the presence of an inhomogeneous condition, in which the grain boundary material differs from the bulk properties. This may be the result of segregation or may be due to intergranular precipitation. These conditions may also be modified by enhanced diffusion effects operating within the grain boundary or by the selective absorption of certain solutes such as hydrogen.

The overall effect of this preferential dissolution is that great damage to the structure can occur with only slight corrosive damage occurring to the main body of the crystal. Because dissolution is confined to such small regions, the actual weight losses are small, penetration rates are high and destruction can be quite rapid. The most common form of intergranular attack occurs with austenitic stainless steels.

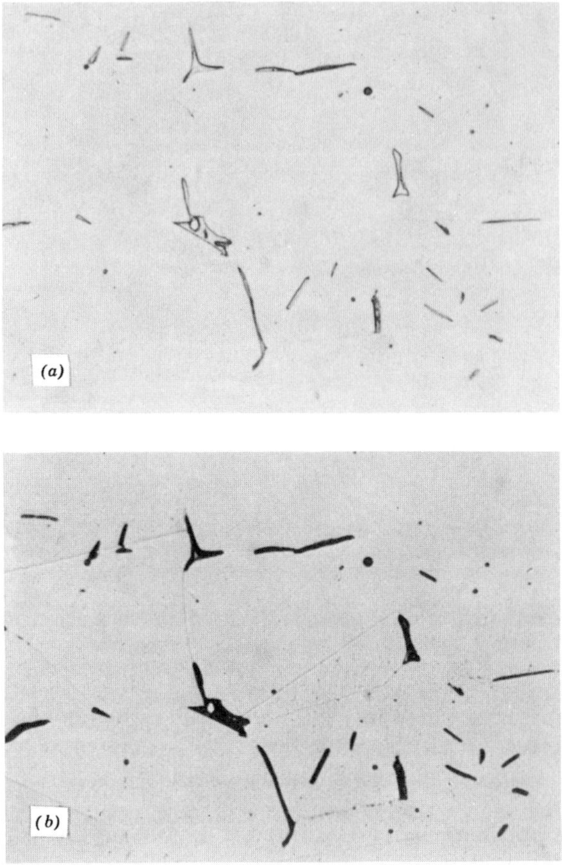

Fig. 9. Selective attack of ferrite in an austenite matrix. (a) The structure of a defective rod end, lightly etched, to show the micro appearance of ferrite phase. (b) The same area, after heavier etching, showing the ferrite is present along grain boundaries and along planes within the grains. Original Mag. 500X. (Reduced approximately 25 percent for reproduction). (Courtesy D. Moore).

Intergranular attack on stainless steels is generally attributed to the depletion of dissolved chromium within the grain boundary. This is directly related to the precipitation of chromium carbides during sensitizing heat treatment. When unmodified 18–8 stainless steels are heated in the temperature range of 950° – 1500°F or cooled slowly through it, the chromium-rich carbide which is relatively insoluble in this range, precipitates. The degree of sensitization is directly related to the temperature and duration of exposure. At the low end of the range, circa 1000°F, times measured in hours are needed to effect a sensitized condition. The precipitation results in depletion of chromium in the grain boundary region.

The effect of this depletion is that the grain boundaries are now compositionally different from the bulk of the grain. Higher corrosion rates result from the interaction of several factors, specifically, the poorer corrosion resistance of the low chromium alloy formed, possible galvanic effects between the grain boundary and interior and an unfavorable cathode/anode ratio.

Sensitization may result from any of the following:

a. Annealing or stress relieving heat treatments.
b. Exposure at some time during the course of service, prior to exposure.
c. Fabrication procedures, generally welding or flame cutting in which the adjacent zones may pass through the temperature range. Figure 10 illustrates the microstructural appearance of an austenitic stainless steel which has suffered intergranular attack.

(a)

(b)

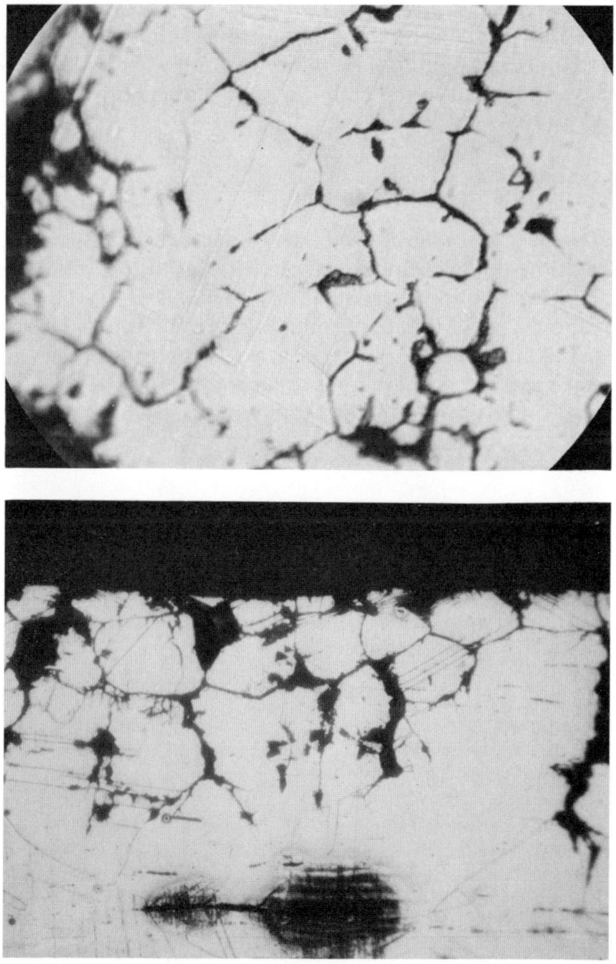

Fig. 10. Microstructure of austenitic stainless steel which has suffered intergranular attack. (a) surface (b) cross section. Original Mag. 100X. (Reduced approximately 15 percent for reproduction).

EXFOLIATION

Exfoliation or lamellar corrosion is a special form of intergranular attack which affects primarily aluminum and magnesium alloys. It is markedly directional and is characterized by attack of the elongated grains on a plane parallel to the rolled or extruded surface. This results in a characteristic delamination or stratification of the surface structure. Figure 11 shows a general view of a component which has suffered from exfoliation. One can have grain boundary attack, however, in susceptible alloys without exfoliation as Lifka and Sprowls [9] have clearly shown in Fig. 12.

The overall susceptibility of an alloy appears to be related to both the cooling rate and compositional differences though the exact relationship is not clear.

It is known that certain environmental factors increase the tendency toward exfoliation. These are: the presence of chloride or bromide ions in the environment, higher temperatures, and an acidic condition and intermittent wetting and drying. This latter condition is believed to result in the creation of insoluble corrosion products which exert a stress normal to the surface.

CAVITATION DAMAGE

Cavitation damage is a well known type of degradation associated with rapid movement of liquid near the liquid metal interface. Stresses in the liquid environment induced by the rapid fluid movement create momentary cavities in large numbers. The duration of each cavity is short and its collapse produces a shock wave which impinges on the metal surface. The most likely effect of the shock wave is that it produces first a compressive stress on the surface and upon reflection produces a tensile stress normal to the surface. Because of the high rate of formation and decay of the cavities, damage is produced rapidly.

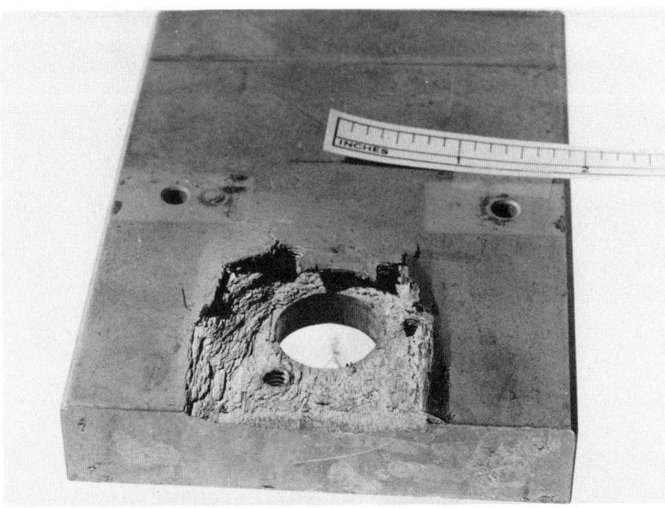

Fig. 11. General view of a 7075–T6 aluminum forging showing exfoliation damage.

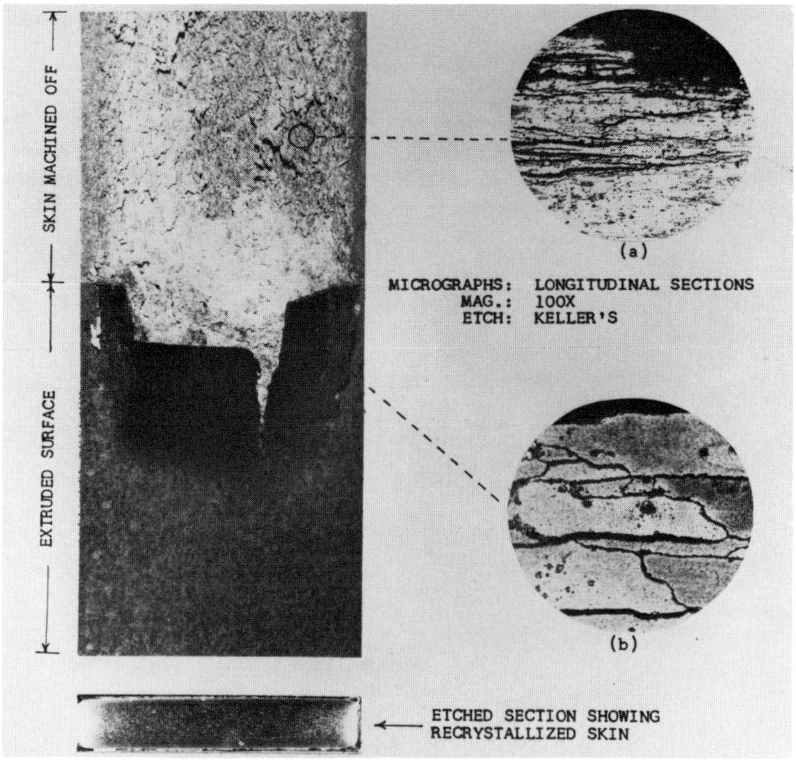

Fig. 12. Bar of 2024–T351 exposed for three years. (a) Fibrous bar interior exhibits exfoliation. (b) Recrystallized surface shows intergranular attack but no exfoliation. (Courtesy J. Rinnovatore).

Cavitation damage superficially resembles pitting but the surface appears considerably rougher with many closely spaced pits. The general appearance of a ship's screw exhibiting cavitation damage is shown in Fig. 13. Micro-sections may show evidence of plastic deformation and honeycombing as in Figs. 14 and 15.

EROSION CORROSION

Erosion corrosion is exemplified by an increase in the corrosion rate caused by relative motion between the surface and the environment.

Many alloys possessing good corrosion depend on the fact that insoluble surface films are created and maintained. While increases in velocity may occasionally decrease the rate of degradation because of kinetic factors, the general effect is that film maintenance is impared with increasing corrosion and increased corrosion rates result. Similarly, turbulence can have an adverse effect on the protecting film. The work of Venzcel [10] indicates that erosion corrosion occurs when the Reynolds number required to produce turbulence is exceeded. Erosion corrosion presents a distinctive appearance. The surface usually exhibits severe weight loss, the overall appearance often presenting a carved almost sculptured impression resulting from the abrasive action of the fluid envir-

Fig. 13. Ship screw exhibiting cavitation.

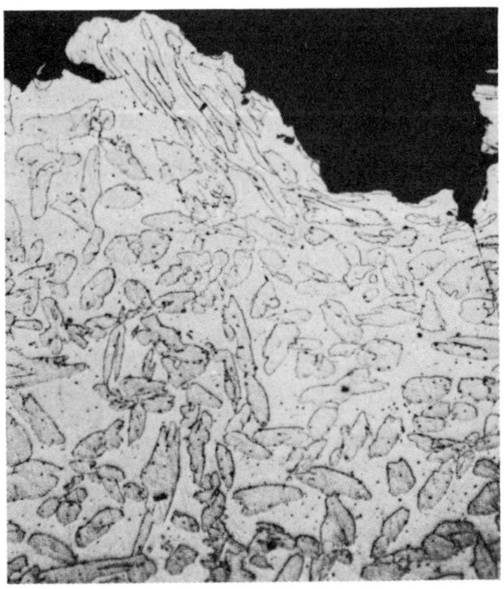

Fig. 14. Microstructure of surface which has undergone cavitation damage. (Courtesy of International Nickel).

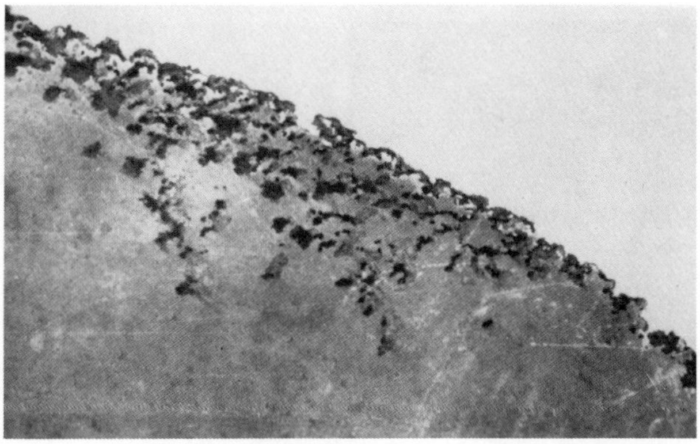

Fig. 15. Honeycombing present as a result of cavitation.

onment. At elevated temperatures, the problem is compounded by the reduced strength at elevated temperatures and the accelerated attack due to temperature. Figure 16 shows a metallographic section taken normal to the surface of an eroded tube of Inconel X. The marked destruction occurring at the surface is evident together with oxygen penetration of both the grains and grain boundaries.

Figure 17 illustrates the progressive thinning caused by erosion corrosion in a tubular section. The study also illustrates the fact that a failure may have multiple causes. In this case, progressive thinning of the wall occurred until the point where the hoop stress generated by the internal pressure exceeded the yield strength of the material and the section failed by overstress as shown in Fig. 17.

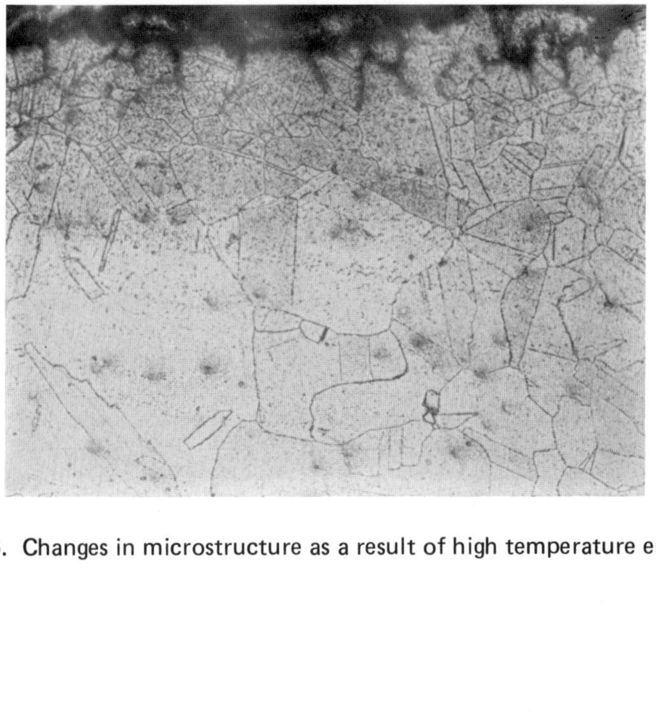

Fig. 16. Changes in microstructure as a result of high temperature erosion.

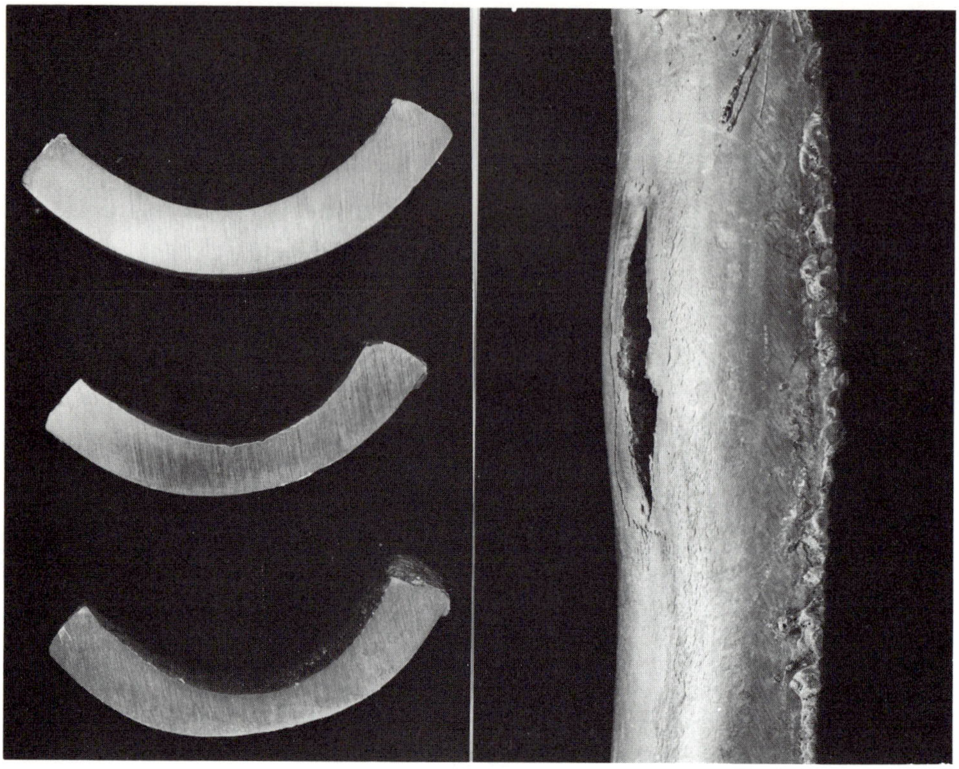

Fig. 17. Progressive thinning due to erosion culminating in final failure due to overstress.

STRESS CORROSION

Premature failure of alloys by stress corrosion has been recognized for many years. The ASTM held a symposium on the season cracking of brass [11] almost 60 years ago. The problem still exists and occasional failures are still seen. Such the interaction of mechanical tensile stresses in a chemically aggressive environment causes cracking that is impossible to predict by independent measurements or corrosion testing. Four basic requirements are necessary to cause stress corrosion cracking; a susceptible alloy, an aggressive environment, applied or residual stresses and time.

Stress corrosion has been observed in many diverse materials and environments. The current literature contains numerous references to fracture of titanium, brass, aluminum, magnesium and various steel alloys [12–20]. With such a variety of distinctly different conditions of fracture, it is difficult to find common features. In work by Hoar and Hines [21], Hines [22], and Parkins [23], the following characteristics consistent with this type of fracture have been observed:

a. There must be simultaneous action of stress and corrosion; alternate application will not produce similar results. The fracture surfaces are macroscopically brittle with little signs of ductile tearing.

b. The induction period necessary to produce crack initiation and pitting is relatively long compared with actual crack propagation. Stress appears to play little part in the induction period; corrosion is the primary driving force. Embrittlement of this region surrounding the crack tip may also be a significant factor.
c. The mode of cracking can be intergranular or transgranular, but is predominantly one or the other. No generalizations can be made regarding the effect of changes in heat treatment or compositions in the mode of cracking but one type of cracking appears to occur more readily for a given alloy.
d. The rate of attack is quite rapid at the crack tip; much less at the sides. The crack propagation process appears to be self catalyzing with the much larger rate being sustained only at the advancing crack tip.
e. Conditions for cracking are specific as to alloy and environment. Specific ions are usually necessary to promote cracking conditions. Although many environments may produce similar corrosion rates, susceptibility to stress corrosion of different alloys may be widely divergent. Certain corrodents giving relatively violent reaction with an alloy do not cause cracking. The question is thus raised of the contributing interaction of corrodent anions to failure.

Tensile stresses at the corroding surface are essential to stress corrosion cracking; however, these tensile stresses may be residual or applied. Both are detrimental to service life which is dependent on the magnitude of stresses. Residual stresses may be more of a problem, however, since they are often not apparent.

Stress corrosion failures generally exhibit little ductility and have the macroscopic appearance of a brittle fracture. There may be multiple cracks originating from the surface but failure usually results from the progression of a single crack on a plane normal to the main tensile stress.

In austenitic stainless steels, cracks are usually transgranular, frequently but not always associated with a specific crystallographic plane. These alloys also exhibit intergranular cracking on certain media notably caustic solutions and highly oxygenated chloride solutions.

Intergranular cracking is the predominant mode of failure for martensitic stainless steels. However, transgranular cracking has been observed in these alloys when tempered below 850°F.

In high strength steels, the crack path is intergranular as shown in Fig. 18. It is widely believed that stress corrosion cracking in these materials is related to a hydrogen embrittlement mechanism as a result of hydrogen being generated at the crack tip as a corrosion product. From a metallographic viewpoint, there is a great deal of similarity between the two. Figure 19 illustrates the appearance of an electron fractograph of a stress corrosion failure in a high strength steel.

The form of the cracks in brasses may be either intergranular or transgranular, depending on several variables. Gilbert has described the crack morphology in several media for α and β brasses as well as for aluminum bronzes.

Stress corrosion cracking in aluminum alloys is characteristically intergranular. Al–Cu, Al–Zn–Mg and Al–Mg alloys are most often affected.

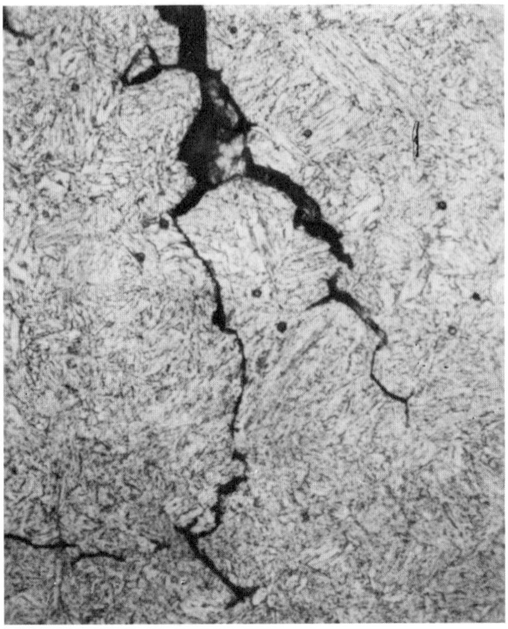

Fig. 18. Intergranular path of a stress corrosion failure in a high strength steel (4355 + V).

Fig. 19. Electron fractograph of stress corrosion failure of a high strength steel.

Of the magnesium alloys, Mg–Al and Mg–Al–Zn are susceptible to stress corrosion damage; attack is usually transgranular.

CORROSION FATIGUE

Corrosion fatigue may be considered as a special case of general fatigue, with certain modifying effects resulting from the environment. The time or cycles required for fatigue crack initiation can be markedly reduced by corrosion processes that create pits or other surface damage. In addition, the fatigue crack propagation rate can be increased by a corrosive environment. Such behavior has been noted in many studies, even when the corrodent was relatively innocuous, such as moist air.

In corrosion fatigue, the total test time is an important factor. Since corrosion is a time-dependent phenomenon, fatigue tests employing lower frequencies exhibit greater degradation than do tests at higher frequencies, where the overall exposure time at stress is greater. Increased temperature also tends to increase crack propagation rate and consequently, to reduce the fatigue life.

A component that has suffered corrosion fatigue damage may or may not exhibit general corrosion on the fracture surface, depending on the immersion duration. Similarly, an ordinary fatigue crack can corrode after the fact, so that immediate identification as corrosion fatigue, based simply on the presence of corrosion products, is not always possible. However, it is known that a corrosion fatigue specimen often exhibits a different macroscopic fracture appearance than one tested in air as shown in Fig. 20. The bars broken in air were bright and lustrous whereas those tested in a corrosive environment exhibited a dull surface with evidence of corrosion product.

Another indication of corrosion fatigue is the presence of a number of cracks, rather than one. These cracks are usually perpendicular to the principal tensile stresses and originate at the surface where the stresses are at a maximum. In general, these fatigue cracks are transgranular in nature; however, some intergranular cracking is found, for example, in lead and aluminum systems that have been embrittled.

Fig. 20. Macroscopic appearance of corrosion fatigue. Specimen on the left tested in 0.9% saline, the specimen on the right tested in air.

Figure 21 is a photograph of the fracture surface of a orthopedic plate which has failed in fatigue. Note the presence of beach markings and the slight discoloration of the surface. Figure 22 is a view of an intermedullary rod which has also failed via fatigue after being implanted for 5 years. Faint beach marks are evident on the fracture surface.

On a microscopic level, striations are often found by examination with electron fractographic methods. These striations are often less pronounced than those found in conventional fatigue, which is probably the result of environmental action on the already fractured surface, but may be related to a change in mechanism as well. Figures 23 and 24 illustrate striations found on the fracture surface using electron fractography.

Fig. 21. Corrosion fatigue in an orthopedic surgical implant.

Fig. 22. "Beach marks" present on fracture surface of a fatigue intermedullary rod.

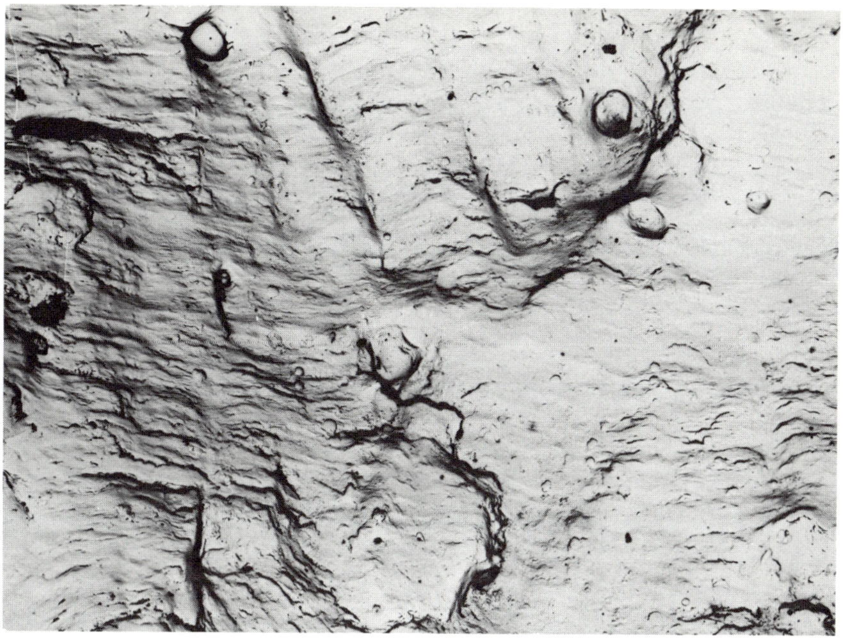

Fig. 23. Electron fractograph showing fatigue striations on the surface of the intermedullary rod, near origin.

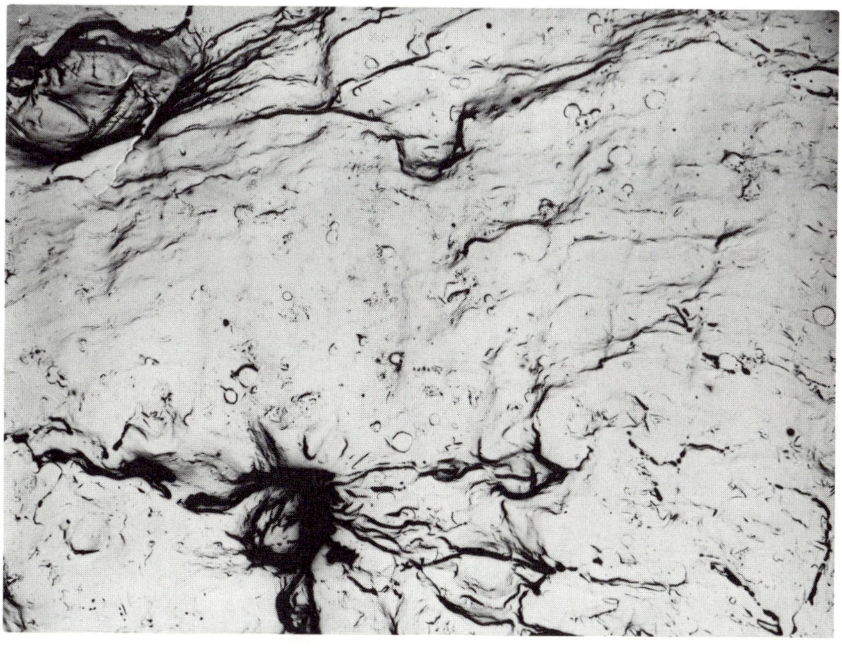

Fig. 24. Same as Fig. 23, further from origin.

REFERENCES

1. Fontana, M.G. and Greene, N.D., *Corrosion Engineering,* McGraw-Hill, N.Y., p. 50 (1967).
2. Greene, N.D. and Fontana, M.G., *Corrosion,* Vol. 15, 25t, (1959).
3. Fontana, M.G. and Greene, N.D., *op. cit.,* p. 57.
4. Espy, H.L., *Metal Progress,* p. 109–115, (Sept. 1964).
5. Shirley, H.T., *Corrosion,* edited by L.L. Schreir, Wiley and Sons, p. 3.62 (1963).
6. Ugiansky, G.M. and Ellinger, G.A., *Corrosion,* Vol. 24, No. 5, p. 134, (1968).
7. Clark, W.D., *J. Inst. Metals,* Vol. 73, p. 263 (1947).
8. Brooks, W.B., *Corrosion,* Vol. 24, No. 6, p. 171 (June, 1968).
9. Lifka, B.W. and Sprowls, D.O., *Corrosion,* Vol. 22, p. 7–15, (Jan. 1966).
10. Venzcel, J., Knutsson, L., and Wranglen, *Corrosion Science,* Vol. 4, p. 1, (1964).
11. *Topical Discussion on Season and Corrosion Cracking of Brass,* Proceedings, Am. Soc. Testing Matls., Vol. XVIII, Part II, p. 147–219 (1918).
12. Newcomer, R., Tourkakis, H.C., and Turner, H.C., *Corrosion,* Vol. 21, #10, p. 307–315 (October 1965).
13. Peterson, M.H., Brown, B.F., Newbegin, R.L., and Groover, R.E. *Corrosion,* Vol. 23, #5, p. 142–148, (May 1967).
14. Johnson, H.H. and Leja, J., *Corrosion,* Vol. 22, #6, p. 178–189 (June 1966).
15. Champion, F.A., "The Assessment of the Susceptibility of Aluminum Alloys to Stress Corrosion", *Symposium on Stress Corrosion Cracking of Alloys,* (STP64) ASTM, p. 358–378 (1944).
16. Dean, S.W. and Copson, H.R., *Corrosion,* Vol. 21, #3, p. 95–101 (March 1965).
17. Johnson, H.H. and Willner, A.M., *Applied Materials Research,* Vol. 4, #1, p. 34–40 (January 1965).
18. Van Der Sluys, W.A., *Transactions of the ASME: Journal of Basic Engineering,* p. 28–34 (March 1967).
19. Henthorne, M., and Parkins, R.N., *Corrosion Science,* Vol. 6, p. 357–369 (1966).
20. Hancock, G.G. and Johnson, H.H., *Materials Research and Standards,* Vol. 6, #9, p. 431–435 (Sept. 1966).
21. Hoar, T.P. and Hines, J.G., *J. Iron Steel Inst.,* 182, p. 124 (Feb. 1956), and 182, p. 166 (Oct. 1956). (2 part)
22. Hines, J.G., *Corrosion,* ed. by L.L. Shreir, Vol. 1, John Wiley and Sons, p. 8.3 (1963).
23. Parkins, R.N., *Stress Corrosion Cracking and Embrittlement,* ed. by W.D. Robertson, John Wiley & Sons, p. 140–157 (1966).

FAILURE ANALYSIS OF AEROSPACE COMPONENTS

T.W. Heaslip*

INTRODUCTION

After the shock and confusion of an air disaster, investigator specialists are charged with picking up the straws and trying to piece together the catastrophic sequence of events that led to the tragic loss of human and material resources. If the machine itself failed unexpectedly and prematurely during the lead event or at some subsequent stage, then invariably engineering specialists will be drawn into the investigation team. If material problems are involved or suspected, then the Materials/Metallurgical Engineering Specialist will be required to perform failure investigations of specific structures, components or parts. In Canada this is one of the functions carried out by the Aviation Safety Engineering Facility in the Department of Transport.

The use of Scanning Electron Microscopy (SEM), X-ray energy dispersive techniques, and metallography are all key stages in the analysis of failures in this laboratory. The SEM capability is unique in that the Engineering Facility has one of the few real-time 3-dimensional image generation systems used in failure analysis in the world (Fig. 1).

This technique takes out the guesswork on topography analysis which is required when using 2-dimensional projection. It removes the ambiguities, especially those you are not even aware exist.

The technologically current X-ray spectrometric analysis capability (Fig. 2) allows the qualitative and semiquantitative chemical analysis of bulk materials, specific zones of parts, grains, phases, and residue on fracture faces and adjacent surfaces. This capability is an important advance employed to ensure a complete investigation of chemical composition, segregation effects, foreign materials, environmental and other effects which leave chemical evidence.

In addition, the projection metallograph is sufficiently sophisticated to perform microstructure analysis at standard and continuously variable magnifications up to 3200X with projection onto a large screen allowing the simultaneous observation by a small number of investigative participants. If the observing group is large there is the capability to also display the image on a color television monitor (Fig. 3). The use of metallography ensures that the metallurgical evidence is thoroughly analyzed.

CASE HISTORIES

To demonstrate the use of these tools and the essential nature of such techniques in comprehensive failure analysis, two recent incidents have been chosen in which the

---

*Aviation Safety Bureau, Transport Canada, Ottawa, Ontario, CANADA.

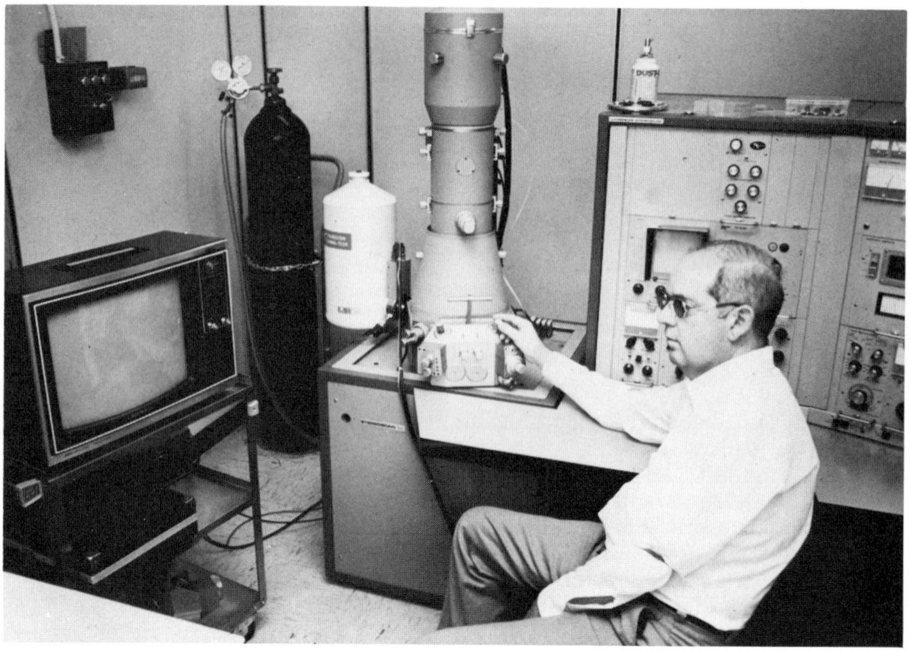

Fig. 1. Scanning Electron Microscope (SEM) with 3–D T.V. display.

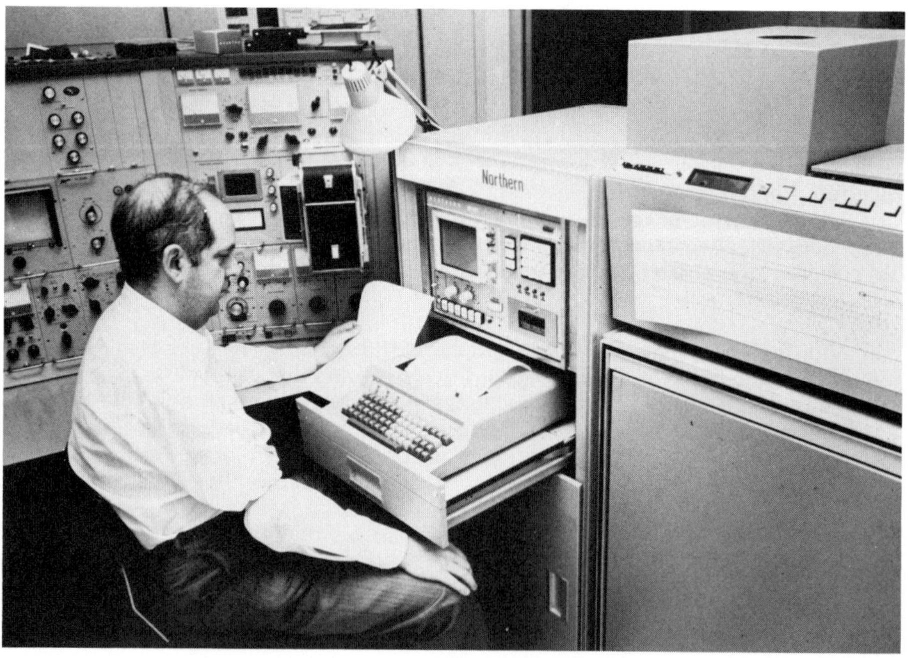

Fig. 2. X-ray spectrometric chemical analysis system.

pertinent areas of investigation will be described.

UNDERCARRIAGE TRUNNION FAILURE

An unusual noise was heard on a Transport aircraft during a right turn onto a runway just prior to take-off from a Canadian airport. After a normal take-off, when the gear was retracted, the red in-transit light for the left hand main undercarriage failed to go out. When the gear was selected down on approach at destination, a sharp bang was heard. The crew was unable to determine if any problem existed with the gear, so a landing was attempted. After touch down the aircraft rolled for a few seconds and then the left wing settled onto the runway. The left hand main wheel assembly separated and punctured the left hand inboard fuel bay and the aircraft slewed off the runway igniting the spilled fuel (Fig. 4).

The accident investigators found that a trunnion arm shaft had failed prematurely on the left hand main gear. Figure 5 shows the main landing gear (MLG) assembly and gives the location of the trunnion and the shaft failure.

Trunnion Examination

The trunnion arms (Fig. 6) were given to Materials Engineering specialists to analyse the component fracture and to determine the reasons for failure.

Figure 7 shows the fracture through the shaft of the main trunnion arm forging. Fortunately one fracture face of the separated shaft was relatively protected from post-separation damage during the accident. This face, shown in Fig. 8, was ultrasonically solvent cleaned to remove much of the fire residues.

Distinctive progressive crack growth markings were found originating from a single initiation area. These were initially suspected as being *fatigue* beach markings. A prominent coarse-surfaced band was noted at mid-diameter (g, Fig. 8). The last 9% of the fracture opposite the crack initiation zone (e, Fig. 8) was angled with respect to the relatively flat fracture surface, and appeared characteristic of final overload rupture.

Scanning Electron Microscope Analysis

A detailed scanning electron microscopic (SEM) evaluation was carried out on the shaft fracture face. The fracture mechanism in the origin area was found to be entirely intergranular in nature (Fig. 9).

Subsequent stages of crack growth were also intergranular (Figs. 10 & 11) and typical of stress corrosion cracking (SCC). In later stages an increasing percentage of local ductile tearing and cleavage was noted intermixed with the SCC intergranular fracture mode. The diametral band (g, Fig. 8) was primarily cleavage with a small amount of ductile dimples (Fig. 12). Final rupture (e, Fig. 8) displayed a mixture of ductile tearing (microvoid coalescence) and some cleavage as shown in Fig. 13.

Metallography Analysis

Metallurgical evaluation was carried out on sections cut from the broken trunnion arm shaft. The axial sections cut from the fracture face to the free, unstressed end of the trunnion arm shaft showed an irregular but repetitive pattern of intergranular cracks in the steel under the chromium plating, averaging round 0.020" deep (Fig. 14). Of significance was the fact that some of these cracks did not exhibit continuity into and through the plating, which indicates they were not due to propagation of plating cracks

Fig. 3. Projection metallograph with TV display.

Fig. 4. Crashed transport aircraft.

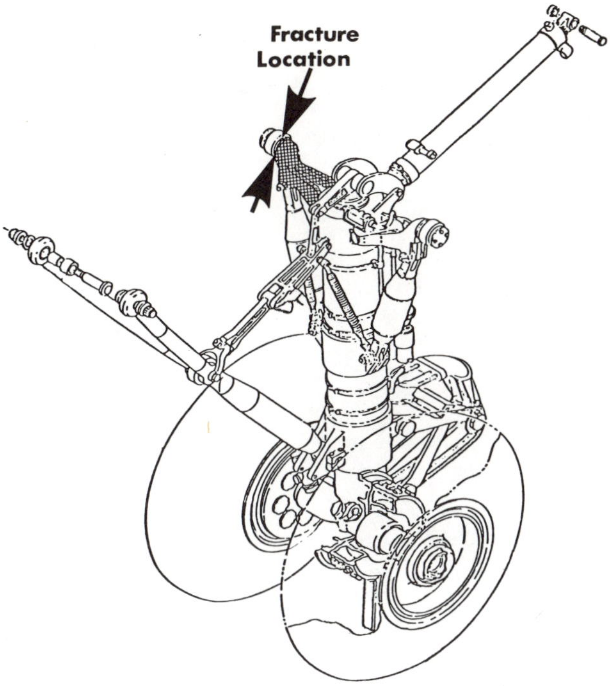

Fig. 5. Main landing gear (MLG) drawing showing trunnion arm location.

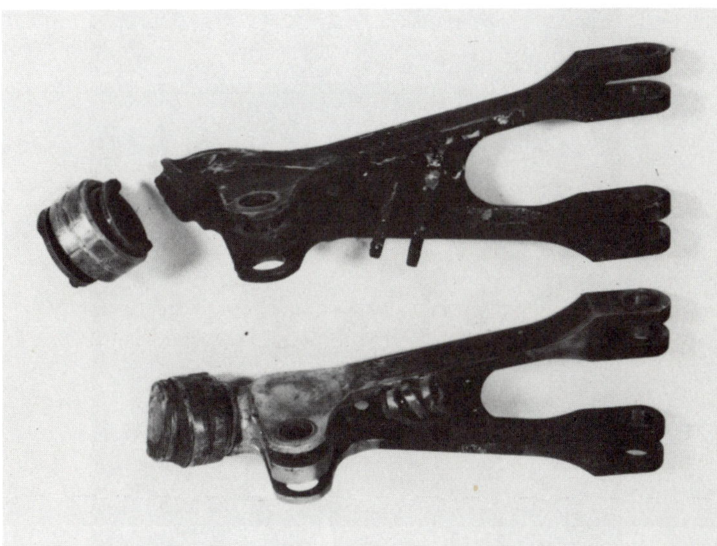

Fig. 6. Both trunnion arms from left MLG. Failure occurred through shaft of the right hand arm.

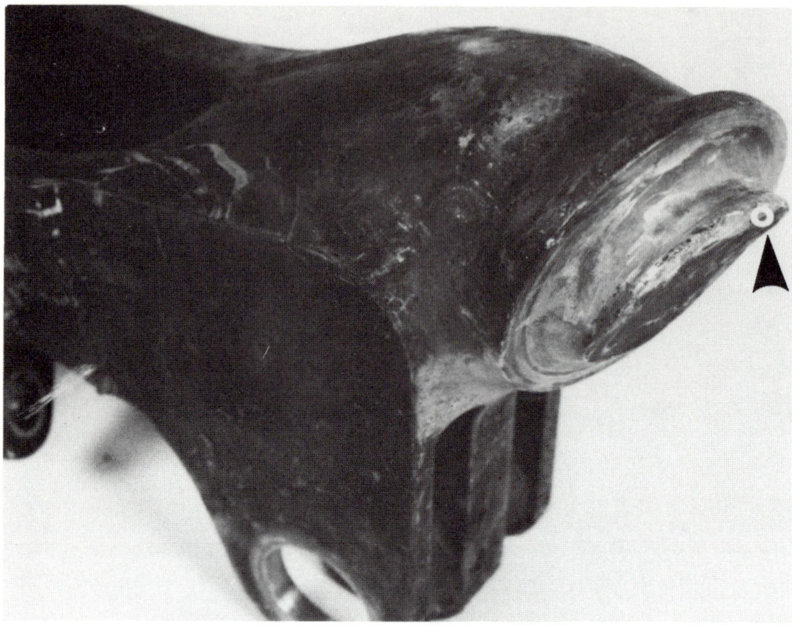

Fig. 7. Side view oт fracture. Note smeared origin 'O'.

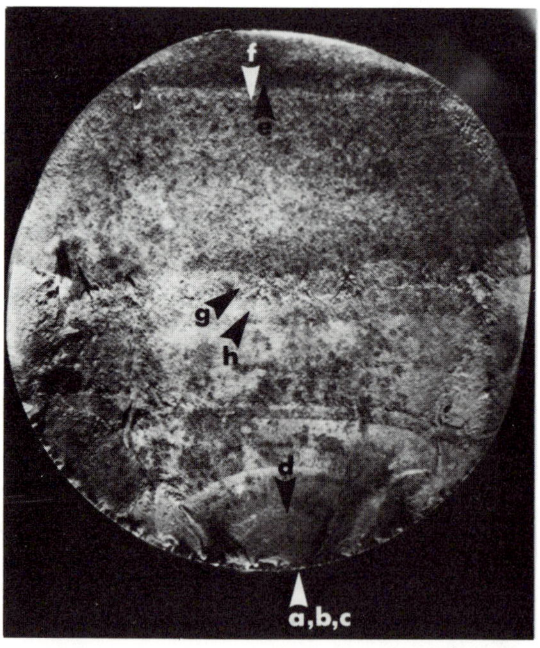

Fig. 8. Fracture face. Approximately 2" diameter.

Fig. 9. SEM image of origin. Note chrome plate zone at bottom.

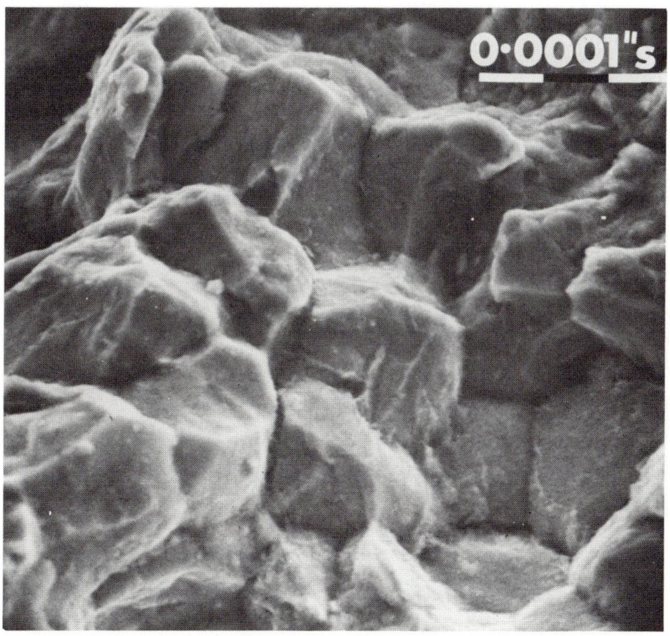

Fig. 10. SEM image of intergranular area 'a'.

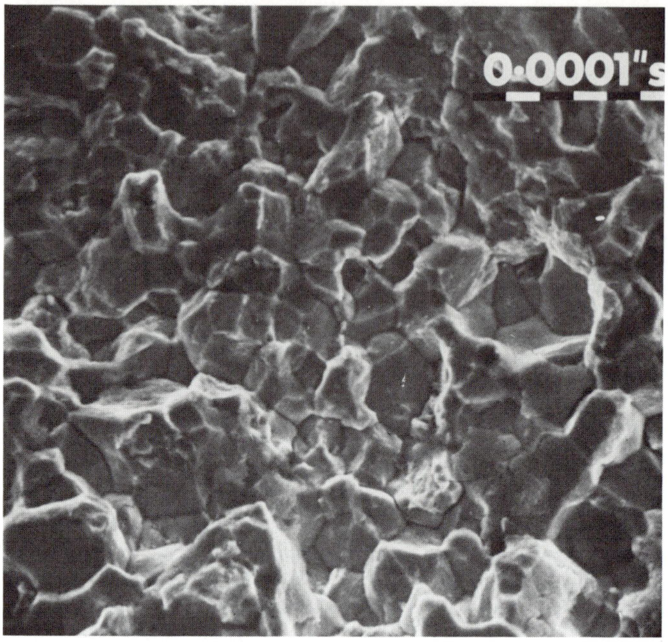

Fig. 11. SEM image of intergranular area 'd'.

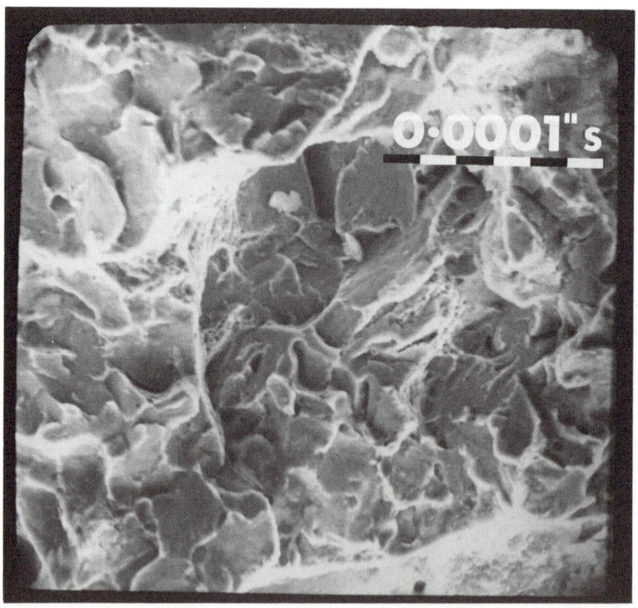

Fig. 12. SEM image of overload band, primarily cleavage with some dimples (Area 'g').

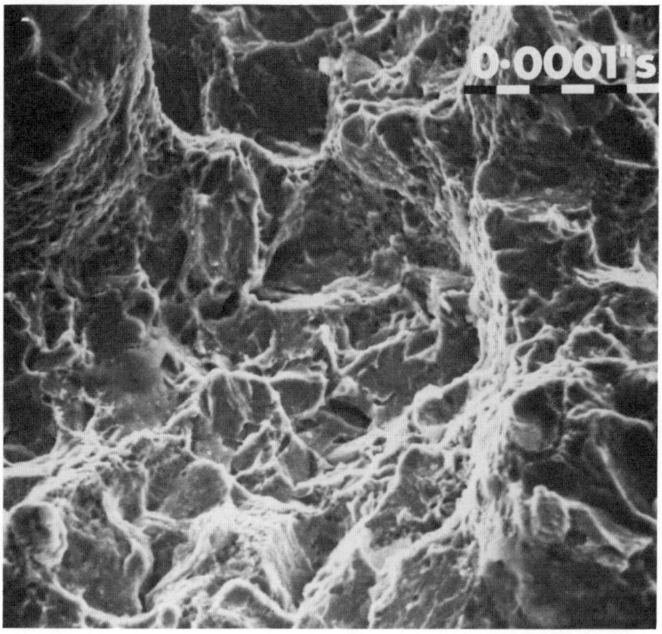

Fig. 13. SEM image of overload area, primarily dimples.

Fig. 14. Typical micro-crack under plating.

on into the steel. It is apparent that the cracks were either present in the steel prior to addition of the plating, or had initiated in the steel after plating. The irregular crack spacing was of the order 0.010" – 0.020".

The general core microstructure as shown in Fig. 14 was typical of a fine-grained, through-hardened and tempered low-alloy steel (4340-type low-alloy steel). The core hardness was uniform, giving an equivalent strength of 260–280,000 psi strength level. There was no evidence of any transformed (martensitic) structure at the surface under the plating, except in the smeared area immediately adjacent to the fracture. Instantaneous rupture and separation energy release has been proven to be capable of generating sufficient local heating to transform steels. The transformation and overtempering seen adjacent the fracture in the smeared area was considered due to fracture energy release only. It was noted that exposure to the post-accident fire had heated the part sufficient to blister the paint but not to temper the steel. These factors were all established by microhardness traverses.

A section of the defective trunnion arm shaft adjacent to the microsections used for metallurgical evaluation was electro-chemically stripped of chromium plating to allow evaluation of the steel substrate. Figure 15 shows a typical region of the steel surface. A large network of long, crosslinked but primarily circumferential cracks were found on this section. Also the surface had evidently not been shot peened prior to plating.

X–Ray Spectrometric Analysis

The alloying elements of the steel were then checked using a fingerprinting technique on the X-ray energy-dispersive equipment by comparing the unfailed trunnion shaft with the failed shaft. The spectrometric chemical analysis display showed the appropriate alloying elements (Ni, Cr, Mo) in the correct proportions for a 4340 low-alloy steel. The plating was confirmed as chromium and no chemical abnormalities were disclosed. Residue (such as corrosion and environmental products) on the fracture face could not be analyzed because the fire and foam effects had to be cleaned from the face to allow topographic analysis.

Analysis of Fracture Mechanism

At this stage it was decided that the basic mechanism of failure must have been intergranular stress corrosion cracking (SCC). Fatigue was discounted because no striations were found during the topography analysis and fatigue is usually transgranular. Also the progressive zone was extremely deep for a high strength notch sensitive steel. The number of stress cycles from side loads on landings and turns during taxiing could only have been in the order of 1000's because of the short time in service (to be discussed under TRUNNION). Therefore a fatigue mechanism would have been high-stress, low-cycle but the extensive cracking precludes this possibility. Failure was associated with very low stress levels. A stress analysis was then performed to determine the stress conditions.

Stress Analysis

An analysis of the loading conditions showed that a 1G static load at gross weight on the aircraft should produce a maximum stress of 40,000 psi in the trunnion arm shaft section. However, the threshold stress for SCC in a 260,000 psi steel is approximately 90,000 psi. Since the 40,000 psi maximum stress is based on a smooth cylindrical shaft section, it is evident some initial sharp stress concentration must have been present to initiate stress corrosion cracking.

Of course, the multitude of intergranular *pre-cracks* under the chromium plating would have acted as severe stress concentrators. These pre-cracks even existed in areas close to the free end of the shaft, which could not have had any significant stress developed by the weight of the aircraft while resting or taxiing and the only other additional continuous stresses could have been residual stresses. The trunnion arm shaft pre-cracks were determined to be typical of 'grinding' cracks introduced into the steel due to severe local overheating during the machining undersize operation at overhaul. Figure 16 shows the crack network displayed under black light on the chrome-plated shaft.

'Grinding' cracks can easily be developed in such high-strength steels if care is not taken during any metal removal operation. Normally such local overheating leaves a shallow transformed martensitic surface layer as evidence of overheating. However, if cracking was developed by an abusive initial coarse cut, a subsequent fine finish cut could remove all evidence of the shallow transformed surface layer but leave the residual crack roots in the steel. It was concluded that the precracks were produced by improperly controlled metal removal procedures prior to hard chromium plating. The *pre-cracks* were not typical of hydrogen embrittlement or stress corrosion.

Trunnion History

The undercarriage assembly had been overhauled one and a half years prior to the accident. As a part of this overhaul the trunnions had been reworked. Subsequently the undercarriage assembly was installed in the aircraft and was subjected to exactly one year of service before the trunnion failed. Stress corrosion crack propagation developed during that year except for the diametral band, which was a band of rapid crack growth under a momentary high load condition. The design of the gear assembly was such that *side loads* would be most responsible for the development of high stresses. Therefore the local jump forward of the crack front in mid-growth was probably the result of a fast taxi turn at near gross weight or a high side load landing.

The investigation next centered on the source of the defectively reworked trunnion. The pre-cracks were apparently introduced during the rework cycle. The standard rework package at overhaul called for machining undersize of the shaft prior to hard chromium plating back to size. The question is "How did the defective component pass through the quality control checks designed specifically to prevent such defects?" At the overhaul teardown of the undercarriage assembly, it was apparently established that the trunnion arm shafts were sufficiently corroded under the steel bushing to warrant having the shafts machined undersize, built up with hard chromium plating, and then remachined to original dimensions. (There is no wear problem in this part since the bushing is shrunk onto the shaft; therefore, only corrosion damage would necessitate rework).

The overhaul agency's processing sequence included two operations, which were carried out at different stages in the process, specifically designed to ensure the detection of 'grinding' defects after the machining undersize operation; these were:—

Fig. 15. Cracking network on surface (arrows).

Fig. 16. Black light fluorescent crack network.

1) A nital etch of the bearing shaft diameter to reveal evidence of any "grinding burns", and
2) A magnetic particle inspection to reveal any evidence of "grinding" cracks (prior to plating).

It was evident that these checks could not have been carried out properly, if at all. However, the identification markings on the failed trunnion were not consistent with the overhauler's work statement sheets; whereas the other unfailed trunnion was unambiguously and correctly marked according to the work statement sheets. Destructive tests were carried out on the two trunnion shafts for direct comparison. Sample sections were subjected to electrochemical removal of the chromium plating which was stopped before the plating was completely removed. Figure 17 shows that the two trunnion arm shafts had been processed in an entirely dissimilar manner for every stage which left a measureable characteristic.

For instance, the failed shaft was *ground* undersize while the unfailed shaft appears to have been *turned* undersize with a single point tool. The failed shaft was grit blasted with no peening whereas the unfailed shaft was shot peened. The fine crack network in the plating on the failed shaft was typical of a high-speed, low-crack bath while the large crack network in the unfailed shaft plating was typical of a chromic/sulphuric conventional bath. Also the end of the failed shaft was detail painted while the unfailed was masked and spray painted. The differences are so great as to strongly suggest they were not processed in the same plant, despite the fact that they were supposedly overhauled together. The unfailed trunnion conformed to the overhaul facility's work statement procedures. The failed trunnion was therefore apparently a *bogus* part.

Conclusions of the Failure Investigation

- The right hand trunnion arm of the left hand main gear failed due to SCC through the shaft until normal operating loads ruptured the remaining section.
- The SCC initiated from pre-existing 'grinding' cracks located in the trunnion arm shaft under the chromium plated surface.
- Abusive removal of the previously corroded shaft surface during rework led to local overheating and the development of the 'grinding' cracks.
- The trunnion was apparently a bogus part, the source of which was undetermined.

Investigations continue to determine where the defective rework operation was performed and how the part came to be installed on the subject undercarriage assembly.

THIRD–STAGE TURBINE FAILURE

A jet helicopter was involved in an accident on a mountain side following an engine failure. After normal start-up and take-off, at an altitude of approximately 20 feet and just prior to entering translational flight, moderate vibration coupled with abnormal engine noise was experienced. The vibration rapidly increased in intensity and an immediate landing was attempted. The helicopter impacted (Fig. 18) sloping ground, and was substantially damaged when it rolled sideways and was also subjected to a post-crash fire, which was quickly extinguished.

Fig. 17. Processing comparison of failed shaft (on right) against unfailed (on left).

Fig. 18. Helicopter crash on a mountain side.

Subsequent examination of the engine disclosed that it was still rotatable, although some stiffness was evident. The engine was stripped at the operator's facility and it was discovered that a section of the third-stage turbine wheel circumferential shroud had separated (Fig. 19), taking with it about 0.4 inch of one vane.

The failed components were initially analysed by an outside consultant agency for the insurance company. The fractures were not, however, allowed to be cut or physically altered in any way during analysis. The parts were then forwarded to the Safety Bureau Facility for investigation.

Initial Analysis

Figure 20 shows the fracture faces A, B and C in the third-stage turbine wheel. The fractures A and C displayed some isolated areas of shrinkage porosity on the blade portion of the fractures but the balance of the fracture faces (in the shroud) revealed no evidence of being primary failure locations. However, the consultant agency established, using a scanning electron microscope, that on fracture 'b' large areas of shrinkage voids existed near the leading edge of the blade section, and a large zone of interdendritic separation existed from the void zones to near the trailing edge. It was considered the interdendritic area was likely associated with extensive microporosity created during casting. The analysis considered why such an apparently defective cast turbine wheel would have operated 807 hours before failure occurred. Creep and thermal fatigue were considered as possible mechanisms of progressive failure. However, no indications of a fatigue mechanism were found on fractographic analysis. Also, extensive secondary cracking was noted, which is not consistent with a thermal fatigue mechanism. Therefore, this analysis concluded that the failure of the wheel was the result of a "cavitation creep" mechanism which culminated in a stress rupture failure of the turbine wheel. The creep strength, it was decided, had been severely reduced due to the presence of extensive shrinkage porosity found in the component. The cause of failure was, therefore, seen as basically due to a *manufacturing deficiency.*

Aviation Safety Bureau Analysis

With this background Materials Engineering specialists in the Safety Bureau examined the turbine remnants. It was found that the wheel shroud fractures A and C displayed minor shrinkage porosity only. They were determined to be secondary overload failures which occurred after the failure in the blade B. Therefore, detailed examination and analysis was concentrated on turbine blade B fracture face and on sample metallographic sections.

Scanning Electron Microscope Analysis

It was obvious that the leading edge zone in blade B (Fig. 20) did contain large areas of gross shrinkage voids and the balance of the fracture was indeed largely of an interdendritic separation nature. However, fine details were masked by a surface film of corrosion and combustion products.

Extensive prior experience with preparation of fractures for scanning electron microscope fractographic analysis has shown it is essential to initially examine the fractures without attempting other than degreasing cleaning to remove volatile agents which may

Fig. 19. Turbine wheel failure.

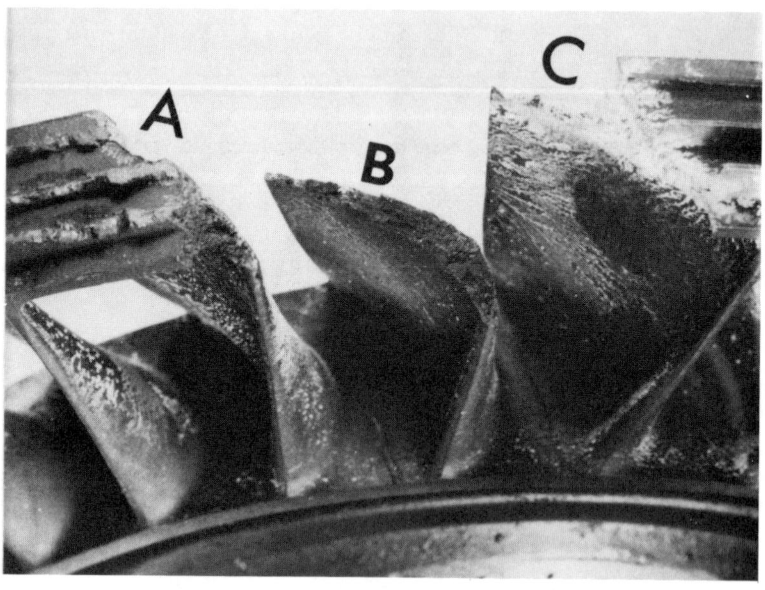

Fig. 20. Blade fractures A, B and C.

contaminate the vacuum chamber of the SEM. Analysis of the non-volatile surface contaminants may be critical to the failure analysis process. In this case, however, it was felt the surface film of corrosion and combustion products was detrimental to the analysis of the failure. Therefore an attempt was made to remove as much residue as possible without causing damage to the fracture surface proper. Solvent and detergent ultrasonic cleaning of this type of nickel-base superalloy does not cause damage to the metal surface. The blade fracture was processed through a program of detergent and solvent ultrasonic cleaning, then remounted and re-examined in the SEM. It was found that significant amounts of the surface soils and combustion products had been effectively removed, and new details were now available for analysis.

Figure 21 shows the cleaned blade fracture ready for SEM analysis. The gross shrinkage voids near the leading edge are readily visible at low optical magnifications.

Figure 22 clearly shows one of these zones in detail. The existence of extensive free standing dendrites indicates that these areas were interconnected voids created during solidification of the casting, and hence were present prior to fracture and are not a result of any in-service failure mechanism.

The cleaning procedures allowed more detailed examination of the relatively flat transgranular zones adjacent to the major voids (Fig. 23). Where these were largely featureless on as-received examination, they now were found to contain many areas of a coarsely striated nature (Fig. 24). These striations showed a general pattern as follows;

> Firstly, they were in general oriented roughly perpendicular to a path from the gross void zones towards the nearest free surface of the blade or towards the blade trailing edge; secondly, the spacing between the striations tended to increase with distance from the void zone; and thirdly, secondary fissuring followed the striations (Fig. 25). No significant fine details were noted in the interdendritic zone of fracture, but as examination progressed towards the trailing edge, zones typical of shear lips were noted increasingly at the free surfaces of the blade. Towards the trailing edge, zones of blocky cleavage-like fracture were also increasingly apparent (Fig. 26).

Metallographic Analysis

The polished and etched blade 'B' metallographic section cut parallel to and below the fracture face was subjected to metallurgical evaluation. The sample showed a normal coarse-grained cast structure with neither macro- nor micro-porosity present (Fig. 27).

Examination of another microsection cut from the adjacent blade at the span location corresponding with the primary fracture, showed similar general characteristics. There was, however, an area of minor porosity and segregation near the leading edge (Fig. 28) where there was an increase in the carbide content and depletion of the dispersed gamma prime phase near the pores (Fig. 29). The normal blade microstructure is shown in Fig. 30.

X—Ray Spectrometric Analysis

An extensive study was carried out to determine the significance of the area of segregation around the pores shown in Fig. 29. X—ray analysis showed a slight depletion

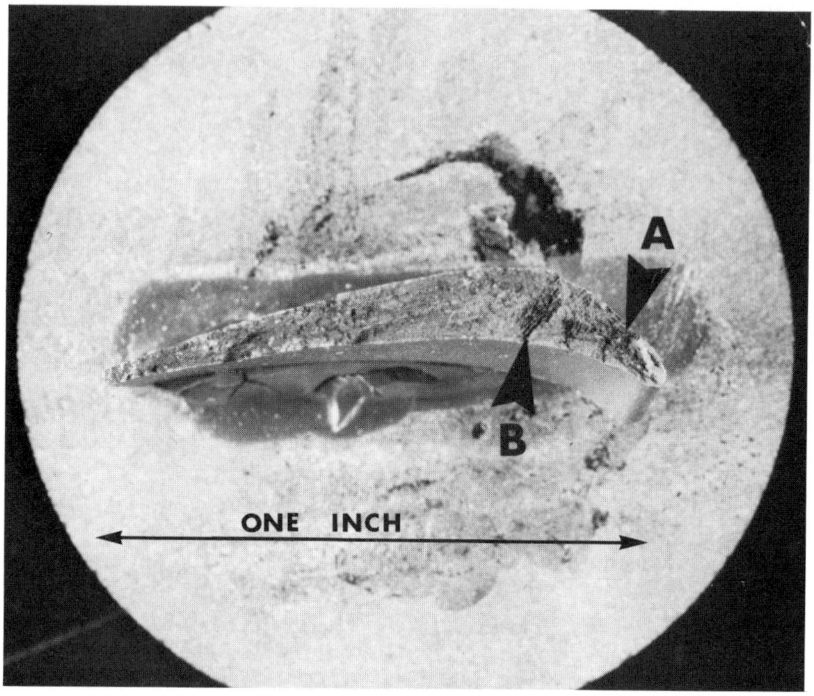

Fig. 21. Cleaned blade fracture. Note shrinkage porosity zones A and B.

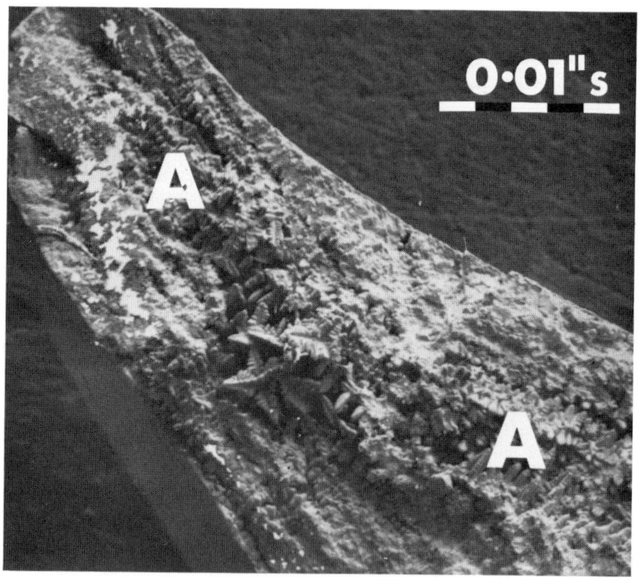

Fig. 22. Gross shrinkage voids 'A' with dendrites.

Fig. 23. Flat zones 'F' and 'X' around voids 'A' and 'B'.

Fig. 24. Typical striated zones in area 'X'.

Fig. 25. Secondary fissuring along striations.

Fig. 26. Blocky cleavage.

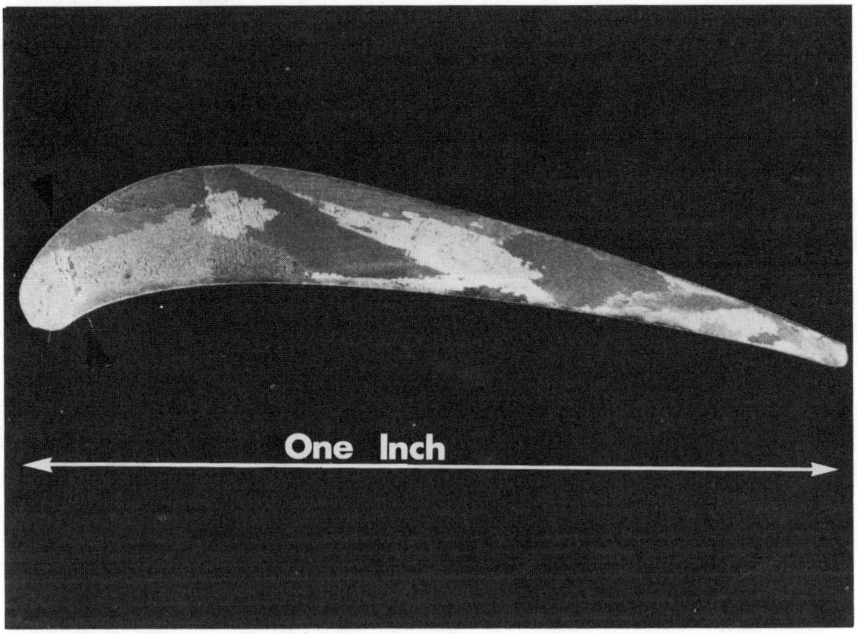

Fig. 27. Typical blade microsection.

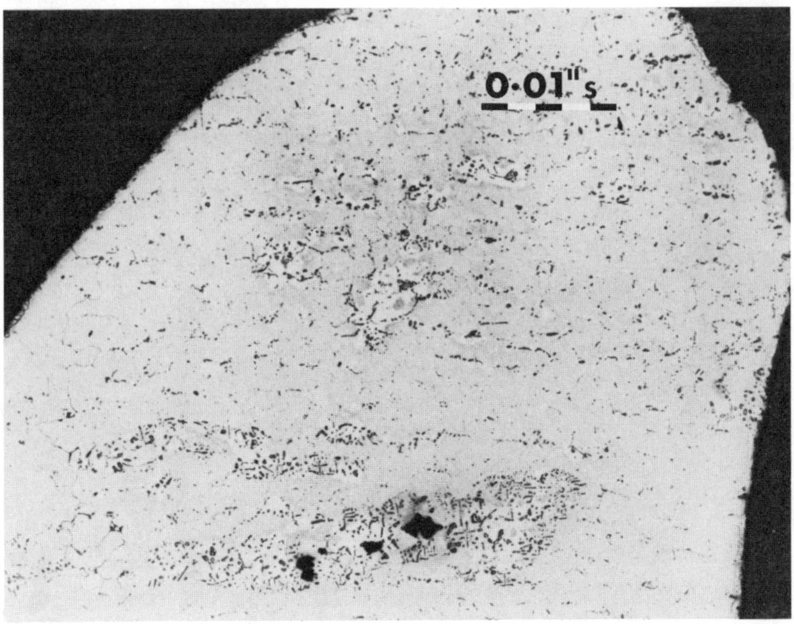

Fig. 28. Minor porosity in leading edge.

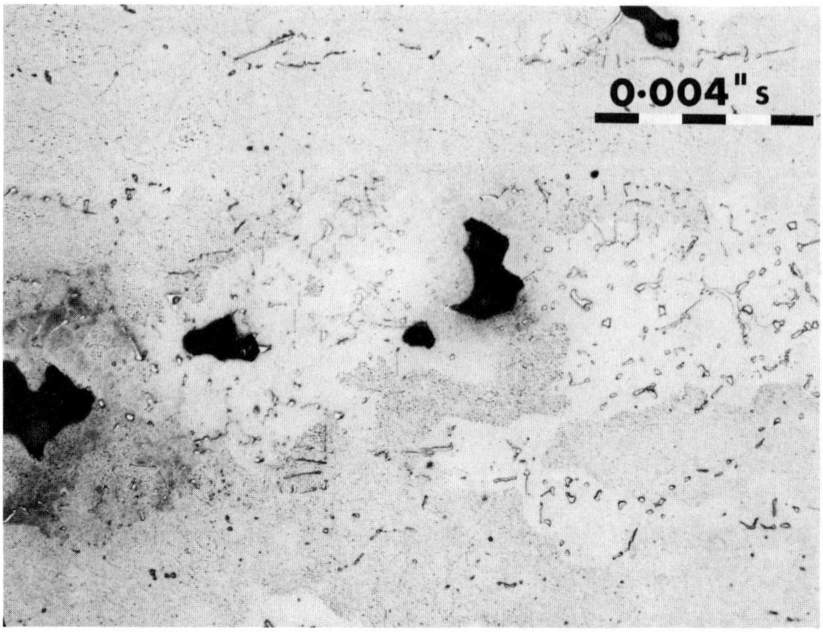

Fig. 29. Enlargement of the pores.

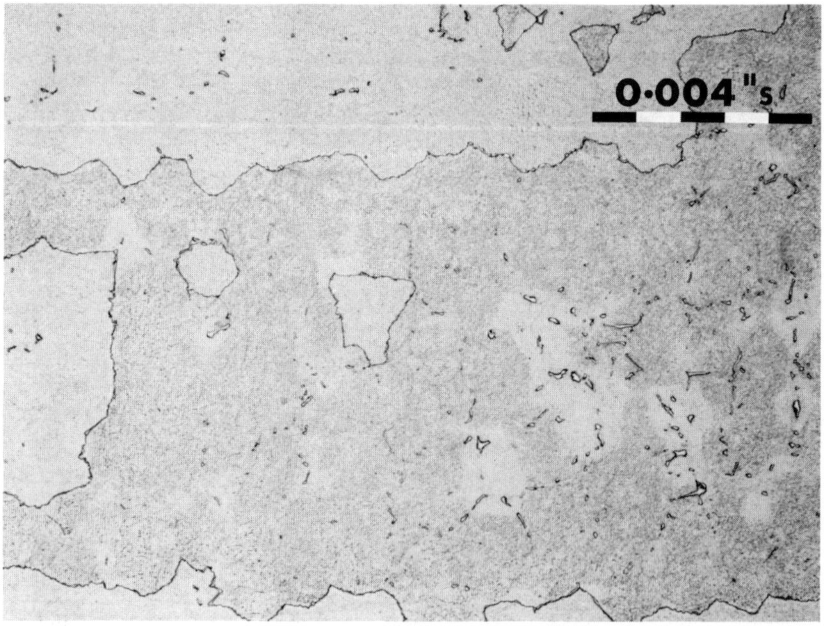

Fig. 30. Typical blade microstructure.

in nickel, and the niobium and molybdenum carbide concentration was also slightly higher around the pores. There was no other detectable variation in chemistry. The general composition conformed to the manufacturer's specified 713-type nickel alloy requirements for Cr, Mo, Cb, Ti and Al. The variations around the pores were consistent with these being the last zones to solidify. The enrichment and depletion of certain elements was therefore considered to be typical and not serious by the Bureau specialists and the manufacturer.

Failure Analysis — Comparison

It is clear that there were very significant areas of difference in the two investigations. As a result of the removal by cleaning of much of the post-fracture corrosion and combustion products, the Bureau analysis disclosed bands containing striation patterns (Fig. 24) around the major shrinkage void zones. The pattern is evidence of the blade section having been subjected to very high-stress, low-cycle fatigue crack growth. It was determined with the manufacturer that the adjacent zone of interdendritic separation is characteristic of overload failure at operating temperature. Figure 31 shows in diagrammatic format the results of detailed fractographic analysis.

In essence, this indicates the turbine wheel was suddenly subjected to a major change in stress sufficient to initiate fatigue crack growth from the internal casting flaws present in the leading edge area of the blade. Since the cyclic stress level was so high, the blade failed in overload soon after this change in stress state occurred, after quite limited but very rapid progressive (fatigue) crack growth. The topographic evidence suggests the cyclic crack growth to be limited to the low thousands or less order of magnitude. Given the speed of rotation (35,000 rpm) this may represent less than a minute from initiation to rupture. Prior to this the blade had operated without any progressive crack growth for some 807 hours.

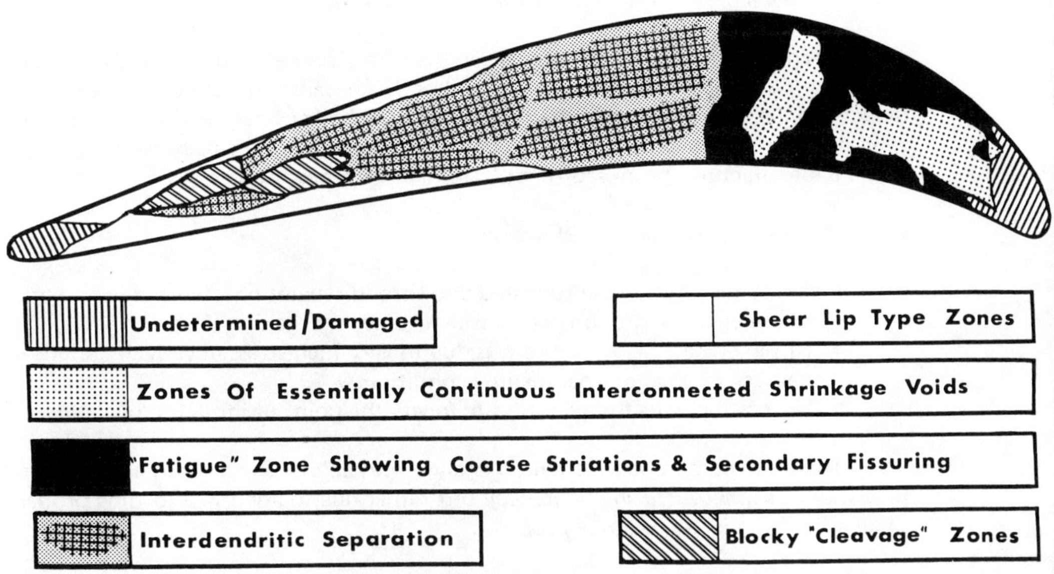

Fig. 31. Identification of fracture mode zones.

The previous investigation however, concluded the blade must have failed as a result of a "cavitation creep" mechanism; the creep strength having been severely reduced by the confirmed presence of extensive shrinkage porosity in the leading edge area. This conclusion was apparently based on the analysis of the interdendritic fracture zone being associated with extensive microporosity created during casting and serving as nucleation sites for cavitation creep failure. However, such interdendritic separation can also be characteristic of overload failure at operating temperatures. No positive evidence of microporosity was found in the microsection cut from the primary failed blade or from an adjacent blade at the span location corresponding to the primary fracture.

The significance of the Bureau's findings is that the sudden initiation of progressive failure due to a sudden increase in cyclic stresses imposed on the turbine blade indicates something other than the existence of the casting defect alone was responsible for initiating this change in stress state. Neither the manufacturer nor the Bureau feel that failure occurred solely due to the presence of this shrinkage. The conclusion of this investigation was that the casting defect served only to locate the initiation point of failure of the wheel when it was subjected to the sudden change in loading just before failure (perhaps a minute or less) after 807 hours of trouble-free service.

A possible causal factor for the change in loading was established which would not have left any positive evidence. The shroud knife edge gas seals operate necessarily with very little clearance. If these edges began to contact the surrounding turbine case, this may have transmitted a changed stress pattern to the blade which failed. Any evidence of such an initiating event would, of course, be completely obliterated by the subsequent severe shroud/casing contact caused by shroud fragment separation and the resulting severely unbalanced turbine wheel. If such contact was to occur it would most likely occur on start-up when thermal expansion of the blades would precede thermal expansion of the casing, and hence tend to minimize tip clearance below steady state levels. This failure sequence could have been caused by a pilot induced overspeed/overtemp of the engine during a previous flight resulting in decreased clearance.

At any rate as a result of this failure the manufacturer reinspected the radiographic records of 637 wheels. It was considered that 37 contained questionable levels of internal defects similar to the shrinkage found here, and 20 had entered service. All 20 owners were notified that these wheels were considered defective and a *possible* safety hazard, and the manufacturer proposed to replace them with new wheels.

Conclusions of the Bureau Investigation

- A turbine wheel blade ruptured and shroud separation occurred causing the severe engine vibration and power turbine damage.
- Turbine wheel blade failure was due to very high-stress, low-cycle fatigue growth because of an abnormal loading condition.
- The casting defect only served to *locate* the point of initiation of failure.

Therefore comprehensive failure analysis was able to show that failure was not due primarily to a *manufacturing deficiency* but rather due to abnormal loading probably originating from an *operational problem.*

SUMMARY

This discussion has described failure investigations in detail to show how complex such analyses can be. Metallography, scanning electron microscopy, and X-ray energy dispersive techniques are essential interdependent tools in performing definitive comprehensive failure analyses. Those who have been involved in investigational work for many years know the tremendous influence these advanced tools have had on improving the results of such analyses. Now by combining the use of these techniques much of the guess work is gone. We find their use is absolutely essential in the performance of effective material failure investigations.

FAILURE ANALYSIS OF ARMAMENT HARDWARE

L.E. Samuels* and I.R. Lamborn*

INTRODUCTION

In the broadest sense, armaments can be said to include the whole spectrum of equipments and associated hardware in the military inventory and, as such, many have commercial counterparts ranging from the commonplace to the sophisticated. However, armaments are perhaps more commonly recognized as those items specifically identifiable with military usage – guns, ammunition, missiles and armoured vehicles, to name a few.

A common factor with all hardware, whether for commercial or specific military usage, is that materials are selected and designed and engineered into components to meet required standards of performance in service, hopefully under predicted conditions of stress and environment. Another common factor is that unintentional failures almost inevitably occur in service. They occur for many reasons but principally because of faults in the choice of material, in the design or manufacture of the component, or because the real service conditions of stress and environment are more severe than or different from what they were predicted to be. Consequently, the approach to the analysis of failures of armament hardware is much the same as for any other engineering hardware; for example, the same techniques of metallography, fractography, chemical analysis, electron probe microanalysis, stress analysis, fracture mechanics, and experimental simulation to confirm the failure diagnosis are used. There are no basic techniques peculiar to failure analysis of armaments hardware, although it is unlikely that some of the experimental simulations used to support a failure diagnosis would find a parallel in commercial failure analysis.

While there is some danger in generalization, there are however, certain features of the design of armament hardware that distinguish it from commercial equivalents where these exist. In general there is a greater emphasis on reliability and high performance under more demanding service conditions. In particular, many equipments and associated components are (or may be) subject to very high and complex loading applied at very high rates which would not normally be experienced by general engineering hardware.

It is in this area of high rate loading that failure problems and associated metallurgical phenomena are encountered that can be regarded as being either unique to some armament hardware or at least much more common than in commercial hardware. In fact, some hardware is designed to fail at high loading rates, or cause failure of other

*Materials Research Laboratories, Department of Defence, Melbourne, AUSTRALIA.

hardware with which it interacts. For example, a high explosive (HE) shell is intended to fragment on detonation of the HE filling. An armour piercing projectile is intended to defeat armour: conversely, armour is intended to defeat an armour-piercing projectile. It is fair to comment that, in the absence of reliable data on mechanical properties and deformation and fracture behavior of metals at the high rates of loading associated with these applications, the design and development of such hardware has proceeded on a largely empirical basis in which conventional metallographic failure diagnosis techniques have had a major role. In the authors' opinion it is also fair comment that the techniques are an important tool in the future identification and control of the material factors necessary for the optimum performance of these items of hardware.

Failure of armament hardware can thus be considered in two categories — those that are unintentional, and those that are intended. The purpose of this paper is to demonstrate some features of the two categories of failure, using examples. As an example of unintentional failure, it will be shown how conventional post-mortem metallography and fractography can be used for analysis of premature explosions of shell within gun tubes, yielding data which, when properly interpreted, can lead to conclusions that cannot be reached with confidence by other means. As an example of the contribution that failure diagnosis techniques can make to the control of intentional failures, the development of shear deformation in concentrated bands at high loading rates and some characteristics of these bands will be considered.

UNINTENTIONAL FAILURE: ANALYSIS OF PREMATURE SHELL FUNCTIONING EVENTS

Background and Causes

A shell can burst prematurely either in a gun tube on firing (commonly referred to as an *in-bore premature*) or at some point in its subsequent ballistic trajectory. Fortunately because of rigid quality assurance controls in manufacture and assembly, such an event is uncommon, but it can be extremely damaging to equipment and hazardous to friendly personnel. Consequently a great deal of effort is expended in attempting to diagnose the event, whenever the conditions are suitable, so that remedial action can be taken to reduce the probability of a repetition. Unfortunately, the inherently complex nature of prematures frequently militates against the required confidence in diagnosis, and anyone who has been involved in an investigation of a premature will be familiar with the inevitable unsupported speculation that arises as to the cause.

There can be many potential causes of a premature depending on the circumstances under which it occurred. A shell consists essentially of the outer steel casing, the rotating band, the high explosive filling, the fuze which can be a highly complicated mechanism with many component parts, and the associated exploding train. Faults or malfunctions in any of these components of a shell or in their assembly are potential causes of prematures. There can also be external causes, such as the presence of a foreign body in the gun tube on firing the round.

The violence of the event can vary greatly, and two extremes of the damage that can result from in-bore prematures are illustrated in Figs. 1 and 2. Figure 1 shows the damage to a naval gun that was mounted on an artillery carriage for range proof firing purposes. Figure 2 is a section of a gun tube in which damage was confined to bulging and other superficial damage features on the bore surface.

Without going into the physics and chemistry of high explosives, it can be said broadly that the most violent prematures in terms of damage to the gun and associated equipment (as illustrated in Fig. 1) can usually be attributed to full detonation of the

Fig. 1. General view of damage to gun following a violent in-bore shell premature.

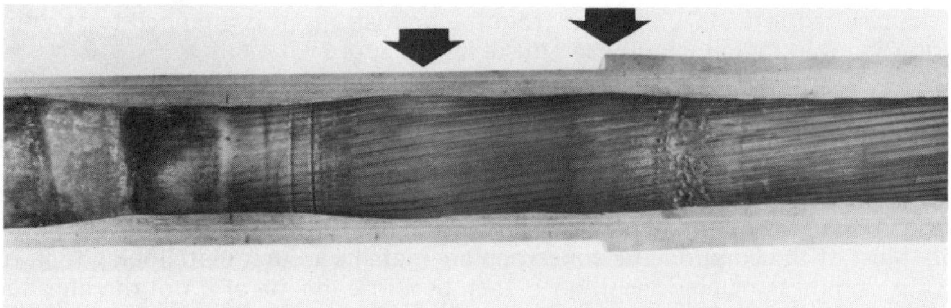

Fig. 2. Longitudinal section of a gun tube showing two bulges and other damage to bore resulting from a low-order in-bore event.

high explosive filling of the shell. The reaction time then is of the order of a few microseconds*. Less violent events, as illustrated in Fig. 2 for example, are usually associated with much longer reaction times, and lower peak pressures, such as those resulting from the ignition and subsequent burning of the explosive filling. However, burning rates can vary widely depending on circumstances and can if fast enough, lead to damage approaching that of a full detonation. Burning can also build up to detonation. On the other hand, burning can cease without even the charge being fully consumed.

This presents a problem in post mortem premature diagnosis. It is helpful if not essential to be able to establish whether detonation was initiated at the outset or whether ignition of the filling and burning was involved if the basic cause of the event is to be

*Highly dynamic stress patterns with components of compression (up to about 300 MPa x 10^2) tension and shear are set up in a body on detonation of a high explosive filling. Strain rates up to about 10^6 s^{-1} might be experiences, and shock-interaction effects might be involved depending on the geometry of the system. While the stress patterns at a given point in time, and the deformation and fracture behaviour of metals in those patterns are extremely difficult to quantify, metallographic techniques have been invaluable in providing qualitative descriptions of explosively-induced phenomena [1].

identified and the correct remedial action taken. The causes can be different in the two cases. Again as examples, premature fuze functioning would be expected to lead to full detonation, whereas adiabatic compression during firing of air entrapped in pockets in the filling or at the filling/shell casing interface could lead to ignition and burning. Remedial action would differ.

Failure analysis of these events is therefore a multidisciplinary exercise involving military personnel, engineers, explosives chemists, metallurgists, and possibly others, and each has a necessary role in a properly managed investigation. The particular role of the metallurgist and of metallography, with which this paper is concerned, is thought by some to be confined to checking for evidence of pre-existing material and manufacturing defects which could have led to the event. While this is indeed an essential part of the exercise, greater understanding of metal behaviour under explosive loading in recent years has substantially expanded the role of the metallurgist and hence his value in diagnosis.

The Approach to Diagnosis

The traditional and conventional approach to premature diagnosis has been, broadly, to attempt to deduce its cause from examination of damage and recovered debris, and evidence from any other source, and then to simulate the hypothesized event and to compare the resultant effects with those from the premature. It is intended to examine the contribution that metallography can make in this approach.

Occasionally, more particularly when the premature occurs under closely controlled proof firing conditions, there might be data available on the prior history of the particular shell involved which could prove relevant to the diagnosis. More often, however, the sole source of material evidence from which the basic cause can be deduced resides in the gun damage (in the case of an in-bore premature) and debris that can be recovered after the event. Most of the components concerned are metallic so that metallography in its broadest sense, and its associated techniques, then becomes the major if not the only tool available for the recognition and interpretation of the evidence.

It is to be recognized at the outset of course that metallography alone cannot be expected to necessarily identify the root cause of the premature. There are, however, three main aims of metallographic examination which, if satisfied, will substantially assist in the diagnosis and point the direction of associated investigations. The three aims are to determine if:

a) detonation or ignition and burning of the HE filling was the prime cause,

b) initiation of detonation or ignition of the filling was at the rear end or the forward end of the shell,

c) material or manufacturing defects present in any of the metal components contributed.

Recovery of Fragments

An exhaustive effort has to be made to recover as much debris as possible (fragments of the gun, shell casing, fuze, unconsumed HE filling or propellant, etc.) to ensure that no relevant information is lost. More often than not, however, the debris is spread over a very wide area and the recovery is incomplete. In the case of the premature illustrated in Fig. 1, for example, most of the barrel and jacket fragments, a few fragments from the base-end of the shell (Fig. 3), but none from the fuze were recovered. As the firing leading to the bulged barrel shown in Fig. 2 took place over the sea, no shell debris was recovered from that event. It is perhaps no coincidence that the cause was established positively in the first case but no agreement was reached on the cause of the second incident.

Fig. 3. Fragments of shell recovered from the event shown in Fig. 1 comprising eleven fragments from the shell base, two from the rotating band (left end of centre row) and the base plate (bottom – sector removed for metallographic examination).

There is always the possibility in the case of an incomplete recovery that the wrong fragments are recovered which, in the absence of adequate background data for their proper interpretation, can lead to an invalid conclusion. For example, many of the fragments from a forward-end-ignited unconfined shell can have fracture features similar to those of a rear-end-ignited shell, Fig. 4. It is only when sufficient fragments are recovered to construct the fracture pattern along virtually the full length of the shell that differences in location of the ignition point become apparent, and a confident diagnosis made. The possibility of this kind of situation is a further compelling reason for ensuring the maximum possible recovery.

Procedure

Because the circumstances of prematures vary so widely, and consequently no generally-applicable procedures for their investigation can be defined, it is intended to use the in-bore premature shown in Fig. 1 as an example to illustrate the essential features of evidence that can be obtained by metallographic examination.

Usually, the bulk of evidence will come from critical visual and macroscopic metallography, although microscopic metallography and other supporting techniques can have an essential role depending on circumstances. Perhaps the major problem to be faced is to sort out from a cornucopia of observations the relevant from the irrelevant, and the primary damage from the secondary and adventitious. It is in this area that well-founded experience has a strong influence in making the right judgment.

It is convenient to consider the example in two stages: firstly, study of the gun damage and, secondly, study of the shell debris.

The Gun Damage

The bore surface of the gun tube usually provides essential information. There are frequently well-recognized "signature" markings of features of the rear end of the

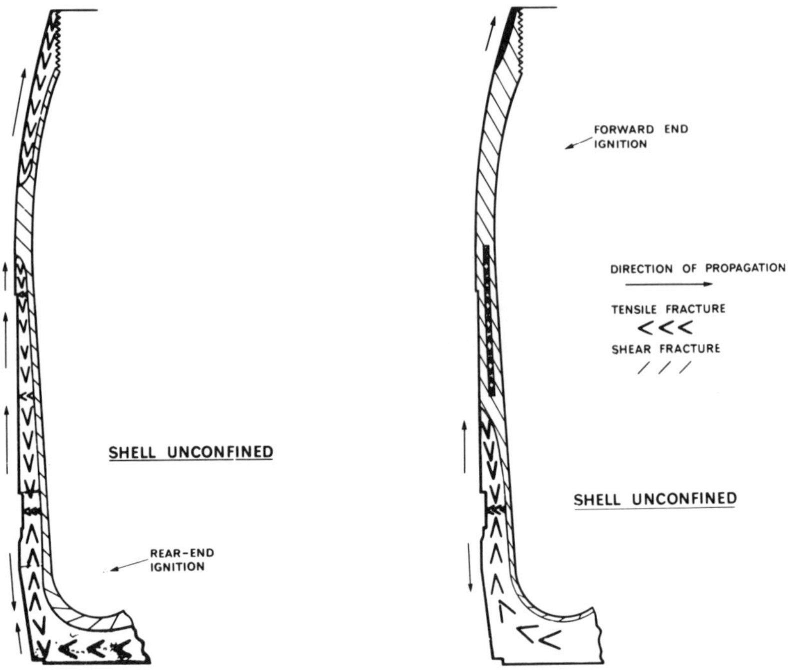

Fig. 4. Longitudinal fracture patterns of a typical shell, unconfined, following ignition of the HE filling near the base (left) and at the forward end (right).

shell — in the example, of the shell-base, the cannelure, the rear edge of the rotating band seat, the groove in the rotating band, and the forward edge of the rotating band seat, Fig. 5. These markings are due primarily to propellant gas wash or erosion of the bore surface caused by turbulence adjacent to the relevant shell features, and their existence implies that the shell was stationary at some instant during the premature. Further, in the example being considered, careful measurement of the distance of the markings from the breech face showed that they were located over an inch *to the rear* of the position that the corresponding shell features would have been with the shell in the rammed position before firing. (The rammed position was known in this case from actual measurement before firing). That is, the shell had moved rearwards against the propellant pressure. The origin of a fifth marking (Fig. 5) could not be identified with certainty, but this did not prevent a confident diagnosis. (The most reasonable explanation was that it was the signature of a shell base feature formed at a later point in time when the shell moved axially and canted, probably when the barrel bulged).

A bulge had developed forward from the signature markings at a position coinciding with the shell bourrelet region. This was associated with some flattening of the rifling lands and opening-out of the craze-cracking normally present in gun-barrel-bore surfaces. Bulging is of course a normal expectation, and there were no conclusions of specific usefulness to this investigation to be gained from its presence. By comparison, there has never been agreement on the interpretation of the *two* bulges in the barrel illustrated in Fig. 2.

Further forward from the bulge, and adjacent to the position of the inner end of the fuze hole with the shell in the rearward position, there is a circumferential band of

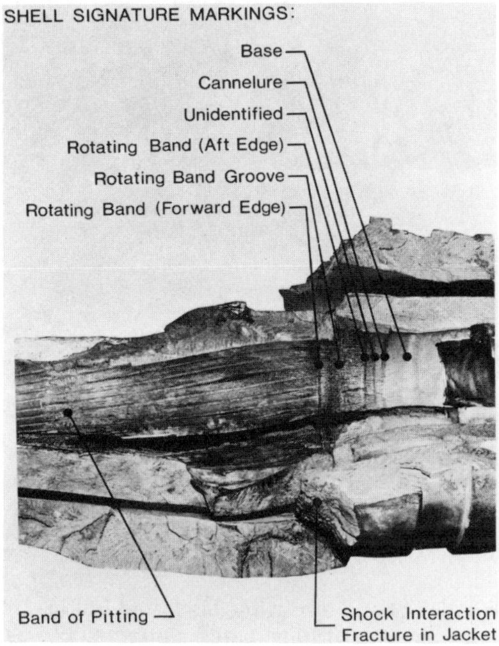

Fig. 5. Part of damaged gun tube and jacket from the event illustrated in Fig. 1, showing features of damage to bore (shell signature markings) and fracture surfaces.

pitting, marked in Fig. 5. This feature is not uncommon with in-bore prematures and appears to be more extensive when the premature is due to a detonation rather than to burning. Until recently there has been controversy as to whether the effect is due to impact of fragments from the shell or to an explosive-gas erosion effect. Careful metallographic examination with associated electron-probe microanalysis has now produced incontrovertible evidence that at least in some cases it is due to impact by fragments of the shell (Fig. 6), although some doubt remains as to the mechanism responsible for the generation of these fragments.

When examining the fracture surfaces of the gun tube and jacket (and for that matter, of the shell casing), it must be recognized that the fracture mode, which of course is determined by the stress distribution in the wall, is influenced both by the nature of the event and the extent to which it is confined. For example, whereas an unconfined hollow cylindrical body might burst with essentially shear fracture of the full wall thickness on detonation of a high explosive filling, the fractures could be essentially brittle cleavage if ignition and burning of the HE were involved. However, if the same body were confined in a sleeve with a wall of sufficient thickness, the fractures could be shear on burning of the HE. If the filling of the confined body detonated, much more extensive shear fracture of the body would be expected because of the higher compressive stress components. Also, depending on its wall thickness, failure of the confining sleeve could be by shear or by brittle cleavage fracture. In the real gun/shell situation, further effects are likely to be introduced by gaps between the jacket and tube,

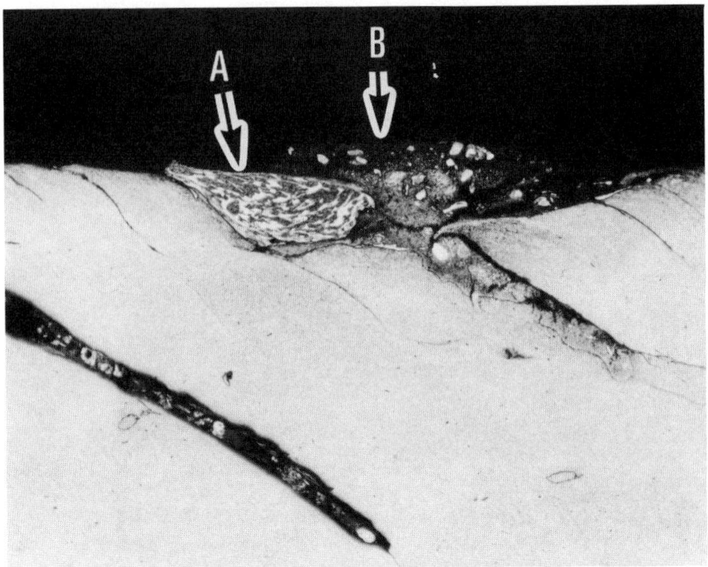

Fig. 6. Section of gun tube at band of pitting illustrated in Fig. 1. A is a fragment of shell body. B is an associated area of tube material affected by gas wash. Etch in nital. Original magnification 100X. (Reduced approximately 20 percent for reproduction).

and tube and shell. Without further elaboration, it suffices that a degree of subjectivity can be associated with deciding from inspection of the fracture surfaces of the gun tube and jacket whether detonation or burning of the HE filling was involved.

In the case of the longitudinal fracture surfaces of the gun tube in Fig. 1, varying lengths of shear fracture of the full wall thickness were developed around the circumference adjacent to the position of maximum bulge. The remaining parts of each longitudinal fracture surface were essentially brittle cleavage with a shear lip at the bore surface over most of the burst length. The chevron patterns of the brittle fracture regions indicated that, while there were a number of origins, brittle fracture had propagated both forward and rearward from the full wall shear length. That is, initial full wall shear characteristic of high compressive loading in the immediate vicinity of the burst had been followed by brittle fracture propagating both forward and rearward. The gun tube transverse fractures and the longitudinal and forward-end transverse fractures of the jacket were in all essentially brittle cleavage.

Although these observations clearly indicated that a high-order explosive event had occurred, none of these features in isolation was sufficiently conclusive to indicate its specific nature.

However, such evidence was found in the transverse fracture at the breech end of the jacket (Fig. 5). The major part of this fracture consisted of an outer annulus, near normal to the outer surface of the jacket, of brittle fracture which had been initiated at the outer extremity of an inner region of normal tensile rupture which had a mixed ductile/cleavage fracture mode. The inner region was convex-forward, angled at about 40 degrees to the longitudinal axis and was located in the jacket at a position a few inches forward from that corresponding with the position of the shell in its original rammed position.

This fracture could be predicted as likely to result from the reinforcement of tensile waves reflected from the outer surface of the system (the jacket) and from the base-end of the shell *when detonation of the HE filling is initiated at the nose end of the shell* (see section on *Model Simulation*) [2].

A transverse section of the jacket cut just forward from the transverse fracture at the breech end showed, in addition to a pattern of radial longitudinal cracks within the wall, an irregular circumferential crack (see Fig. 7). This crack extended for a few inches towards the forward end of the jacket. Fractographic examination of the crack surface after breaking open showed it to have a mixed ductile/brittle cleavage tensile mode. Thus it could be identified as a scab or spall-fracture characteristic of shock reflection from the outer surface of the jacket following a detonation of the HE filling [2]. By invoking certain simplifications of the system, an available computer programme predicted spall fracture in the jacket at approximately this distance from the jacket surface following a detonation of the HE filling of the shell.

Both gun tube and jacket had been manufactured from the correct steels in satisfactorily heat treated conditions, and no pre-existing defects likely to have contributed to the event were identified.

These observations summarize the important evidence from the gun damage relevant to the diagnosis of this premature. As with other investigations of this kind, much data was also gathered relating to damage to the cartridge case, effects on the gun bore surface of vaporization or melting of brass components, secondary damage to fracture surfaces from impact of fragments of unknown origin, and other features. None of these data were of direct material assistance in the diagnosis.

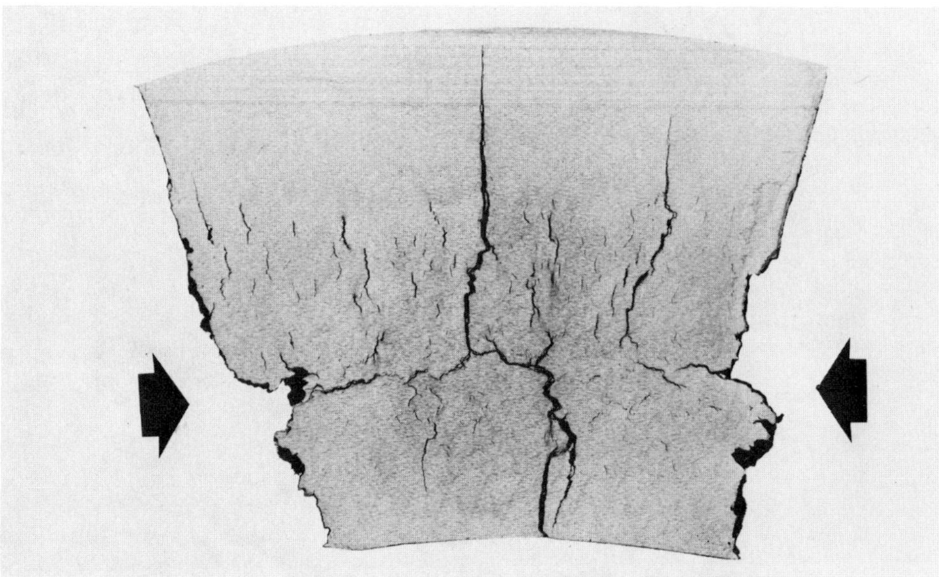

Fig. 7. Section of jacket forward from transverse fracture at breech end, after magnetic particle inspection. Showing radial cracks, and circumferential spall crack (indicated by arrow). Magnification 1X.

The Recovered Shell Fragments

The shell fragments recovered, Fig. 3, comprised the base-plate (which was found in the cartridge case), parts of the copper rotating band and a few fragments from the rear end of the shell casing. There were no fragments from forward of the rotating band seat or from the fuze. While the fragments recovered were exhaustively examined, only those features directly relevant to the diagnosis will be summarized.

Notwithstanding extensive hot gas erosion of the fracture surfaces, it was evident that some of the smaller fragments identified as originating from the top of the base-plate recess and from between the top edge of the recess and the base of the shell had resulted from reflected shock reinforcement. Fractures and fragments of this kind could be predicted from consideration of the reinforcement of tensile waves reflected from the outer surface, the top of the base-plate recess and the base of the shell when detonation of the HE filling was initiated at the forward end of the shell. Fragments with identical characteristics are in fact obtained from an identical shell detonated in the manner indicated.

The mild-steel base-plate and the medium-carbon steel shell casing both had pearlite/ferrite microstructures, and shock twins (Neumann bands) were present in the ferrite of both materials, Fig. 8. While the presence of these mechanical twins, formed under explosive loading, is commonly associated with a detonation event for diagnostic purposes, there is evidence to suggest that the correlation in isolation should be regarded with some reservation. It appears, for example, that twins can form at very low pressures [3]. Further, they have been identified in association with what would normally be termed a burning event [4]. Further study of the loading conditions leading to shock twin formation is warranted.

Fig. 8. Mechanical twins (Neumann bands) in the mild steel base plate. Etched in nital. Original magnification 500X. (Reduced approximately 15 percent for reproduction).

The Diagnosis

Taken together, five critical observations leave little room for doubt in the diagnosis of this premature within the framework of the aims set out in the section on *The Approach to Diagnosis.* The five observations are:

(i) The event initiated at or very close to shot start. This is indicative of a very short reaction time of the HE filling compatible with a detonation.

(ii) Spall fracture in the jacket is indicative of a detonation.

(iii) The presence of shock twins in the ferrite phase in fragments of the shell casing and base plate are indicative of a high order event, probably a detonation.

(iv) The transverse fracture at the breech end of the jacket resulted from reflected shock interaction, indicative of a detonation. Further, the location of this fracture was compatible with initiation of the detonation at the forward end of the shell filling, rather than at the rear end.

(v) The fracture features of the recovered rear-end fragments of the shell casing are characteristic of reflected shock reinforcement, again indicative of initiation of detonation at the forward end of the filling.

This evidence clearly points to initiation of a detonation of the high-explosive filling near the forward end of the shell at or very close to shot start (i.e., the first forward movement of the shell on firing). An interesting result was that the pressure so generated overmatched that of the propellant at that time, resulting in rearward movement of the shell.

While no fragments were recovered from the forward part of the shell, there was no evidence to suggest that material or manufacturing defects in the gun or the shell casing had contributed to the event.

Verification of Diagnosis – Event Simulation

Full Scale Simulation

As indicated in the section on *The Approach to Diagnosis,* the need for confidence in the diagnosis of a premature usually leads to experimental simulation of the event so that the resultant damage features can be compared. This can be an expensive and time-consuming exercise, since usually it involves static trials in which shell are burst (i.e., detonated or ignited) under conditions matching those of the premature event as hypothesized in the diagnosis.

In the case of the example being considered, two shell were prepared for detonation, one by initiation at the forward end through modification of the normal fuze, and the other by initiation at the rear end close to the base of the shell filling cavity. These shell were then inserted in lengths of gun tubes at a position corresponding with the rammed position of the prematured shell, and detonated.

There are, of course, certain criticisms of the use of static firings of this kind for the simulation of a shell being fired, one of the more obvious perhaps being that there is no cartridge case or propellant and associated gas pressure behind the shell. A consequence of this is, for example, that there might be no signatures of shell base features on the gun tube bore as described above. If they are found, they are much less distinct and possibly are formed by a different mechanism. They were present, if somewhat indistinctly, in the present trials.

The important observations made after the trial in this instance were that, with the forward-end initiated shell, transverse fracture at the breech end of the tube with similar characteristics to that in the jacket of the premature at the equivalent position, and shock-interaction fractures at the base of the shell identical to those of the prematured shell occurred. On the other hand, these fractures did not occur with the base-end initiated shell. The observations thus confirmed the diagnosis that the premature had resulted from detonation of the HE filling initiated at the forward end of the shell.

By way of comment, it was noted that both trial shell moved rearward from their original position by roughly the same distance as that observed for the prematured shell. Consequently, rearward movement of the prematured shell is not, in itself, evidence of forward end initiation of the HE filling.

Model Simulation

A feature of shock interaction effects, which constituted the major source of evidence for the diagnosis, is that they are determined primarily by the geometry of the system rather than by its size. It should therefore be possible to simulate the essential features of the gun/shell system of the premature in a scaled-down model for use in investigating shock interaction fractures. If use of the model for the purpose could be validated, it would have considerable economic advantages in terms of both cost and time over the full scale approach as described above. The main criticism is probably that, in a reduced-scale model, the reduced pressure impulse on detonation of the smaller diameter HE charge could influence resultant metal deformation/fracture effects.

Figure 9 is a drawing of a simplified model reduced approximately 4 to 1 from full scale, used to simulate the subject prematured gun/shell system. Dimension B represents the combined wall thickness of the shell, gun tube and jacket; dimension C is the combined thickness of the gun tube and jacket; and dimension A is the thickness of the shell-base. The shell cavity and gun chamber (for the cartridge case) are indicated. The shell cavity is filled to the level shown with the same HE used in the premature shell.

When detonation of the HE filling is initiated at the top of the model (simulating nose-end initiation in a real shell) it can be predicted qualitatively that the reinforcement of tensile waves reflected from the outer cylindrical wall and from the lower (rear) surface of the shell-base of the model, will lead to fracture located approximately as shown schematically in Fig. 9. While fracture does not result at this location when the filling is initiated at the base of the simulated shell cavity, a classic "cone-fracture" can be predicted at the top end of the model. This is also a shock-interaction effect, but as it is a consequence of reflection of tensile waves from the top face of the model, it is irrelevant for the purposes of the study.

Figure 10, showing the results of initiation of detonation at the nose-end and base-end of the simulated shell cavity respectively, confirm that transverse shock-reinforcement fractures occur at the predicted locations. That resulting from nose-end initiation has characteristics similar to that of the transverse fracture at the breech end of the jacket involved in the premature. It is noteworthy that the radial annulus at the outer surface of this fracture was approximately one quarter the width of the annulus in the corresponding premature fracture. The convex-forward part of the fracture extended virtually to the shell cavity surface. This represents a difference when compared with the equivalent fracture of the premature which extended to the inner wall of the jacket only. The difference can be attributed to effects of the interfaces between the jacket and gun tube, and between the gun tube and the shell in the real gun/shell system. The circumferential spall fracture in the jacket, and the shock-interaction fractures in the shell-base were not reproduced in the model. There was extensive twinning in the ferrite phase of the mild steel of the model.

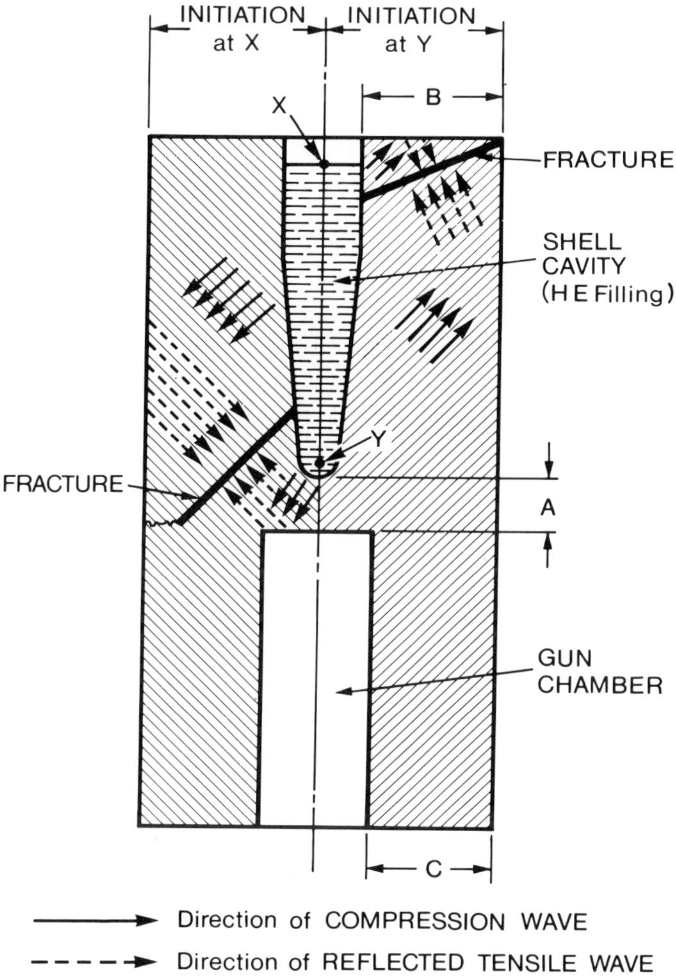

Fig. 9. Section of simulator, schematic, showing shock reflections and approximate locations of fractures resulting from initiation of the HE at the top (left) and at the bottom (right) of the simulated shell cavity.

It can be concluded from these tests, therefore, that small scale models can be used to reproduce at least some of the effects in full size gun/shell systems on detonation of the HE filling. They are therefore potentially useful supplementary tools for the diagnosis of in-bore prematures. However, it is evident that the simple model discussed above needs refinement, particularly to simulate the interfaces between the components of the full-scale system, before its real value can be assessed.

The Role of Metallography

The role of metallography in concluding that the premature in the example given was due to initiation of detonation of the shell filling at the nose end at or near to shot start should be evident. It is equally evident that the investigation does not end there as the specific cause remains to be identified. However, the metallographic examination

Fig. 10. Premature simulators after initiation of HE filling at (A) top and (B) bottom of simulated shell cavity.

has pointed the way for the investigations of experts in other disciplines to identify the cause; it provides the essential first step in the total investigation. It is sufficient to say that, in the example given, a fuze fault was ultimately established to be the cause of the premature.

It should not be inferred that conclusions from metallographic examination can always be reached with the same confidence as in the example discussed. It is frequently not possible to reach a conclusion, and the premature illustrated in Fig. 2 is a good example of this. The essential point to be recognized in a successful diagnosis is not just that the necessary evidence must be available in the fragments recovered (which is not likely under active service conditions) but that the investigator must have the competence to be able to recognize it and that the necessary background data must be available to enable him to interpret it correctly.

INTENTIONAL FAILURES: THE ROLE OF CONCENTRATED SHEAR BANDS

Background

There are a number of situations where military hardware is intended to fail at the last phase of its operational life. For example, the casing of some types of ammunition is expected to fracture under the influence of an explosive charge detonated on impact and the resultant fragments, propelled by the explosive charge, are the parts of the weapon that actually attack the target. The objective then is to induce a number of fractures in the casing which interact to produce fragments of a size that has maximum effectiveness in defeating the target. The optimum size of fragment varies markedly with different targets, and fragment size is determined by metallurgical as well as explosive factors. At the same time, an overriding requirement is that the launch safety of the round not be compromised, i.e., the probability of unintentional failure during firing must not be increased by any metallurgical changes made to influence fragmentation behavior.

A second important example is the interaction between metallic targets and projectiles. Failure of the projectiles is intended by those using the armour, and failure of the armour is intended by those using the projectile. A third example is the type of device often known as an explosive bolt. This is a fastener which is required to hold components together during one phase of its life but to fracture and so enable the components to separate when an explosive charge which it contains is fired. It must perform both functions reliably and, particularly as space for the explosive charge usually is limited, this introduces difficult competing requirements in design.

The types of hardware covered by these illustrations have in the past largely been designed by highly empirical methods. It is only recently that studies of the factors which control the initiation and propagation of the fracture processes involved have commenced. These phenomena turn out to be complicated, are often peculiar to the high strain rates involved, and are far from being well understood yet. Only one facet will be pursued here, and this is that failure commonly results from shear being concentrated in narrow bands. This is a phenomenon of considerable metallographic interest.

Characteristics of Concentrated-Shear Bands

As an illustration, when a steel plate is attacked by a hardened-steel projectile with an ogive penetrating end, normal plastic deformation occurs over a considerable volume of the plate, Fig. 11. Fracture occurs by a fairly normal plastic mode. However, an aluminum alloy plate attacked by a similar projectile but with a flat penetrating end may show the sequence of deformation and fracture events illustrated in Fig. 12 [5]. Deformation is now localized in a narrow band which is initiated at the corners of the projectile end early in the penetration sequence (Fig. 12(a)); the band propagates across the thickness of the plate faster than the indentation velocity (Fig. 12(b)), and eventually reaches the back face (Fig. 12(c)). Cracks then initiate in the band at the back face and propagate back along the bands until eventually they reach the corner of the projectile indentation, at which stage a plug separates and is ejected. The armour then has been defeated, Fig. 12 (d). The energy absorbed by this fracture mechanism is a good deal less than that for the purely plastic failure mode.

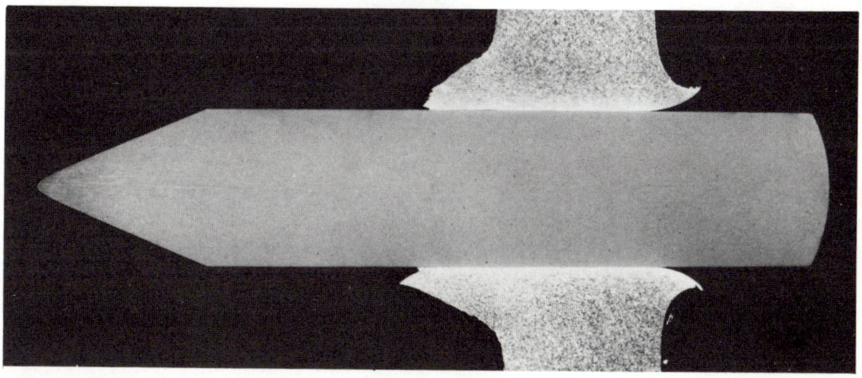

Fig. 11. Section of a mild-steel plate after penetration by a conically-pointed projectile. Widespread plastic deformation of the plate material has occurred. Etched in nital. Magnification 3X.

Moreover, situations may arise where the rate of energy input per unit volume of shear band may be such that the heat is generated at a faster rate than it can be conducted away. The temperature of the material in the shear band will then rise. If the strength loss due to this temperature rise is greater than the increase due to strain harden-

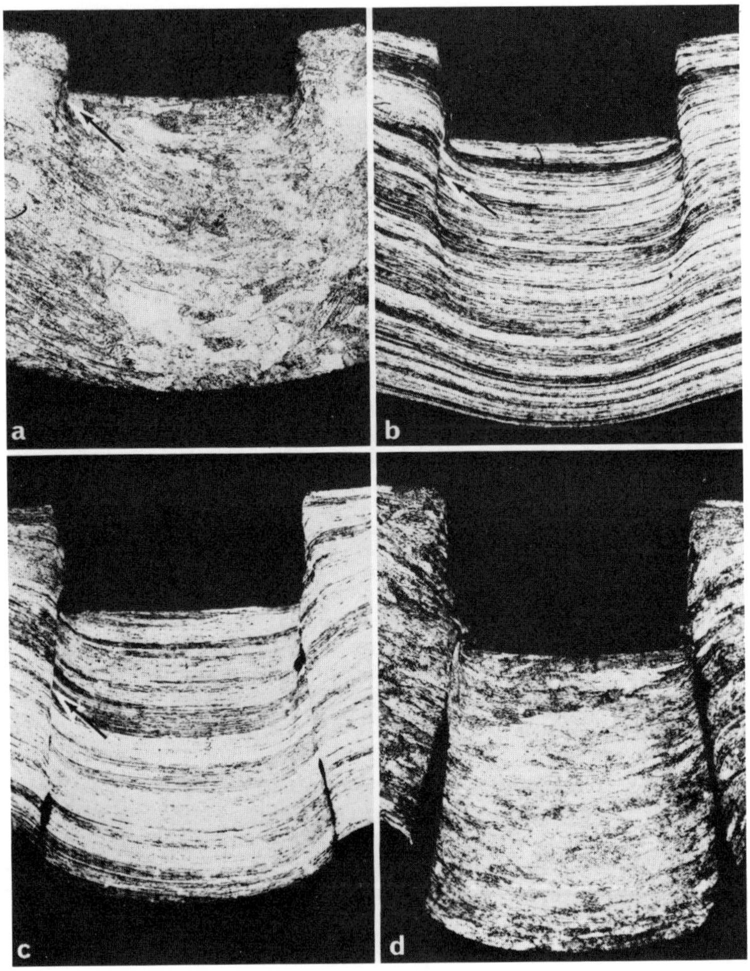

Fig. 12. Stages in the penetration of an aluminum alloy (2014–T6) plate by a flat-ended projectile. Showing development of shear bands in which the plastic deformation is largely confined. Sections etched in mixed acid reagent. (Taken from reference 5). Original magnification 5X. (Reduced approximately 15 percent for reproduction).

ing, unstable catastrophic plastic deformation will ensue and the energy required to induce fracture will be further reduced. The phenomenon, which was first proposed in a classic paper by Zener and Holloman [6], is known as adiabatic shear (although the process is not strictly adiabatic in the thermodynamic sense). Adiabatic shear frequently is regarded as a phenomenon in itself, but more likely it is a special case of concentrated shear as is implied in the above description, the transition between the two being illustrated in Fig. 13.

The classical examples of shear bands purported to be adiabatic shear bands are found in quenched-and-tempered steels subjected to high strain rates by a variety of means. Narrow light-etching bands may be detected by ordinary techniques of optical metallography (Fig. 14), and it has always been assumed that the light-etching regions are untempered

Fig. 13. Transition from wide shear band (a) to narrow white-etching band (c) in a quenched steel tempered to various hardness, viz.: (a) 230 HV; (b) 350 HV; and (c) 420 HV. All specimens subjected to impact compression. Etched in nital. (Taken from reference 17). Original magnification 100X. (Reduced approximately 20 percent for reproduction).

Fig. 14. Quenched-and-tempered steel subjected to impact compression. Showing white-etching bands on shear planes, known as adiabatic shear bands. Fractures have developed in many of the bands, but not in all. Etched in nital. Original magnification 40X. (Reduced approximately 20 percent for reproduction).

martensite. It is hypothesized that adiabatic heating has been sufficient to cause austenitization in the band and that this austenite subsequently transformed to martensite due to rapid quenching by the bulk of the specimen. We shall return to this point later.

In any event, it is found that the development of shear bands correlates with a reduction in the resistance of armour to penetration (Fig. 15), or with an increased tendency for a penetrator to break up on impact. The formation of the bands thus may be crucial in determining the resistance of materials to fracture under high-strain-rate loading. Studies of these bands, particularly studies aimed at establishing the factors which determine whether they form or not, are therefore an essential prerequisite to furthering improvement in the resistance of many items of military hardware to failure. The presence of shear bands in a failed component is also an important indication in post-mortem diagnosis of the circumstances of failure.

Shear bands develop within the bulk of a specimen, and thus can only be studied in sections. Consequently, optical metallography is a most useful method of establishing the presence and distribution of the bands, and for determining several important characteristics. The two basic parameters that need to be known are the shear strain and the temperature developed in the band.

Estimation of Shear Strain by Metallographic Methods

If only partial penetration of a projectile into a plate has occurred, it may be possible to obtain some estimate of the strain from a knowledge of the depth of penetration and the width of the shear band [6]. This method is applicable only if the shear band extends completely through the specimen thickness and is reliable only if the strain is reasonably uniform through the width of the band, which probably is not so (see below).

The strain is more reliably estimated from the displacement of internal markers in the specimen which run nearly perpendicular to the shear band. One example is shown in Fig. 16, where segregation bands in the aluminum alloy provide a suitable marker system. The displacements of the edges of segregation bands can be determined with reasonable accuracy, and the distribution of strain across the band can be determined, Fig. 17.

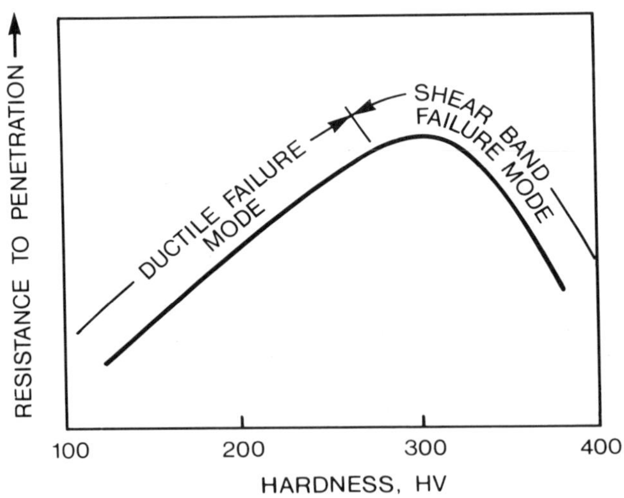

Fig. 15. Diagrammatic representation of the variation with hardness of the resistance to penetration of a quenched-and-tempered steel. The decrease in resistance to penetration at a hardness of about 350 HV is associated with the development of adiabatic shear bands during penetration. (Taken from reference 11).

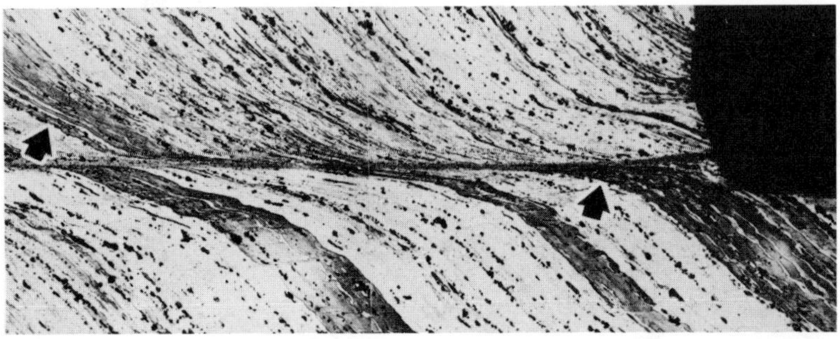

Fig. 16. Detail of the shear band indicated by the arrow in Fig. 12(b). The arrows indicate the displacement of the edge of a segregation band, from which the strain in the shear band can be estimated (Fig. 17). (Taken from reference 5). Magnification 125X.

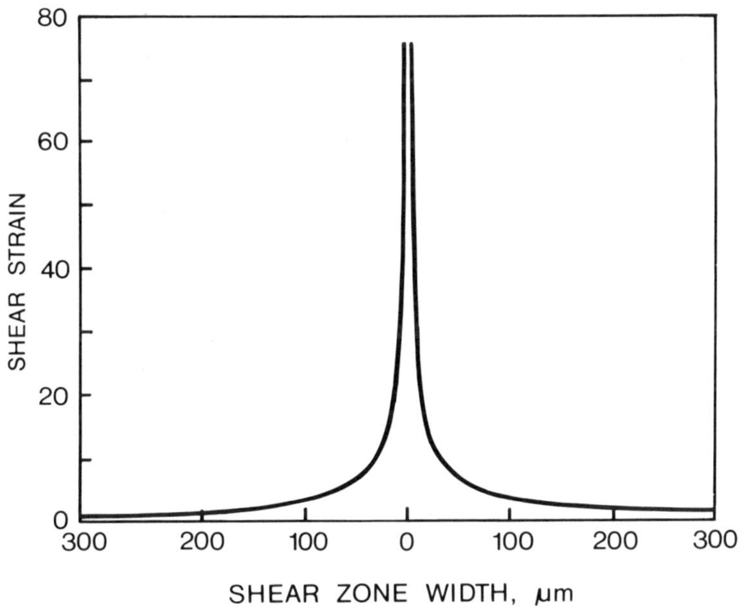

Fig. 17. Estimated distribution of strain across the shear band illustrated in Fig. 16.

Note that much of the shear strain is concentrated in a narrow central region of the band. Another possible marker is a ductile elongated inclusion, such as a manganese sulphide inclusion in steel, if fortuitously it is located across the shear band. With luck, this does occur, Fig. 18(a)[15]. Note that the width of the band in which the shear actually occurred can again be determined positively. In the case illustrated, this was much less than the width of the white-etching bands, so that measurements based on the latter would have underestimated the maximum strain in the band.

Even ignoring these refinements, a number of determinations have indicated that the strains in the bands are very high indeed, being in the range 50 to at least 100.

(A)

(B)

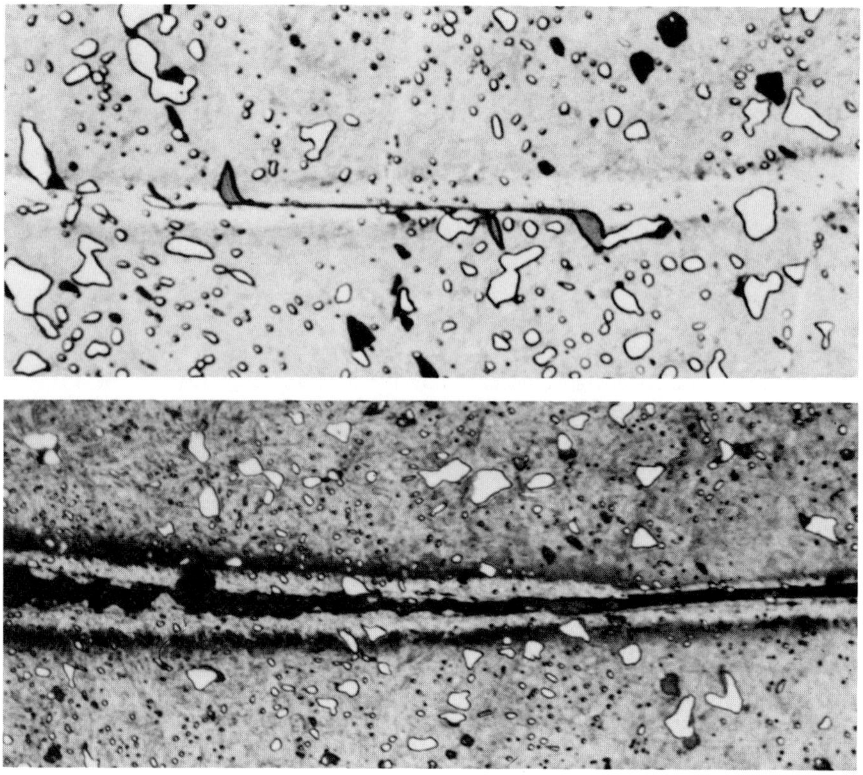

Fig. 18. White-etching shear band in an as-quenched high-carbon steel. Etched in picral. (A) shows shear of a MnS inclusion across the band and (B) shows dark-etching (tempered) regions adjacent to the white-etching (martensitic) band. (Taken from reference 15). Magnification of (A) 1000X and (B) 250X.

Estimation of Temperature by Metallographic Methods

The only methods currently available for estimating this temperature are to try to ascertain whether changes in either hardness or structure have occurred which differ from those to be expected from the plastic deformation in the band alone.

Changes in hardness provide only indirect evidence, and may be difficult to determine reliably when the bands are thin, but the technique should be applicable to a wide range of alloys. Techniques based on temperature-induced structural changes are more direct, but are applicable presumably only to systems in which structural changes are to be expected on heating, and then can only indicate that a certain temperature has been exceeded. Even so, there are difficulties in ascribing a reasonably accurate figure to this temperature because the duration of the heat pulse is so short that values determined under normal quasi-equilibrium conditions are not directly applicable. Special calibration experiments can however, be carried out at heating rates closer to those expected in the shear band [7].

When temperature-induced structural changes do not occur in the alloy under investigation, it might be imagined that some estimate of the temperature reached could be made by observing the defect structures in the band (principally by transmission electron microscopy) and assessing the extent to which structures expected for the strains involved had been relaxed by thermal effects. The difficulty at present is that the defect structures developed by the very large strains in the shear bands are not well understood at all. The information so far available indicates, in fact, that they are of an unusual nature and that the distinction between pure deformation effects and thermal relaxation effects is not straight forward [8,9]. The interpretation of experiments such as those carried out by Craig and Stock [10] on shear bands in 70:30 brass is therefore now considered to be unreliable.

In any event, most work to date has been carried out on steels. The presence of shear bands of white-etching material of the type illustrated in Fig. 14 almost universally has been taken as an indication that the A_3 temperature of the steel had been exceeded. Usually, however, such a conclusion is based on the interpretation of only two pieces of evidence, namely:

1. The band is light etching in optical microscopy, which is assumed to be indicative of untempered martensite. This is scarcely definitive evidence. "White-etching" means only that the band is more highly reflective than the matrix after metallographic etching, and this could be due to any number of causes.
2. The hardness of the band is higher than that of the matrix. This evidence is not conclusive either. For example, it is possible to plastically deform even the hardest steels under these conditions of constraint and this could, in principle, just as well account for the hardness increase.

On the other hand, there are a number of observations, some of which are long standing, which suggest that the white-etching bands may not be composed of a normal untempered martensite. The hardness of the band is usually higher than that of bulk as-quenched martensite of the same steel. Moreover, the band hardness varies with the prior structure of the steel [11]. The results of X-ray diffraction investigations have been very variable, various investigators claiming that the structure of the band material is essentially ferritic, martensitic, or austenitic (see Ref. 11 for a review of this topic). Few of these investigations have been comprehensive and it is not a simple matter with thin bands of this nature to distinguish positively by XRD methods between deformed ferrite and martensite, certainly so in steels with medium and low carbon contents or if large strains are present in the martensite. Next, the microstructure of the material appears to be very fine grained but it has not been possible to resolve the acicular structure expected of martensite. Further, the tempering characteristics of the band material are usually quite different from that of bulk martensite, much higher tempering temperatures being required to produce a dark-etching effect in optical microscopy, this temperature also depending on the prior structure of the steel [11]. Finally, white-etching zones of similar appearance in optical microscopy are found on machined surfaces; it has been shown positively that some are composed of deformed ferrite, that some are martensite of an abnormal structure, and that some are martensite of a normal structure [12].

Clearly, more definitive evidence than any of that outlined above is required to settle this issue. One obvious method of attempting to obtain such evidence is the use of transmission electron microscopy, supported by electron diffraction, and it is surprising that only two investigations of this nature so far have been carried out [13,14]. Both have been no more than exploratory, but reached the conclusion that the bands were martensitic and that the martensite was not of normal structure. It appeared to be very fine grained (crystallite size about 0.1μm diameter) and to contain many dislocations. It was proposed that this structure resulted from deformation during one or other of the transformation stages. Neither paper, however, gives sufficient data to establish that the identification of martensite was quite positive. A further complication arises in some later work by one of these groups (reported by Bedford, Wingrove and Thompson[11]) which showed that bands produced at very high strain rates did have a normal martensitic structure.

Some light is thrown on the problem by a simple experiment carried out by Manion and Stock [15] in which shear bands were formed in a high-carbon steel initially in the quenched and lightly-tempered condition. Optical metallography showed that a dark-etching band was present adjacent to each edge of a white-etching band, Fig. 18(b). It seems reasonable to conclude that these are fully-tempered regions, which should be present adjacent to locally reaustenitized regions. The outer dark-etching bands, however, were not apparent in a similar experiment carried out by Glenn and Leslie [13], although these workers did report a decrease (but only a slight decrease) in hardness in the transition regions adjacent to the white-etching bands. The type of TEM investigation referred to above could profitably be extended to a search for these associated tempered layers.

The present situation therefore is that, on balance, it does seem probable that the white-etching bands are martensitic, and hence indicative of adiabatic heating to above 900 degrees, but this cannot be said to have been conclusively established. If a band is martensitic, its general characteristics are represented diagrammatically in Fig. 19.

Even so, it seems possible that there may be two different types of white-etching adiabatic bands which are not directly distinguishable by optical microscopy. It is also now known that there is a third type of white-etching band resulting from localized plastic deformation, and even a fourth resulting from the diffusion of elements such as carbon and nitrogen into a surface [16]. It is clearly dangerous to jump to conclusions about the nature and cause of any white-etching layer observed by optical microscopy.

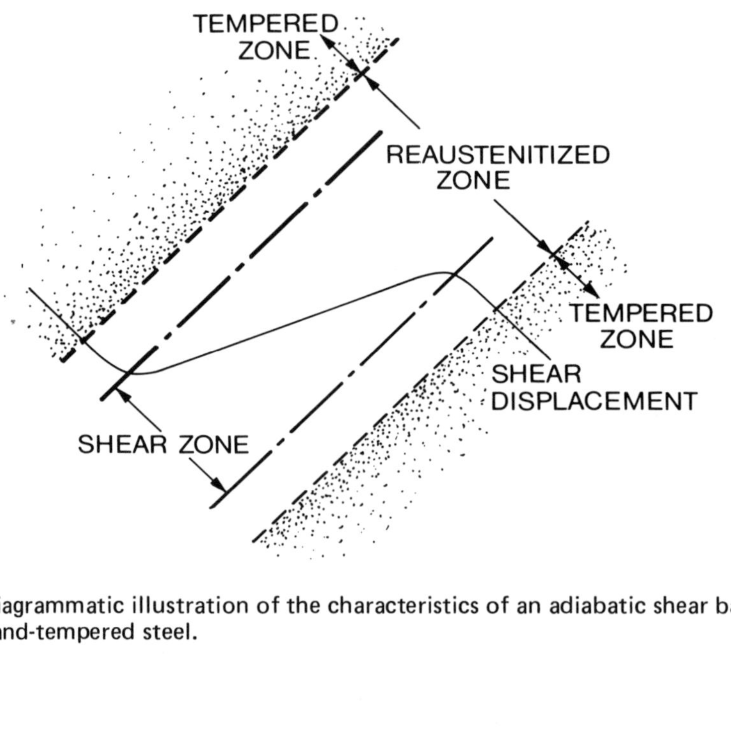

Fig. 19. Diagrammatic illustration of the characteristics of an adiabatic shear band in a quenched-and-tempered steel.

The whole story is a good example of the strengths and weaknesses of metallography. The technique is indispensable for establishing the existence of bands of concentrated shear, has established their critical importance in determining resistance to failure at high strain rates, and has elucidated many of their important features. However, it has not been conclusive in establishing critically if and when adiabatic heating has occurred. This has perhaps been due to a basic problem of metallography, namely, that interpretation very often is subjective. Practitioners, being human, are liable to draw the conclusions that they want to draw. Moreover, it shows that, certainly in difficult cases like the one under discussion, all the weapons in the metallographer's armoury may have to be applied, and to be applied critically and thoroughly, before sound conclusions can be reached.

CONCLUSIONS

Although failure analysis of armament hardware frequently is no different from failure analysis of engineering hardware in general, certain important types of hardware specific to military usage may be subjected to complex loading applied at very high rates which is difficult to quantify. Failure may then be characterized by modes which are not commonly encountered in commercial hardware failures and which are not well understood.

Metallographic techniques can have a major and indispensable role in post-mortem diagnosis of failures of these types and in their rectification. They are also indispensable research tools for clarifying modes of failure associated with such hardware, and for identifying the material factors that affect these modes. It is clear from the examples given, for example, that metallographic techniques will be largely responsible for elucidating the adiabatic shear phenomenon and so providing data necessary to improve the performance of certain military hardware.

However, it is also clear, if the point needs to be made at all, that the value of metallography for these purposes depends entirely on the ability of the metallographer to interpret correctly his observations against the background of available data relevant to the problem at hand. How well he does this can be, and often is in the areas discussed in this paper, dependent in turn on inputs from other technical disciplines — on inter-disciplinary technology transfer.

REFERENCES

1. I.R. Lamborn, A.J. Bedford and B.E. Walsh, Institute of Physics Conf. Ser. No. 21, p. 251, (1974).
2. J.S. Rinehart and J. Pearson, *Behaviour of Metals Under Impulsive Leads,* New York: Dover Publications (1965).
3. R.W. Rohde, *Acta Metallurgica,* 17, p. 353 (1969).
4. N.M. Burman, to be published.
5. A.L. Wingrove, *Met. Trans.,* 4, p. 1829 (1973).
6. C. Zener and J.H. Holloman, *J. Appl. Phys.,* 15, p. 22 (1945).
7. R.A. Huggins, H. Udin and J. Wulff, *Welding Jnl.,* 35, p. 18s (1956).
8. D.M. Turley, *J. Inst. Metals,* 97, p. 237 (1969).
9. D.M. Turley, *Met. Trans.,* 2, p. 3233 (1971).
10. J.V. Craig and T.A.C. Stock, *J. Aust. Inst. Metals,* 15, p. 1 (1970).
11. A.J. Bedford, A.L. Wingrove and K.R.L. Thompson, *J. Aust. Inst. Metals,* 19, p. 61 (1974).
12. D.M. Turley, *Mat. Science and Eng.,* 19, p. 79 (1975).
13. R.C. Glenn and W.C. Leslie, *Met. Trans.,* 2, p. 2945 (1971).
14. A.L. Wingrove, *J. Aust. Inst. Metals,* 16, p. 67 (1971).

15. S. Manion and T.A.C. Stock, *J. Aust. Inst. Metals,* 14, p. 190 (1969).
16. B.D. Grozin and V.F. Iankevich, *Friction and Wear in Machinery,* 15, p. 143 (1962).
17. A.L. Wingrove and G.L. Wulf, *J. Aust. Inst. Metals,* 18, (4), p. 167 (1973).

FAILURE ANALYSIS ASSOCIATED WITH MINING AND HEAVY MECHANICAL EQUIPMENT

Charles R. Morin*, Kenneth F. Packer*, and John E. Slater*

INTRODUCTION

The analysis of failures of mining and heavy mechanical equipment, in general, follows the same guidelines and techniques as applied to other failures. Since these general procedures with examples are discussed elsewhere in detail, this paper will concentrate on an in-depth discussion of the tools and techniques for failure analysis which pertain to the unique problems in mining and heavy mechanical equipment. Mining encompasses many forms such as hard rock, strip, deep tunnel, and, of course, many forms of ore and minerals. Bearing in mind that each type of mining and each mineral will place different burdens on equipment, common requirements for design and performance allow a certain amount of generalization for the purpose of this discussion.

Failure analysis is not treated as an end to itself; the cost of equipment today and the small profit margins of operating companies requires round the clock production averaging 350–360 days per year. Failure analysis is viewed as a constructive procedure intended to first identify failure modes and causes; this information is then fed back to designers and operators to minimize future problems. In this light, the procedure is often referred to as *product performance analysis.*

GENERAL APPROACH

The analysis of failures in mining equipment and heavy machinery is complicated by the size of the equipment, the uncontrolled environment attendant with field use, and field modifications, occasioned by previous repairs. Additional steps required in such a failure analysis therefore, will include factors which are often considered adventitious in simpler systems.

The two most important factors needed to conduct a failure analysis are 1) undamaged samples of the material from the origin area, and 2) a complete history of the machine from design, fabrication, use and maintenance. The investigator must insist to examine the failure as soon after the occurrence as possible. Corrosion and handling damage may obscure or destroy the critical samples needed for the laboratory analysis. It should be made clear that the fractures must be protected from the weather. If necessary, fractures can be oiled and wrapped in plastic sheeting until a qualified analyst arrives to direct the program.

---

*Packer Engineering Associates, Inc., Naperville, Illinois 60540 USA.

The wreckage distribution pattern is also an important source of information for reconstructing the circumstance of failure. Documentation of the wreckage pattern may be accomplished by sketches or photographs and preferably both. Particular attention should be paid to gouges and impact marks, corrosion, deformations of beams and other structural elements, failure mode and direction of fasteners and welds. During this survey the analyst can begin forming a plan of attack and make provisions for tagging, protecting and sampling the critical area. If parts must be moved before the analyst arrives, documentation of the above-mentioned wreckage pattern becomes even more valuable.

After the preliminary plan is established, the investigator should begin collecting the background information required for the failure analysis. Important documents to obtain and review are: 1) the original engineering drawings, 2) drawings of all field modification, 3) operation and service manuals, 4) operators and maintenance log books, and 5) service history of similar machines in the field. Often the manufacturers service engineers or the technical staff of the operating company can be a valuable resource for this information.

UNIQUE PROBLEMS OF MINING AND HEAVY MECHANICAL EQUIPMENT

The material failure problems in the mining and heavy mechanical equipment industries are unique for several reasons. The size, complexity, operating environment, materials of construction are all dictated by economic considerations that require high reliability to amortize the initial cost. For instance, the cost of a single large power shovel for strip mining coal costs in excess of $25,000,000. The depreciation charges alone can exceed $10,000/day. A failure resulting in loss of sales will cost an operating company from $300/hr. for a small machine to $4000/hr. for the largest dragline in operation today.

In addition to significant direct cost of failure, an equally important consideration is safety. The mining industry offers one of the most hazardous occupational exposures in this country. Table 1 collects injury frequency rates and severity measures for selected industries in 1968. These data show that for the industries studied, coal mining has the greatest frequency of injuries. As a class, the extractive industry has the highest injury severity rate as measured by days of lost work per million man-hours exposure. While not all injuries are a result of mechanical failures, an improvement in reliability of mechanical equipment is sure to have an impact on safety.

Size Effects

Probably the most difficult problem to deal with in mining and heavy mechanical equipment is the effect of size. The largest dragline in the Illinois coal field weighs 27,000,000 pounds. Booms are being produced by a power shovel company that are 320 feet long. The size of individual sections is too large to be post-weld heat treated. Field welds are required to complete the boom on-site. Even though critical joints are inspected by ultrasonic, radiographic or magnetic particle testing, one manufacturer reports that the major cause of failure is fatigue cracking from flaws in heat-affected zones and area of incomplete penetration.

The original design is based upon sound welds and allowable stresses on standard types of joints. Much of the welding is not subject to codes (such as API or ASME) and designers will use design limits based upon experience for the "normal" quality welding standards that in-house production usually meets. One mining equipment supplier uses an allowable design stress of 15,000 psi for any welded joint that is not post-weld heat treated or fully inspected.

Characteristically, the mining and mechanical equipment industries solve problems by "beefing-up" areas of recurring problems. This approach results in generally thicker

TABLE I

Injury Frequency Rates and Severities in Selected Industries, 1968*

INDUSTRY	SEVERITY RATE man–days lost per million man–hours exposure	FREQUENCY RATE injuries per million man–hours
Manufacturing:		
food products	1,023	26.8
lumber and wood	2,973	36.1
furniture	945	22.3
stone, clay, glass	1,212	21.6
primary metals	917	15.2
fabricated metals	924	21.1
Mining:		
coal mining	9,859	40.8
metal mining and milling	3,227	21.1
non-metal mining	3,010	22.5
sand and gravel	2,440	20.2
stone quarrying	2,505	16.6
Construction :	1,992	26.9
Transportation:		
passenger	1,220	23.6
motor freight	1,821	31.7
Services:		
auto repair	568	14.5
misc. repair	710	20.1
state hospitals	535	20.0

* From Reference 1, page 109.

and stiffer components. As the section thickness increases three factors can contribute to unexpected failures. First, a normally ductile steel in thick section is more susceptible to brittle fracture. Second, the metallurgical quality of the steel is decreased because of the lower level of reduction from the ingot and the poor response to thermal treatments such as normalizing or quench and temper. Third, the stress gradients and dynamic effects increase as the structure becomes larger and less compliant.

Figure 1 is a macro photograph of a section through a crawler tractor idler wheel. During developmental testing, the fatigue life of the weld joint did not exceed 35,000 cycles. The first design change simply enlarged the sections making the structure much stiffer and the fatigue life reached 150,000 cycles, but the design life called for 250,000 cycles. As the structure was further stiffened, the fatigue life fell to less than 25,000 cycles.

We decided to take the opposite approach, a design shown in Fig. 2 was developed and tested. The fatigue life for the structure, which was actually less stiff than the original, exceeded 300,000 cycles. This design approach should be purposely considered because the first tendency is to assume stiffer is stronger is better.

Brittle Fracture

The effect of section size on the brittle fracture of steel has been extensively studied. The techniques of linear elastic fracture mechanics (LEFM) will apply to any brittle fracture of steel if the following conditions have been met:

1. The failure occurred at a stress below the nominal yield stress.
2. The structure contained a pre-existing sharp flaw small with respect to the machine element dimension.

A basic expression used in LEFM analysis relates a material property (K_{Ic}, the critical stress intensity at fracture, referred to as the fracture toughness) and the simultaneous effect of stress (σ) and flaw size (a). The basic functional relationship is expressed as follows:

$$K_{Ic} = \sigma (\pi a_c)^{1/2}$$

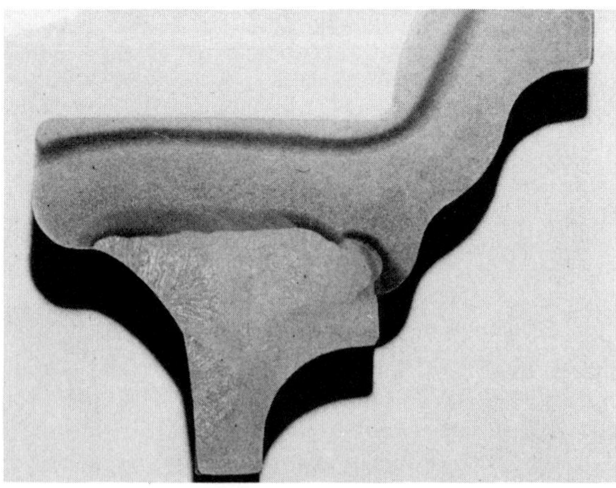

Fig. 1. Macro photograph of an etched cross-section through an idler wheel showing the rim-hub weld joint.

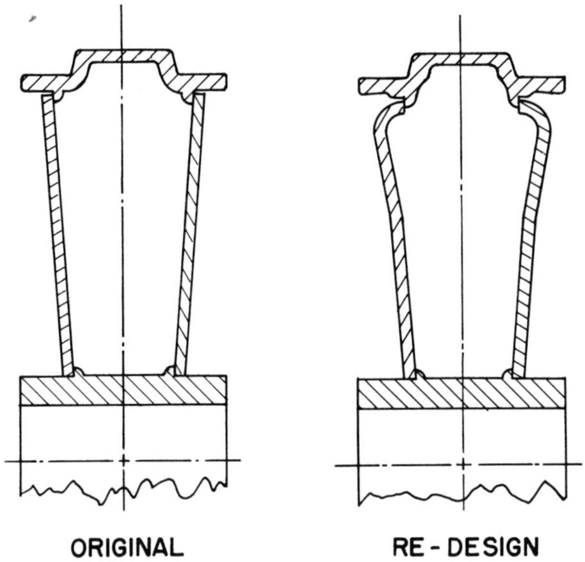

Fig. 2. Sections through the original, "stiff", and final "flexible", wheel designs.

Therefore, if the fracture toughness is known, knowing either the initiating flaw size (a) or the nominal stress at failure (σ) allows calculation of the other. For a detailed discussion of the limitations of LEFM, the reader is referred to the ample published literature. A simple method of checking the validity of assuming the ideal case is provided by ASTM A399 to insure that plane strain conditions are met. The condition is predicated on the requirement that the thickness of the member, B, will provide sufficient restraint at the crack tip to prevent gross yielding. The required thickness is a function of the fracture toughness and nominal yield stress as follows:

$$B \geqslant 2.5 \left(\frac{K_{Ic}}{\sigma Y} \right)^2$$

However, for the purposes of estimating failure stress or critical flaw sizes, the above requirement is often too conservative. If the thickness is at least equal to about $(K_{Ic}/\sigma Y)^2$, assuming ideal behavior will often allow reasonable bounding estimates to be made.

Much of the mining equipment utilizes high strength or hard, abrasion resistant materials. The combined effects of high hardness and strength and heavy sections cause a low tolerance to flaws. Many of these flaws originate in original manufacture, but flaws resulting from service related damage or defective field repairs often initiate fatigue and fracture.

Figure 3 is a photograph of a tooth holder from a 60 cubic year bucket that fractured in service. The metallurgical analysis of the component, which is a cast 4340 type alloy heat treated to 350 BHN, showed that casting defects were present at the fracture origin, Fig. 4 and Fig. 5. Metallography showed that the crack interior was decarburized (Fig. 6). These metallographical conditions and the location prove that the casting defect was a hot tear. However, this tooth holder operated for several months prior to fracture. Tooth loosening and subsequent impact loading was a likely contributing factor. Improved casting quality through more thorough quality control and provisions for a tighter fit has significantly reduced this problem.

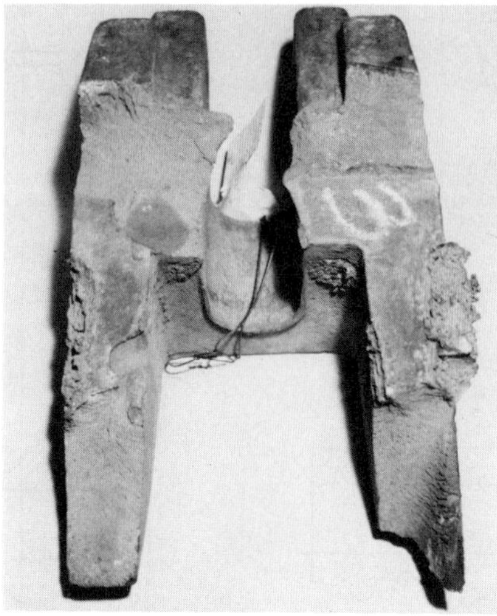

Fig. 3. Fractured tooth holder from a 60 cubic yard bucket.

Fig. 4. Macro photograph of the tooth holder fracture origin.

FATIGUE DAMAGE

LEFM also finds application to fatigue damage analysis of heavy equipment by providing a relatively simple method of determining the remaining useful life in a structural member. It has been established that cyclic stress intensity ($\triangle K$) governs the rate of crack growth according to the following power law:

$$da/dN = C\triangle K^n$$

where:

da/dN = the crack growth per cycle (the in./cycle)
C = empirical constant
K = cyclic stress intensity (KSI $\sqrt{}$ in.)
n = empirical constant (generally about 4)

Fig. 5. Close-up photograph of hot tear as tooth holder fracture origin.

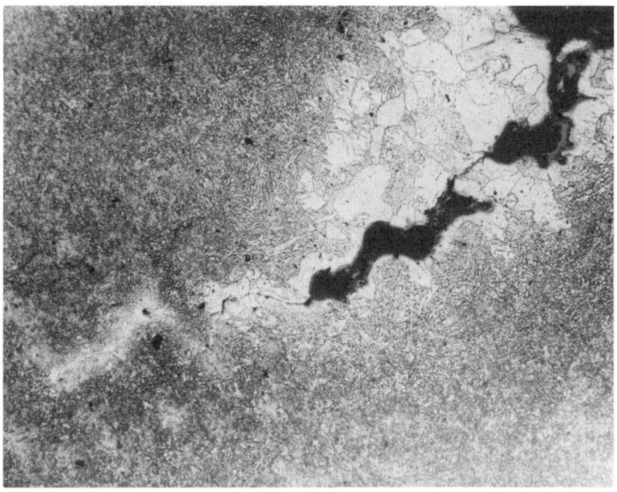

Fig. 6. Metallographic section through hot tear in tooth holder. Original Mag. 50X. (Reduced approximately 35 percent for reproduction).

LEFM analysis of propagating fatigue cracks is valid for ductile materials because the plastic zone ahead of the advancing crack is usually small. Corrections for significant plasticity are not required for cyclic stress ranges below the nominal yield. Therefore, fatigue analysis in the high cycle regime is fairly straight forward. Figure 7 is a typical plot of da/dN as a function of ΔK, which can be integrated to determine the number of cycles necessary to propagate a crack from its initial size to failure. In other cases, LEFM is used to determine the maximum allowable time between inspections to assure that any flaws below detectable limits on the last inspection will not become critical before the next.

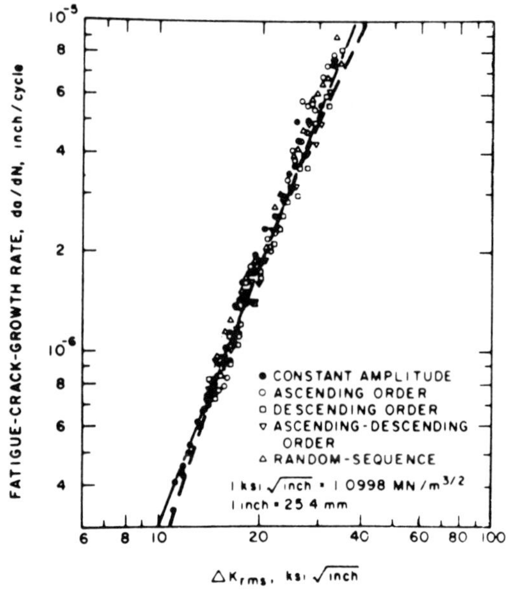

Fig. 7. Typical LEFM characterization of crack growth rate, (Refr. 1).

Figure 8 is a photograph of a large strip mining power shovel that suffered a complete fracture of one rear leg. By the time the failure analysis was started, the origin area had been destroyed by the maintenance crew. Remaining sections of the fracture surface were recovered from a scrap area. These samples were reconstructued into their original location and chevron marks were traced to the forward surface of the leg where the arch plate was welded into the leg. The weights of individual members were estimated and cable tensions were calculated from a free body diagram (Fig. 9). Since the structure is statically indeterminant, both a finite element and a classical superposition stress analysis were performed. The cyclic stress was calculated to be 28,000 psi in the critical area. Assuming a fracture toughness of 45 KSI $\sqrt{}$ in. and a failure stress of 28,000 psi, the flaw size at failure was probably about 0.6 in. deep. The number of stress cycles accumulated was estimated knowing the time to complete one lift, the days of operation per year, and the years of service. The original flaw size in the structure was calculated by solving the crack growth rate expression for a_i. Based upon this analysis, NDT specifications were prepared to detect flaws equal to or greater than 0.1 inch in length in the arch area. Inspecting for and repairing all flaws above the specified size assures a useful life of at least 8.6 years. Of course, periodic inspection and proper maintenance will indefinitely extend the useful life and prevent future costly down-time.

The Problem of Technology Transfer

A major problem in the mining industry relates to the lack of technology transfer to the personnel actually performing design and maintenance functions.

Many failures occur due to improper maintenance and repair procedures. A very costly failure shut down a system that conveys coal on rubber belts 12 miles from the mine to a riverfront loading dock. A well-intentioned maintenance man welded coal deflector shields to the side panels to prevent abrasion at the point where coal dumps onto the belt. After a short period of time, several deflector shields broke free and were carried on the conveyor to a transfer point. One shield measuring ¼ in. x 12 in. x 96 in. became lodged at this point and caused over five miles of reinforced belting to be torn. An examination of the fracture surfaces revealed that the plates were skip-welded to the

Fig. 8. Fractured rear leg of a large strip mining power shovel.

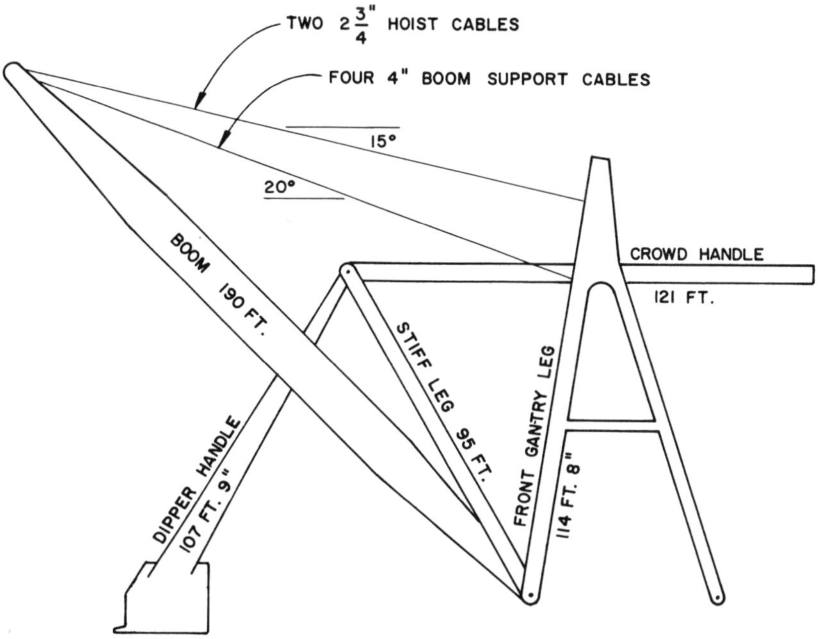

Fig. 9. Power shovel geometry and dimensions.

side panels with 1/8 in. filets. Metallography revealed that the heat-affected zone (HAZ) was completely transformed to martensite (Fig. 10). A microhardness traverse of the HAZ showed that adjacent to the weld the hardness exceeded Rc60. Chemical analysis revealed that the deflector shield steel was of nearly eutectoidal composition. Medium and high carbon steels cannot be welded without special precautions with regard to hy-

Fig. 10. Crack through HAZ of a deflector shield. Original Mag. 200X (Reduced approximately 35 percent for reproduction).

drogen pickup, preheat, interpass temperature, and post-weld heat treatment. The mode of failure was determined to be under bead cracking through the brittle HAZ. The cause of this failure was the selection of an improper welding procedure. Fundamentally, the cause of the failure can be traced to a more general problem in the mining and machinery industries, i.e., the difficulty in supervising and training unsophisticated maintenance personnel who are unaware of basic material behavior. Alban[4] found in a survey of hardened gear failures that most could be traced to human errors.

Wear

Abrasive wear, erosion and corrosion cause a general loss of material from machine elements which necessitates replacement when damaged beyond a useful limit. Wear is combatted by utilizing alloys of complex microstructure or by hardfacing more common steels. One of the first alloys developed by the mining industry specifically to resist abrasive wear was patented by Robert Hadfield in 1883. The ferrous alloy contains 1.2% C and 12.5% Mn and has become known as Hadfield manganese steel. ASTM A128 lists nine grades of carbon-manganese steels in common use today.

Several compositions of cast iron have been developed for abrasion resistance. Chill cast white iron was probably the first alloy utilized for ore grinding. As toughness requirements increased, irons with an austenitic matrix with dispersed carbides were utilized. An alloy like Paraboloy (2.5% C, .75%Mn, 0.6% Si, 19.2%Cr, 1.2%Mo, 0.7%Ni) exhibits a microstructure of Cr_7C_3 type carbides dispersed, in an austenite-martensite matrix when air hardened from 2000°F. This alloy work hardens from in-service and wear proceeds by chip formation and breaking out of microscopic fragments from the surface.

Hardfacing components involves weld depositing a wear-resisting alloy on the surface to be protected. A typical cobalt base alloy like Stellite 6 (3.0%Ni, 2.0%Si, 3.0%Fe, 2.0%Mn, 30%Cr, 1.5%Mo, 4.5%W, 1.0%C) can be applied in the field to build up worn parts. The deposit contain 1400 HV particles in a 400 BHN matrix. Figure 11 is an example of how to take advantage of controlled wear. By hardfacing one side only, the

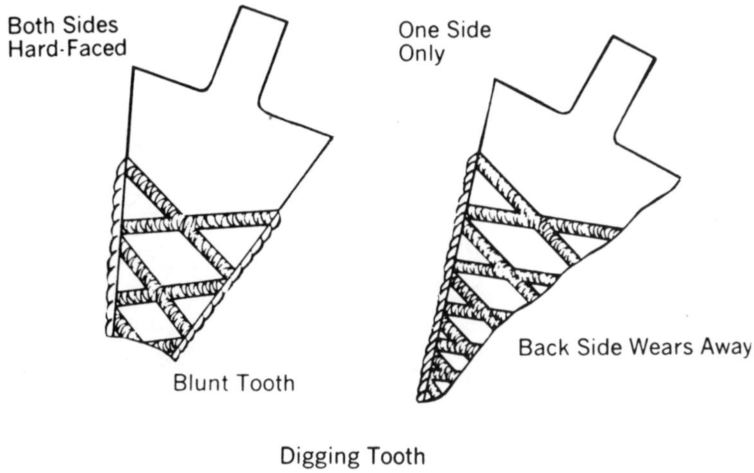

Fig. 11. Example of controlling wear through selective hard facing, (Ref. 2).

digging tooth becomes self-sharpening. The deposit configuration can be tailored to the existing conditions. Figure 12 shows typical hardfacing patterns to combat specific wear problems. In rock mining parallel beads act as rails to guide the flow of aggregates along the tooth. However, granular ores will gouge areas between a rock tooth overlay. Transverse beads will trap a quantity of ore between welds and protect the underlying steel from abrasion. In a situation where both rock and sand are encountered, a herringbone or diamond pattern is a useful compromise.

Erosion-Corrosion

Corrosion of all engineering materials is theoretically possible because of the basic reactivity of the metals with the environment. Corrosion is minimized when a surface barrier, such as paint or the naturally produced oxide film, separates the alloy from direct contact with the corrosive medium. Erosion-corrosion is a phenomenon that occurs when the flow of the environment over the surface of the part strips away the protective surface layer. Pumping of slurries, which are liquids that contain suspended abrasive particles, is a common problem in the extractive industries. Figure 13 is an example of a stainless steel impeller after three weeks service. It is obvious that the naturally occurring chromium oxide passive film was rapidly destroyed in operation. Erosion-corrosion is combatted by utilizing inherently hard and corrosion resistant materials. An iron containing 14.5% silicon has a high resistance to erosion-corrosion in acid slurries which are very corrosive to many other corrosion-resistant alloys.

Often, common structural materials can be coated or lined with highly resistant materials. Carbon steel pipe lined with basalt, stainless steel, ultrahigh molecular weight polyethylene, etc. finds application in specific service conditions. Rubber surfaced parts have been used for many years for ore handling systems.

Erosion-corrosion problems can be solved by changing the alloy, changing the environment, and sometimes by redesigning the system to prevent turbulence or impingement. A failure analysis of an erosion-corrosion problem first requires a complete analysis of the operating parameters including: pH, temperature, flow rate, type of abrasive, and oxygen content. Better materials or designs cannot be evaluated unless the specific service conditions are well understood.

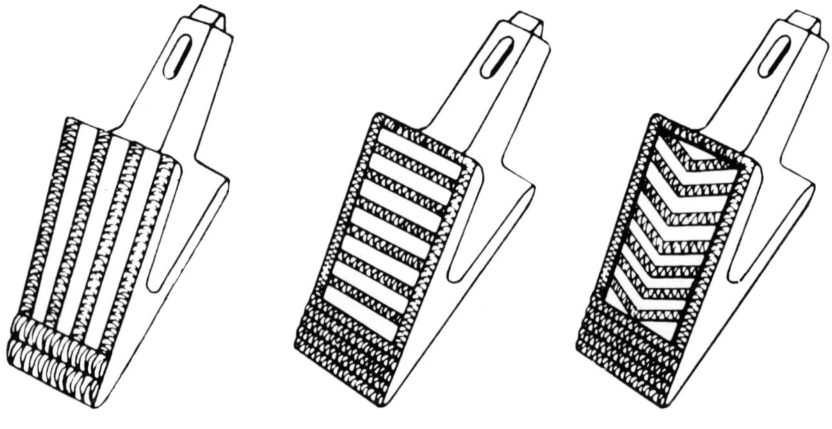

Fig. 12. Hard facing patterns for specific digging applications, (Ref. 2).

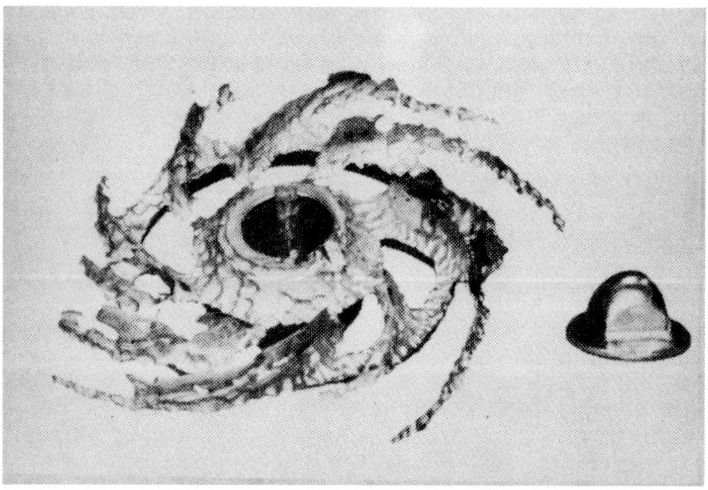

Fig. 13. Stainless steel impeller after three weeks use in acid slurry pumping, (Ref. 3).

Corrosion-Fatigue

The reduction in fatigue strength or service life due to the simultaneous influence of environmental interactions is termed corrosion fatigue, CF. While the detailed mechanism is still subject to dispute, the phenomenological aspects of CF have been well documented. Important considerations are:

1. stress ratio and maximum stress,
2. the specific environment and,
3. the cyclic frequency.

Higher stress ratios and mean stress accelerate the growth of CF cracks. More aggressive environments and lower cyclic frequencies also accelerate damage.

Consider a large wheel rim used for off-highway equipment. The inside radius is subject to cyclic loading on each revolution (Fig. 14). During development, the high mean and cyclic strain readings raised fatigue durability questions. A laboratory fatigue program was conducted on wheel sections cycled between 1100 to 2000μ strain at 800 cpm without failure in up to 12 x 10^6 cycles. However, in field performance evaluation, cracks were noted in the critical radius after 250,000 cycles. Figure 15 is a photograph of the fracture surface after mechanically bending the specimen to open the cracks. Figure 16 is a close-up view of the crack surface having a typical fatigue crack geometry and covered with black oxide. Figure 17 is a photomicrograph of a section through a typical fatigue crack which also shows a corrosion pit on the critical surface. This field failure was determined to be a result of CF. The inside tire environment was found to be 180°F containing substantial water vapor. The light surface rusting and pitting reduced the fatigue life from the infinite life regime to about 6 months. The failure analysis ultimately pointed the way to solving the problem by paint coating the critical radius. Vapor phase corrosion inhibitors are now being evaluated as an additional safeguard.

SUMMARY

While the failure mode of mining and heavy mechanical equipment is similar to other systems, several unique problems influence the cause of failure. The problems of size, abuse, maintenance, wear and corrosion syngergistically affect the reliability of such machines. The role of a failure analyst is to resolve the myriad of possible man-machine-

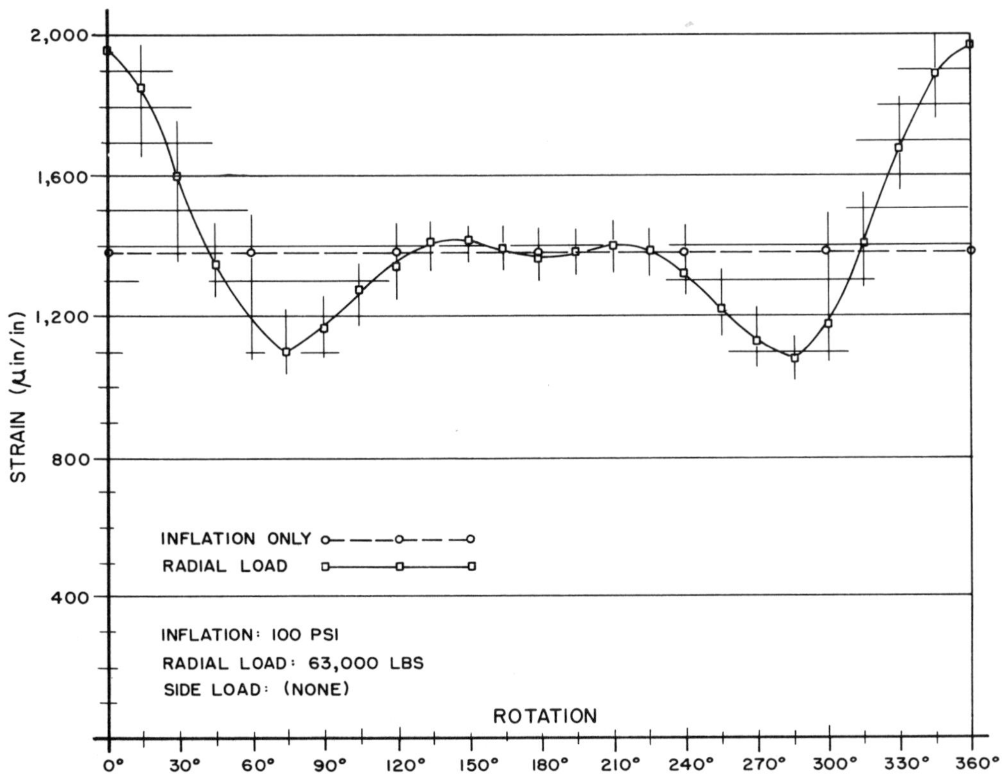

Fig. 14. Cyclic strain amplitude as a function of wheel rotation at critical radius.

Fig. 15. Fracture surface showing CF cracks in critical radius.

Fig. 16. Close-up view of one CF crack shown in Fig. 15, the surface of the critical radius is at the left. Original Mag. 50X. (Reduced approximately 35 percent for reproduction).

environment interactions to define the cause and to provide design input to minimize future problems. The failure analyst and the designer must work together to optimize reliability in a cost-effective way.

REFERENCES

1. T.R. Shives and W.A. Willard, *Mechanical Failure: Definition of the Problem*, NBS, p. 23 (1974).
2. Symposium, Materials for the Mining Industry, Vail, Colorado, pp. 90,91 (July, 1974).

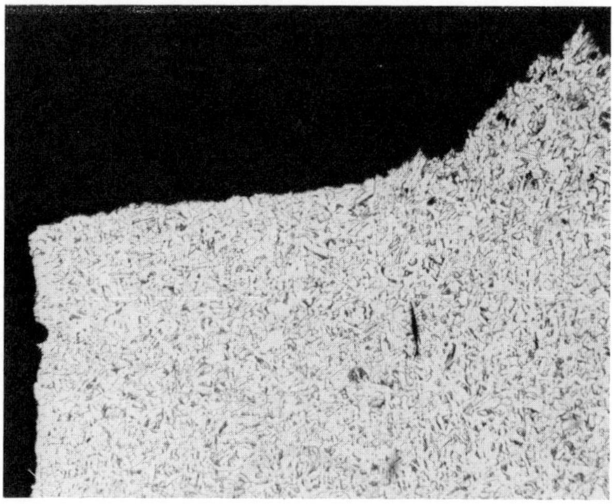

Fig. 17. Photomicrograph of the CF crack shown in Fig. 16, the surface of the critical radius is at the left. Original Mag. 50X. (Reduced approximately 40 percent for reproduction).

3. M.G. Fontana and N.D. Greene, *Corrosion Engineering,* McGraw-Hill, p. 72 (1967).
4. L.E. Alban, "Why Gears Fail", *Metals Progress,* p. 95 (Nov., 1970).

BOILER AND TURBINE COMPONENT FAILURES

R.D. Barer*

On many occasions, problems which superficially look like a simple "it bust" type of failure, have been shown to be much more significant when micro-examination is applied to the parts involved. It is, of course, important that the people who are in first hand contact with such failures should be sufficiently curious, want technical verification, or be required formally to make the evidence available for diagnosis. In this paper several case histories of failures in boilers and related components in which metallography played an unsuspected level of significance will be presented. Also included are a couple of examples which demonstrate the value of visible and macro evidence in the diagnosis.

Case History No. 1 – Boiler Tube Failure. The first case history involves a very simple boiler tube failure [1]. A 2" fire row tube burst in a controlled superheat marine boiler resulting in flooding of the furnace in several feet of water. The incident occurred during ship trials in the first few hours of the ship's life. The tube burst open, as shown in Fig. 1, just below the bend leading to the steam drum.

Visual examination showed the following;

a) A split at the center line of the tube, i.e., at the high heat input zone
b) Thin edges of metal existed at the split, practically forming a knife edge, and "stretcher" marks were present on the inner surface at the break, as shown in Fig. 2, which are indicative of plastic flow.
c) A relative absence of any internal deposits in the area.

This tube and an adjoining one were made available for examination. Radiographic and metallographic examination did not reveal any evidence of defective material. The microstructure of the steel 4 ft. away from the burst and from the neighboring tube showed a normal ferrite–pearlite mix typical of a 0.25% carbon steel.

Small pieces of metal were cut from areas close to the lip of the burst. Metallographic examination of these pieces established the fact that the metal in this area had been heated to temperatures of approximately 1700–1800°F before being rapidly quenched by the tremendous flow of steam and water released by the burst. The microstructure of the steel from this area is entirely martensite as shown in Fig. 3.

Examination of specimens from areas adjoining the burst, shown in Fig. 2, showed increasing amounts of pearlite and bainite, as might be expected, indicating a less drastic

*Defense Research Establishment Pacific, Victoria, British Columbia, CANADA.

Fig. 1. Location of burst tube in tube bank.

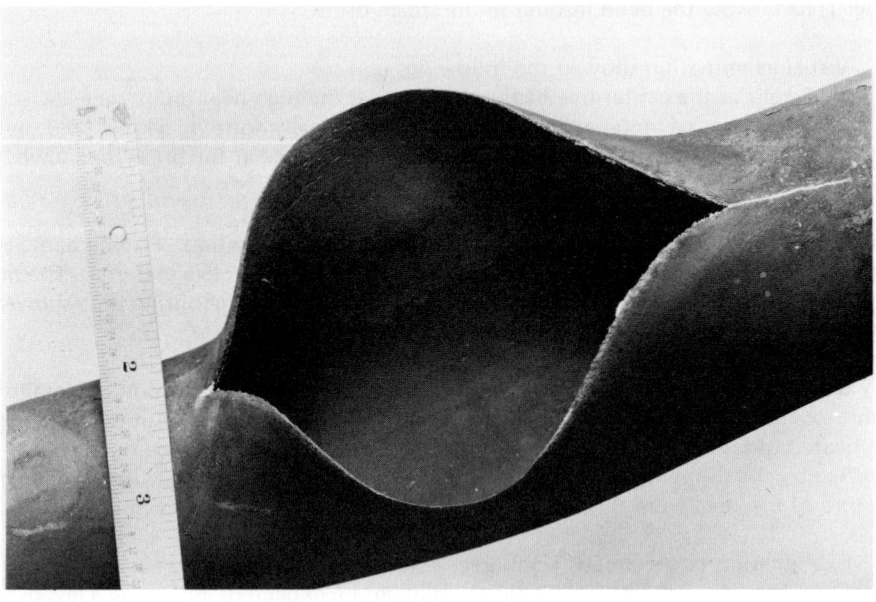

Fig. 2. The burst tube. Note the thin edges and the "stretcher" marks on the inside surface.

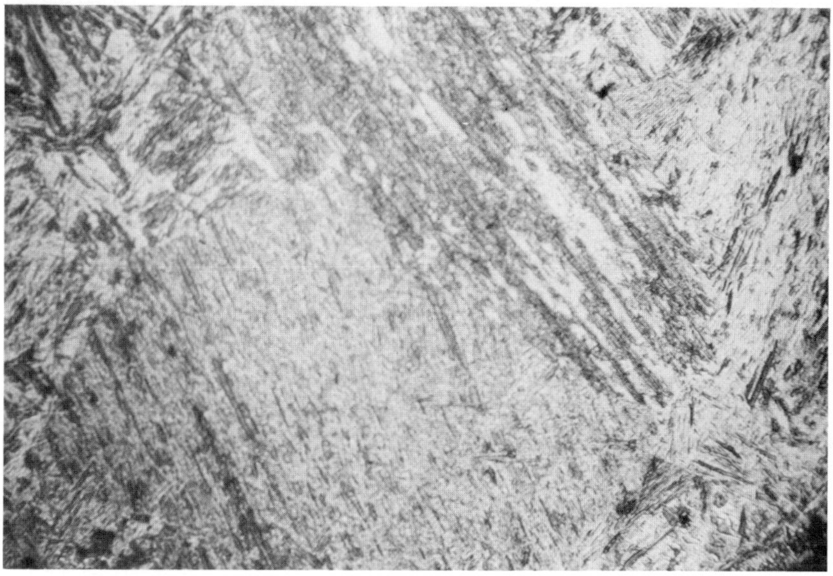

Fig. 3. Typical martensitic microstructure at the burst. Original Mag. 400X. (Reduced approximately 35 percent for reproduction).

quench rate and, possibly, a little lower temperature. The heat-affected zone extended about 2" above and below the rupture, but not fully around to the back. The sequence of events which must have occurred during the failed, as deduced from the above evidence, was:

a) Excessive steam generation occurred because of high heat input at location of the burst, and this was combined with some temporarily poor circulation (cause unknown) which impeded the transfer at the heat.
b) The tube metal at the area where the burst occurred became progressively hotter, eventually reaching 1700–1800°F. This may have occurred in a short period of time.
c) At the high temperature, the steel would have approximately one twentieth of its usual strength and also possess great ductility. The result was gentle bulging until the wall of the tube became too thin to withstand the internal pressure.
d) At the instant of rupture a great flow of water and steam (much of the water would flash to steam) effectively quenched the overheated metal.

It was concluded that failure was by simple overheating, probably related to something inadvertently being left in the boiler which impeded the flow in this one tube. It was considered an unusual event and one not likely to occur again since the generator tubes of this power plant served for many years without incident.

Case History No. 2 – Generator Tube Failure. The second case history[2] also involved a burst generator tube in a ship's power plant, and at a similar location, i.e. just below the upper bend. There was nothing unusual in the operating conditions of the ship at the time when the failure occurred.

The burst tube is shown in Fig. 4. There appeared to be no metal wastage and only a minor amount of stretching. The separated edges had a "blocky" or plate-like appearance. This is a typical appearance of previously cracked metal.

Other tubes in the second row, shown in Fig. 5, also showed some cracking which became more evident when the cracks were split open. The combination of full and partial ruptures suggested a momentary change in operating conditions and may have occurred in the tube.

A section cut from the burst tube near the burst, Fig. 6, shows the presence of other, i.e., secondary, cracks as well as ductility of the metal.

The microstructure of the steel close to the burst is shown in Fig. 7. The intergranular cracking is typical of stress corrosion. Note also the metallurgical evidence of overheating and rapid cooling in this tube as compared with the more common microstructure of a tube from the second row, Fig. 8. This type of cracking was found in many of the sections examined and is indistinguishable from cracking exhibited by the steel of an autoclave exposed to 50% caustic soda at 250°F and 400 psi, as reported by Copson[3].

Further examination showed that the cracking was confined to tubes in the two rows only. Sound tubes were found next to cracked ones. In the defective tubes, the cracking was in a zone approximately centered at two to four feet from the top of the tubes. Similar examinations of tubes in other boilers on the ship revealed no cracking.

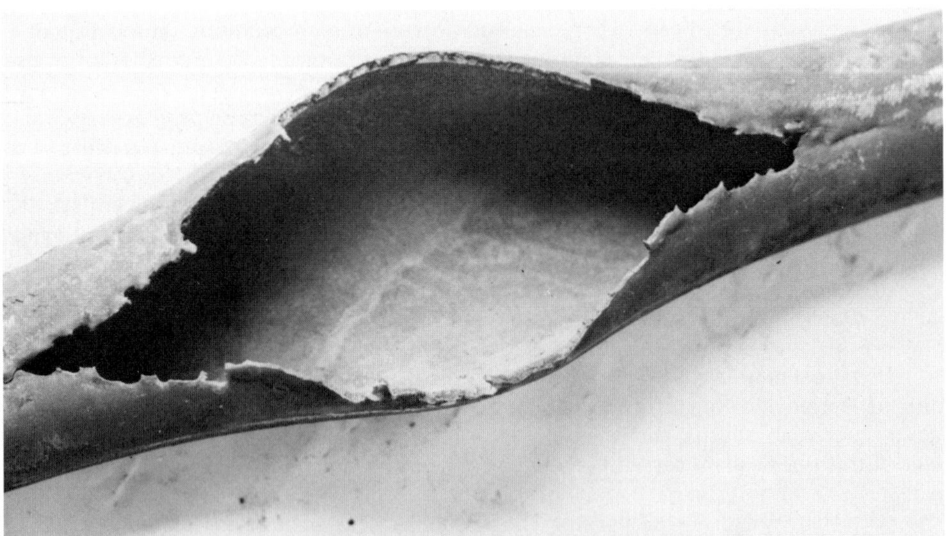

Fig. 4. Burst boiler tube. Note thick walled rupture. (Compare with Fig. 1).

Fig. 5. Exterior of three leaky boiler tubes.

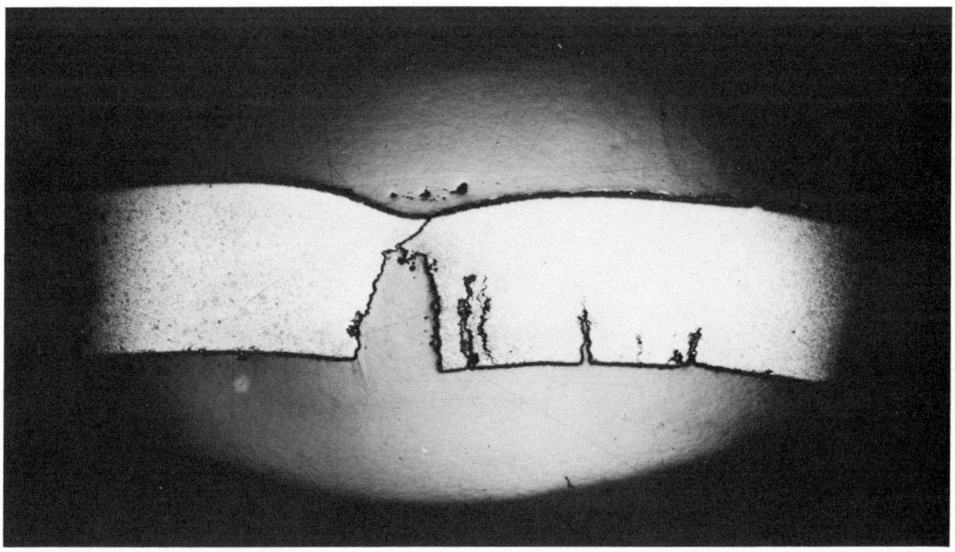

Fig. 6. Section through tube wall near the burst showing ductile necking down and final 45° rupture. Note other cracks. Original Mag. 7X. (Reduced approximately 25 percent for reproduction).

Fig. 7. Fine intergranular cracking in non-fire, inside surface of tube near the burst. Note normalized structure indicative of overheating. Original Mag 200X. (Reduced approximately 33 percent for reproduction).

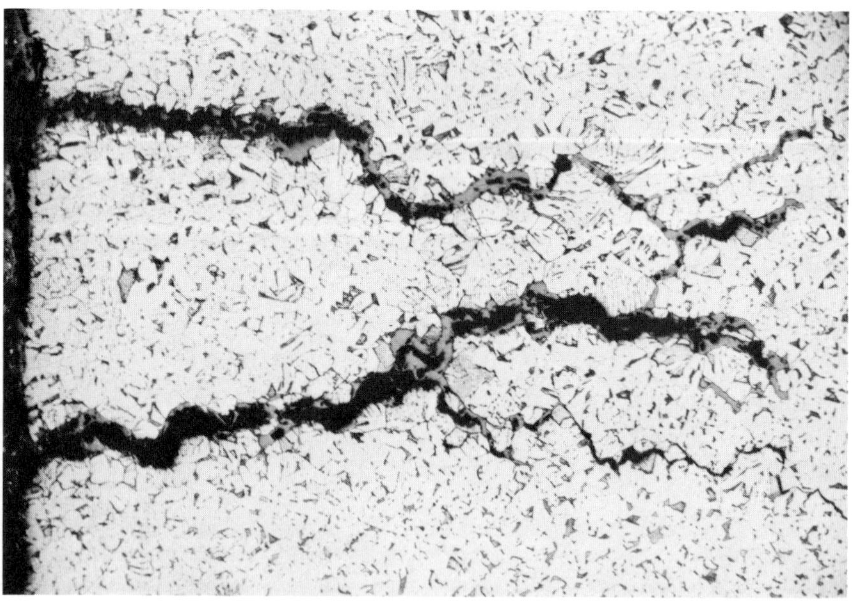

Fig. 8. Section through tube wall showing intergranular cracking typical of caustic cracking. Original Mag. 100X. (Reduced approximately 33 percent for reproduction).

The caustic cracking or stress corrosion cracking had initiated on the water side of the generating tubes. It only remained to establish the presence of stress and a corrosive environment.

The stresses in the boiler tubes were a combination of the normal operating stresses resulting from steam generation at 400 psi and 650°F and the residual stresses. The high residual stresses that resulted from fabrication by extrusion and from bending the tube would be reduced somewhat after several years' operation, but would still be significant in relation to the operating stresses.

The corrosive environment could only be caustic from the boiler water treatment. The usual concentration is commonly 200 ppm, or much less than one percent of that required for aggressive attack. Normal caustic cracking of boiler metal usually occurs where caustic concentrates by leakage, e.g., between seams and around rivets, or by capillary action and subsequent evaporation. No such action could have occurred on these boiler tubes. The boiler solution in this instance had to develop a higher concentration right on the steam generating surface of the boiler tube.

Severe overheating in a localized area could occur and proof of this is given by Davidson and Associates [4]. They used thermocouples to study various conditions and found zones of severe overheating as shown on Fig. 9. They considered that this resulted from "steam blanketing" following flame impingement. Figure 10 indicates how rapidly the caustic content in the concentrating film of boiler water can rise with an increase of temperature above that of the over-all boiler water. A 100°F rise corresponds to 50 percent caustic in weight percent. These facts establish a mechanism to account for this case history of stress corrosion cracking occurring directly in boiler tubes.

Case History No. 3 – Leaky Boiler Tubes. A third case history of failures in boiler tubes [5] involved leaky tubes in an older ship. As a preliminary to more detailed examination, the tubes were checked by radiography which revealed a fairly extensive cracking as shown in Fig. 11.

Sections were cut from affected areas on these tubes for microexamination. The general findings were:

a) The presence of a high incidence of cracks.
b) A great distribution in the depths of the cracks, some perforating the tube wall and others extending only part way through it, as in Fig. 12.
c) The major cracking was on the water side, Fig. 13, as a result of especially severe flame impingement.

A closer look at a typical beginning crack, Fig. 14, provided additional details:

d) The cracks generally did not follow the grain boundaries.
e) There was little, if any, oxide in the larger cracks.
f) There was generally an oxide layer on the tube surface – approximately 0.001" to 0.002" thick – and it contains particles of copper.

As shown in Fig. 14, there appears to be some correlation between the cracks in the oxide layer and the beginnings of the cracks in the metal. This is confirmed in Fig. 15, in which are shown "embryo" cracks, i.e., the earliest stages in corrosion of the tube, the formation of cracks in the oxide, and the apparent pointing-up of the general oxide layer. These appear to be the initiating points of cracks in the tube wall.

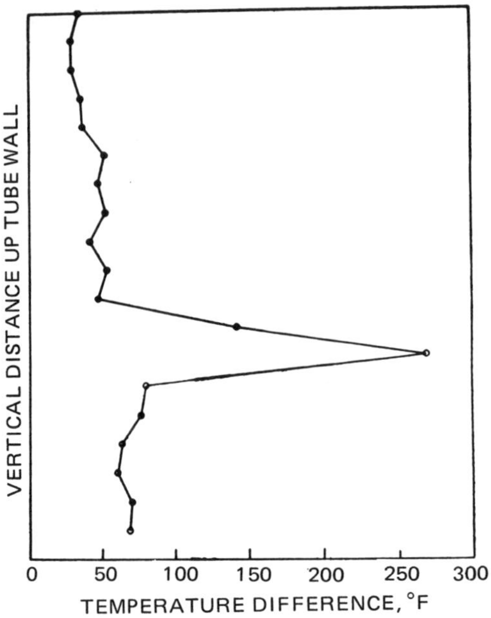

Evidence of a hot spot in a 50 ft. vertical water wall tube.

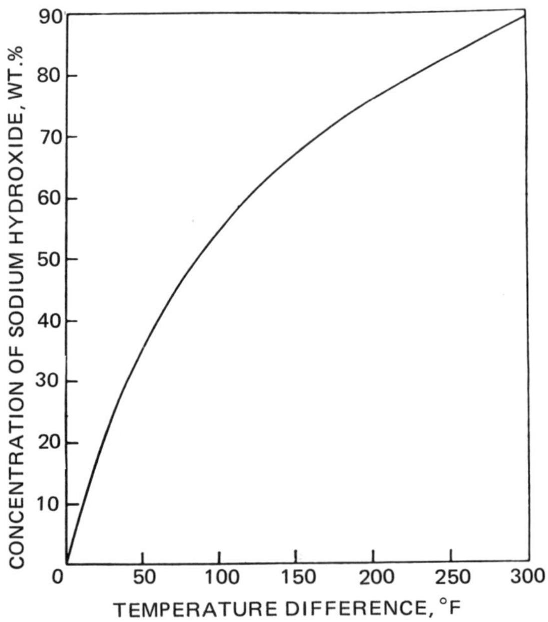

Fig. 10. Caustic content attainable in concentrating film of boiler water. [Based on data from *International Critical Tables,* 3,370 (1928).]

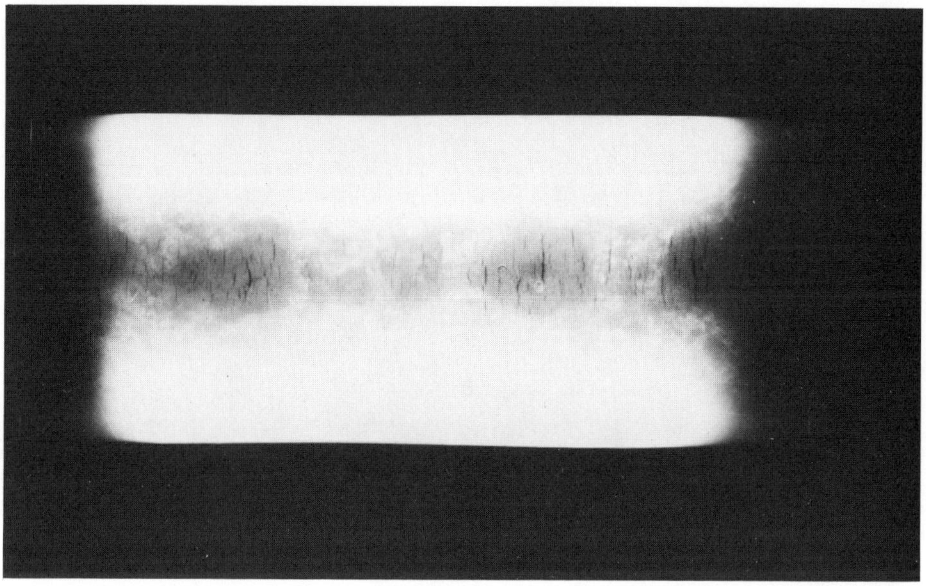

Fig. 11. Radiograph of boiler tube showing many fine transverse cracks.

Fig. 12. Section through tube wall showing transverse cracks. Original Mag. 20X. (Reduced approximately 25 percent for reproduction).

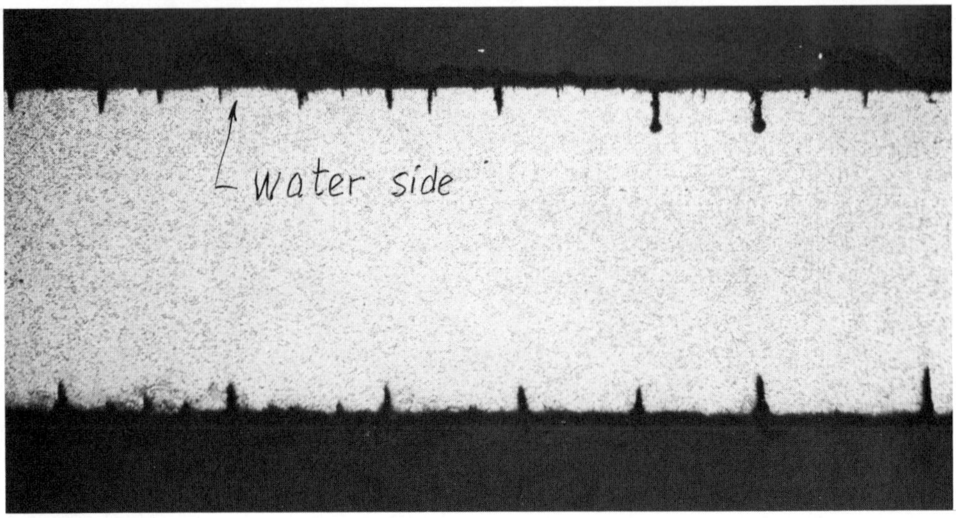

Fig. 13. Section through tube wall showing transverse cracks on both water side and exterior of tube. Original Mag. 20X. (Reduced approximately 25 percent for reproduction).

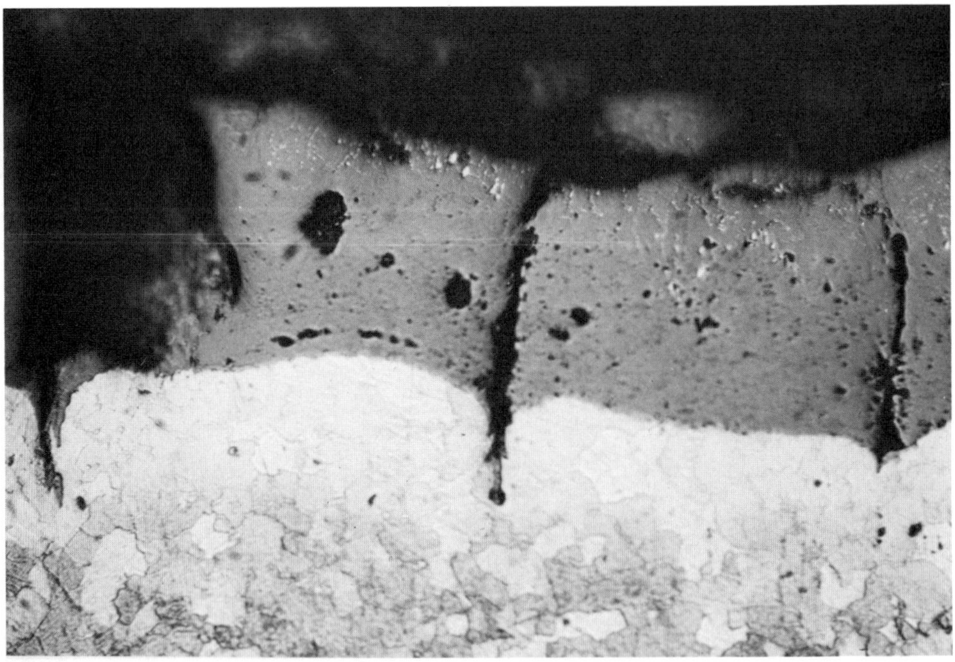

Fig. 14. Close-up of a crack shown in Fig. 13. Note the oxide layer on the tube surface. Original Mag. 100X. (Reduced approximately 25 percent for reproduction).

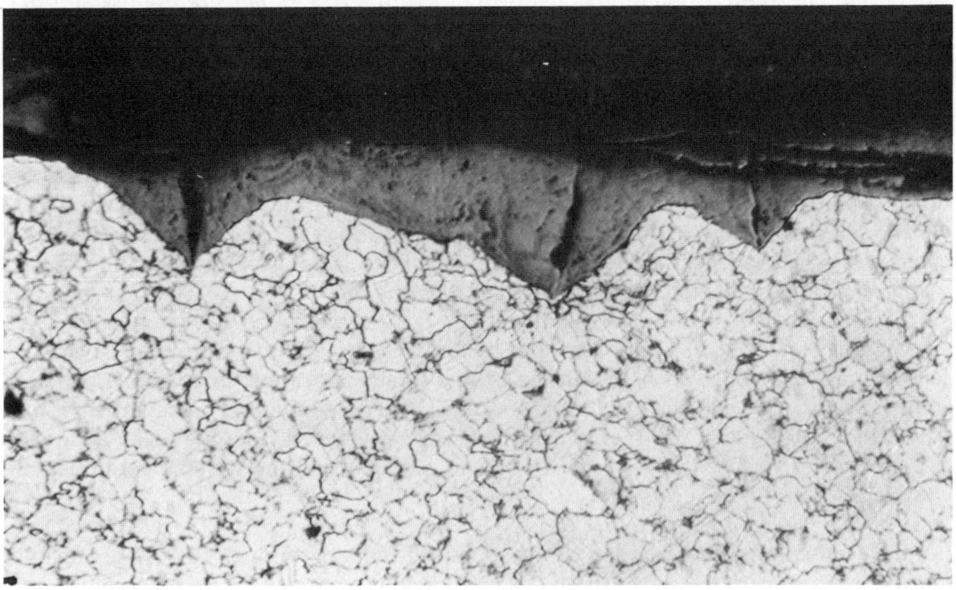

Fig. 15. Early stages of corrosion fatigue cracks. Note cracks in oxide layer. Original Mag. 100X. (Reduced approximately 25 percent for reproduction).

An important factor noted in the metallographic examination was the absence of pearlite, and in its place, small particles of "spheroidized" carbide. The normal structure of the tubes, based on examination of nearby sound tubes, was that of ferrite plus pearlite. A spheroidized structure indicates lengthy exposure to temperatures of 900°F or higher. An indication of the amount of time required is shown by the fact that spheroidization which can be achieved in approximately 15 hours at 1200°F may require almost one thousand times as long at 1000°F.

It was established then, that the defective tubes had been overheated. The task faced then, was determining why some of the tubes overheated while others did not. Similar internal deposits were found on adjacent sound tubes, so that while the deposits may have contributed (by an insulating effect) to overheating, they were not the significant factor. One possible mechanism which could provide high temperatures for the metal for long periods would be some hampering of the circulation of water (and steam). It is possible to visualize a situation where once slightly more than normal steam generation occurred (by slight overheating) causing an increasingly deteriorating situation to follow.

Because of earlier favorable definition of fissures by radiography in the laboratory, it was decided to use radiography as an inspection tool in other ships of this class. These examinations proved successful in finding some other defective tubes, not otherwise suspected.

Calculations of stresses which may be present in a tube which overheats while its neighbors do not, show that quite high stresses can develop. For example, the stress that will develop in an 8 ft. long tube under restraint with a 200°F temperature rise approximates 36,000 psi. Residual stress levels of this magnitude were confirmed with strain gauges. It seems reasonable that normal operational cycling over many years could interact with some part of this stress and result in corrosion fatigue.

The evidence obtained metallurgically indicated that the tubes failed by corrosion fatigue. Overheating was the main factor which changed the non-corrosive nature of the boiler water to a corrosive condition; it also served to slow the flow through the tubes, which in turn was more harmful; and relative overheating of some tubes contributed to cyclic stressing. Recommendations included chemical cleaning and periodic nozzle adjustments to avoid flame impingement.

Case History No. 4 – Rupture of Two Superheater Tubes. Another instance of failure in a ship's boiler involved rupture of two superheater tubes in one boiler while the ship was engaged in exercises thousands of miles from home port. This superheater had been recently re-tubed. Samples of these tubes were removed from the boiler and mailed to the laboratory. The tube ends in the header were plugged and the ship continued on the exercises.

Examination of the samples of the tubes showed a fairly thick (0.02") crystalline-appearing deposit on the steam surface, a considerable amount of scale externally, and thinning to the point of rupture in a relatively small area. This was confirmed by a radiograph of one of the tubes, Fig. 16. Metallographic examination showed no defects or cracking in the steel. Analysis of the deposit revealed the presence of boiler chemicals, predominantly phosphate, which do not belong in the superheater.

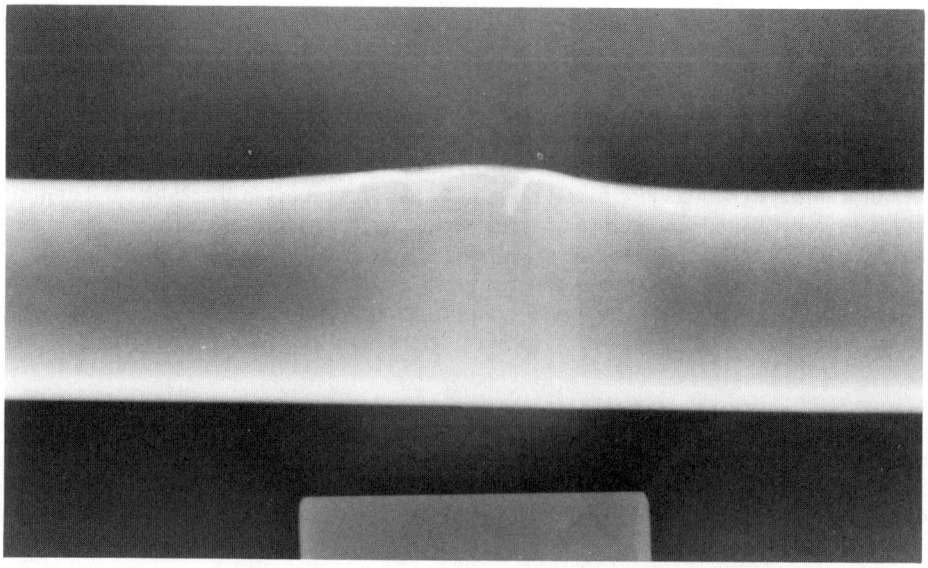

Fig. 16. Radiograph of tube showing the localized thinning of the tube wall.

It was concluded from the laboratory evidence that boiler chemicals had deposited in these tubes and had served as an insulating layer, and that the tube metal had reacted with the environment both internally (corrosion) and externally (scaling). Rupture of the tube wall resulted.

The ship was asked to check into possible causes of "priming" (or carryover) and a small gap was found just above the water line in the flanged joint of the steam line from the drum to the superheater. This was corrected and the ship was allowed to operate at reduced power because of the existing deposit. Subsequently, when the ship arrived in a foreign port, a split was noted in the stub ends of two of the tubes. These were removed and replaced, as were also the other two matching ends. The laboratory at the foreign port examined the stub ends and found that cracking was caused by caustic embrittlement (or SCC). At about the same time, on pressure testing, a leak was detected in another tube, which proved to be a small split just past the header, on the fire side. This too proved to involve caustic embrittlement cracking.

The finding of caustic cracking in the stub ends of the three tubes, as shown in Fig. 17, led to a concern that there could be a condition of generally prevalent cracking. Considerable doubt arose about the future reliability of this power unit. The other (port) superheater had not suffered any priming and its reliability was not in question.

Obviously, in view of the questions regarding operating on a possibly defective power plant, a decision was needed. The facts were carefully reviewed at the home base and the following points made:

a) The failures of the initial two tubes occurred by high temperature corrosion and oxidation.
b) Nearby tubes, exposed to similar conditions, were in some danger of failure, but at reduced output the hazard was reasonable.
c) Cracking in the *stub* ends of the two tubes was by a different mechanism. Cracks had likely resulted from the high residual tensile stresses produced by insertion of the plug.
d) Cracking in the third tube that failed was of some concern. Its exposure to heating was somewhat less, and it was possibly a special case, confined to the one tube.

Acting on this diagnosis and these premises led to a decision to continue the ship's operation. The ship performed with no problems for two months (at reduced output on the one unit) until it could return to home bse.

Case History No. 5 – Main Turbine Failure. This next case history of a failure involved a ship's main turbine. This failure initially appeared to be an isolated case – but metallography played a part in showing it to be far more widespread and serious.

Investigation of the turbine failure showed that two (of three) cover plates, Fig. 18, had been allowed to drop into the turbine by an apparent shear of a bolt, Fig. 19. If failure was by shear, there was an implication that the bolt might have been over-torqued and cracked during installation or that a foreign object had been somehow wedged between the plate and the turbine, causing the high stress necessary to shear the bolt. Either of these would represent one-of-a-kind failure. As shown in Fig. 20, failure had taken place just under the head and not in the weaker neck where shearing would have occurred by over-torqueing.

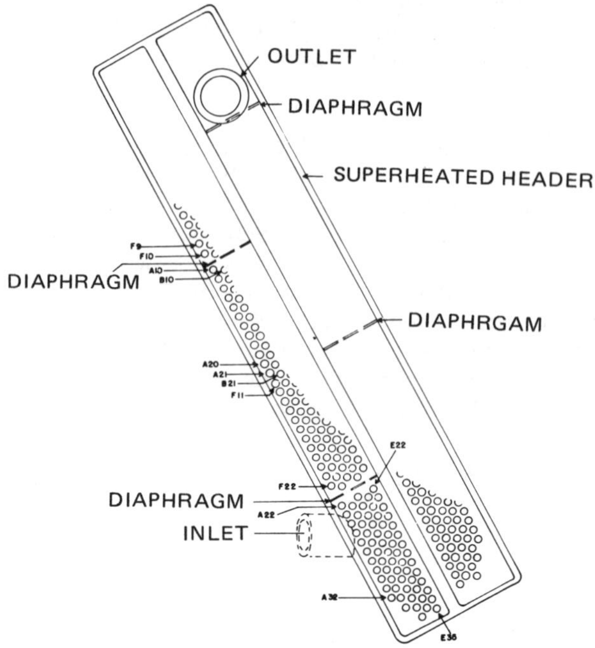

Fig. 17. Arrangement of tubes in the superheater.

Fig. 18. Remaining cover plate and locking bolt location.

Fig. 19. Failed locking bolt. Note the shear appearance.

Fig. 20. The two locking bolts. Note that failure took place just under the head.

Sectioning of the remaining bolt showed evidence of stress corrosion cracks, as in Fig. 21; and some cracking also was found in the failed bolt. It was noted that the crack extended into the undeformed grains of the bolt, showing that cracking was not associated with a shearing mechanism. It was also noted that the metal at the fracture surface had suffered very little deformation over most of the fracture face, so that what appeared to be shearing, was superficial smearing which occurred when the cover plate worked past the bolt after it had all but failed.

The bolts were of Type 303 free machining stainless steel. Apparently these bolts were part of a modification kit which called for "stainless iron", i.e. an iron-chromium alloy. In any case, the Type 300 stainless steel is prone to cracking by stress corrosion in this environment, as chloride contamination can be expected to be present to some degree in a ship.

The diagnosis of stress corrosion cracking and the presence of these stainless steel bolts in other ships with modified nozzle plates, raised the question about this now being a much wider problem. Accordingly, recommendations were made that at the earliest opportunity, other such units should be examined, and preferably the bolts should be replaced with a low alloy steel known to be resistant to SCC in this environment. A steel such as ASTM A193 B16 (1 Cr, 0.5 Mo, 0.3 V) was suitable.

The very next ship examined had a bolt partly failed (part of the head was missing) and one cover plate had dropped approximately ¼ in., Fig. 22. Subsequent checking resulted in the finding of five installations with cracked bolts, saving five main turbine failures. Replacing these contributed considerably to the reliability of operation of the ships.

Case History No. 6 – Overheated Thrust Bearing Failure. The next case study is one of two examples where metallurgical examination was involved but visual or macro examination was more significant. A rather overheated and distorted thrust bearing, part of a main turbine drive train, shown in Fig. 23, was brought to our attention following failure in service. The collar, normally a flat surface, had been "re-shaped" by the severely damaged retaining ring surface, onto which the remains of the thrust pads had been smeared, Fig. 24. The normal tin-base babbitt and bronze-backed thrust pads were largely obliterated off the retaining ring. Large pieces of the bronze thrust pad, Fig. 25, appeared to have "escaped" at an early stage of the failure process. Aside from some tumbling and loss of babbitt (by melting) they had not been too damaged. They did not appear to have been subjected to any special loading.

The severe damage and distortion in this area were final stages in a particular sequence of events. It remained to establish what the events were.

A careful examination of the back of the thrust bearing retaining ring showed an interesting deformation at each of the radial flat notches, Fig. 26. Also the inner surfaces showed signs of light pounding, but the "lip" (arrow in Fig. 26) did not show this pounding, suggesting that it formed at a later stage in the deterioration sequence.

Pieces of steel from the surface of the retaining ring were examined under the microscope. These showed evidence of having been to a high enough temperature ($\sim$1700°F) to result in martensite, and was further evidence of overheating. The subsequent chemical analysis showing the presence of nickel confirmed that some welding had occurred

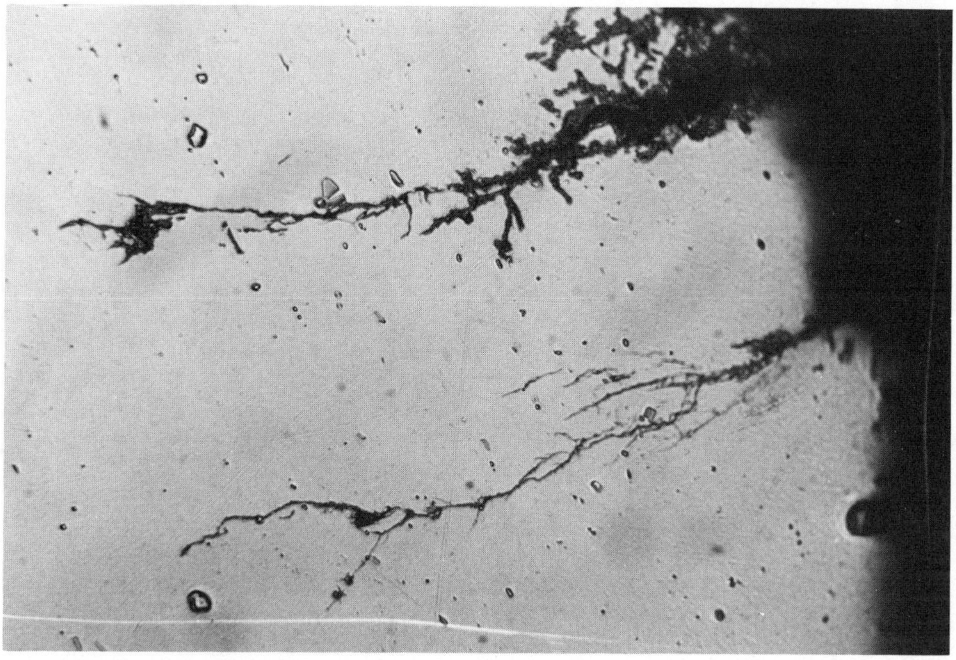

Fig. 21. Branching stress corrosion cracks in locking bolt. Original Mag. 100X. (Reduced approximately 5 percent for reproduction).

Fig. 22. Partly fractured bolt head, and cover plate movement of ¼ in.

Fig. 23. Badly worn gear and bearing assembly.

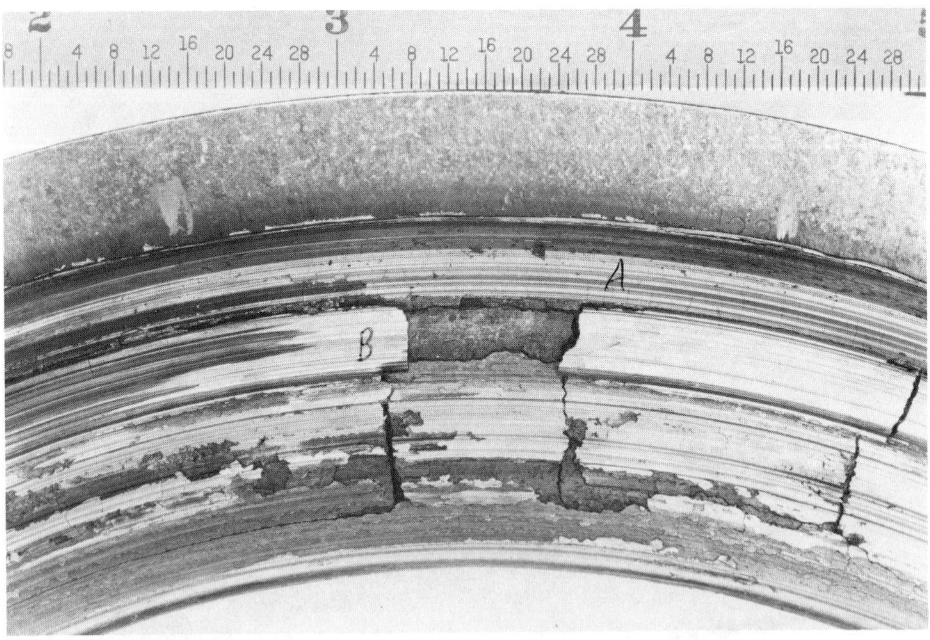

Fig. 24. Smeared face of thrust bearing retaining ring.

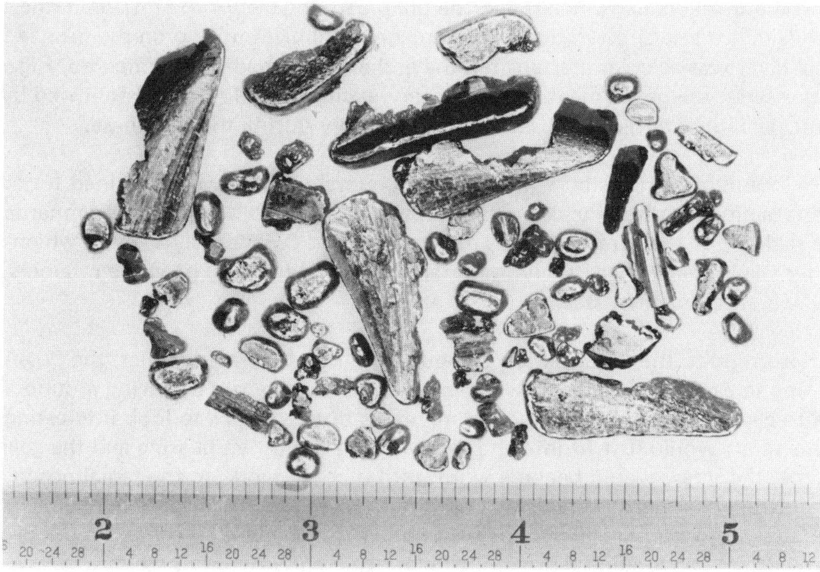

Fig. 25. Pieces of bronze and steel found associated with the failure.

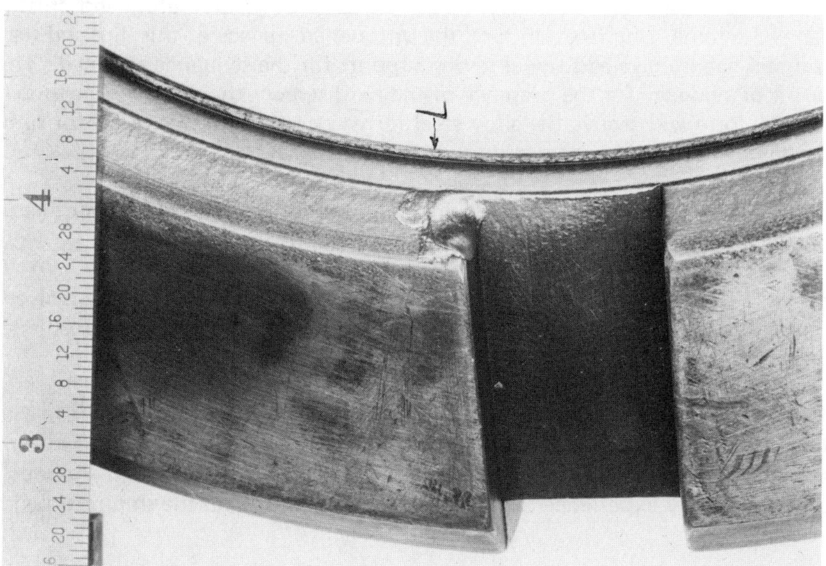

Fig. 26. Deformation in corner of radial notch in thrust bearing retaining ring. Note absence of pounding on "lip" L.

between the thrust collar (3% Ni) and the plain carbon steel of the retaining ring. Metallurgical evidence also showed that the 3% nickel steel had built up on the bronze in stages — the final layer was acicular martensite; the earlier layers had been tempered, Fig. 27. This suggests a sequence of overheating (temporary exclusion of lubricant) followed by quenching (return of lubricant) as well as welding, probably during the hot phase.

In looking for a possible "foreign" item, metal samples were obtained for analysis off the retaining ring face including small fragments lodged between the remnants of the bronze pads. One steel fragment was different, it had 0.35% molybdenum which none of the other steels nearby had. Only the retaining screws (used to hold the retainers in place) were 0.35% molybdenum steel.

Returning to the deformation or pounding that was noted earlier, the possibility of a retaining screw being accidentally dropped and subsequently bouncing around in such a way as to cause the peening and pounding marks noted, begins to look interesting. In fact, this screw would just fit into the gap between the thrust bearing and the gear shaft. With time, the screw would become worn until it could work up the radial notch and over into the thrust bearing side. This would initiate some jamming, some temperature rise, and the destructive process would be well underway. By what mechanism the bronze pieces of thrust pad would be broken off was not readily determined.

In any event, a complete breakdown followed in which the pads did overheat and break up. A fragment of the screw — by this time well mutilated — lodged between the thrust pads and remained there to give strong support to this breakdown sequence.

The lodging of the screw fragment also explains the absence of pounding on the 'lip', which was steel that flowed over under the action of temperature and pressure at the mangled retainer ring face. In fact, the untouched surface of this 'lip' and the peening on the metal nearby give additional strong support for the sequence outlined. There is a further bit of evidence for the presence of the hard screw: the much softer thrust pads could not deform and groove the alloy steel thrust collar. Only the screw (or similar object) could have initiated and contributed to such destruction.

This coherent sequence helped to indicate that failure was the result of an unusual circumstance, alleviating concern about other such units.

Case History No. 7 — Pinion Gear Failure. The final failure problem also involved some metallography, but careful observation and trying to develop a reasonable sequence proved more significant.

During high speed operation a failure of a pinion gear bearing occurred. The failure was detected by a sudden temperature rise to 390°F. There had already been some concern at the somewhat higher temperature of operation, approximately 200°F — about 18–20°F above that normally experienced. The unit was shut down and the ship returned on the other engine.

When the main gearing was opened up, it was found that the bearing had failed by wiping, Fig. 28, and the adjoining bearing had also failed by wiping at some previous period.

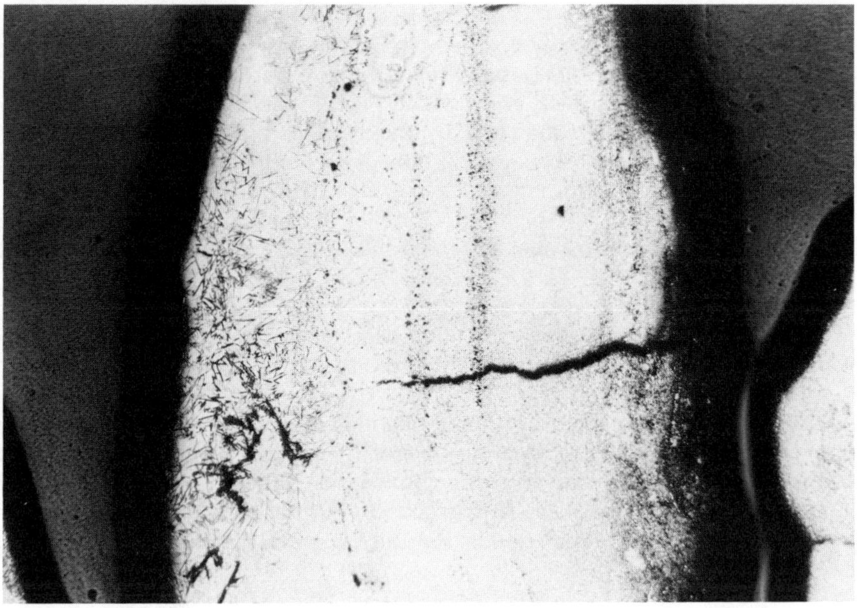

Fig. 27. Smeared high nickel material from thrust collar (B) in Fig. 24 showing acicular martensite at interface (top) and tempered martensite. Original Mag. 100X. (Reduced approximately 30 percent for reproduction).

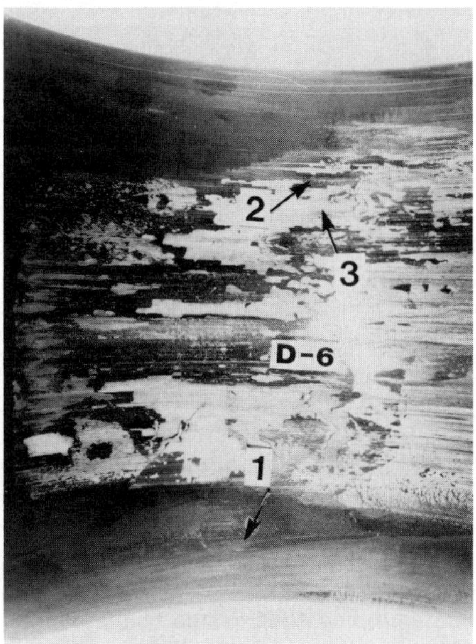

Fig. 28. Wiped bearing. Note non-wiped but stained area (arrow 1) and black oxide area (arrow 2).

All metal surfaces in the gear box had a brownish stain, presumably dating back to the period when an EP gear oil was used. The stain on the steel surfaces was hydrated iron oxide; on the babbitt surface, as shown on Fig. 29, it was tin oxide. The oxide just outside the wiped area, was less than 0.001" thick and represents a corrosive attack. Tin oxide is known to be hard and abrasive and some light scoring of the shaft journals confirmed this. It was noted that in the area of the wipe there were dark areas (arrow 2), from which the black scale could be lifted. Samples of the thin black scale were mounted and examined under the microscope, and the result, Fig. 30, showed tin oxide to a thickness of 0.002".

Engineering personnel also reported that bearing clearance for these two bearings had been somewhat below the desired 0.011" to 0.015".

A fairly usual diagnosis of this type of bearing failure has been to suggest that pieces of the tin oxide layer broke off and caused overheating, etc. However, no evidence was seen of the thinner tin oxide being anything but sound. Other investigators have reported that 0.002" thick tin oxide can also remain sound. Further, this mechanism does not explain how the tin oxide more than doubled in thickness in a specific area of the bearing surface.

Putting the preliminary observations together led to the following concepts: if the bearing clearance is slightly below normal, it would be reasonable to expect that the flow of oil might be less than that considered optimum. This could cause a small temperature rise. Also, the tin oxide (from a previous era) had more than doubled in thickness only in the wiped area, i.e., the area which was a bit hotter. It only remained then to postulate that the tin oxide increased in thickness under the operational "stress" of increasing temperature.

A possible mechanism then includes the following:

a) inadequate (or borderline) clearance for high speed operation, leading to a small temperature increase;
b) small growth of the oxide during periods of high speed operation;
c) the expansion from the additional tin oxide would further restrict the clearance, leading to further temperature increase and oxide growth;
d) eventually the melting point (approximately 430°F) of the babbitt is reached, the metal would flow, and oxide would be displaced, improving the clearance and followed by a resultant temperature drop.

This sequence met all the observed facts and seemed to explain satisfactorily these bearing failures. As a result, priority for prevention would be on ensuring suitable clearance, and stained bearings of adequate clearance were not a problem.

ACKNOWLEDGMENT

The author wishes to acknowledge the work of his associates in Materials Engineering at DREP, who participated in and contributed to these investigations.

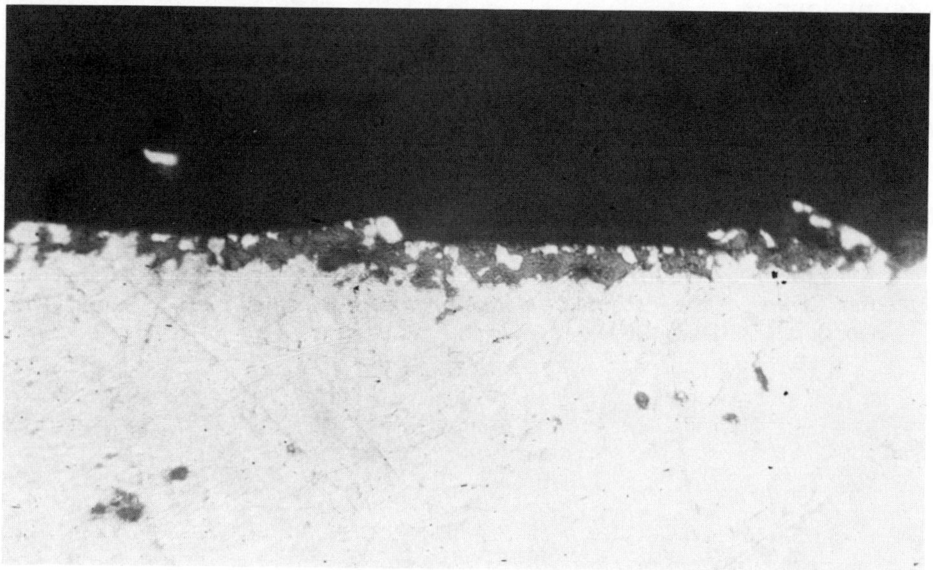

Fig. 29. Microstructure of area indicated by arrow 1 in Fig. 28. Depth of oxide approximately 0.0007 inch. Original Mag. 500X. (Reduced approximately 25 percent for reproduction).

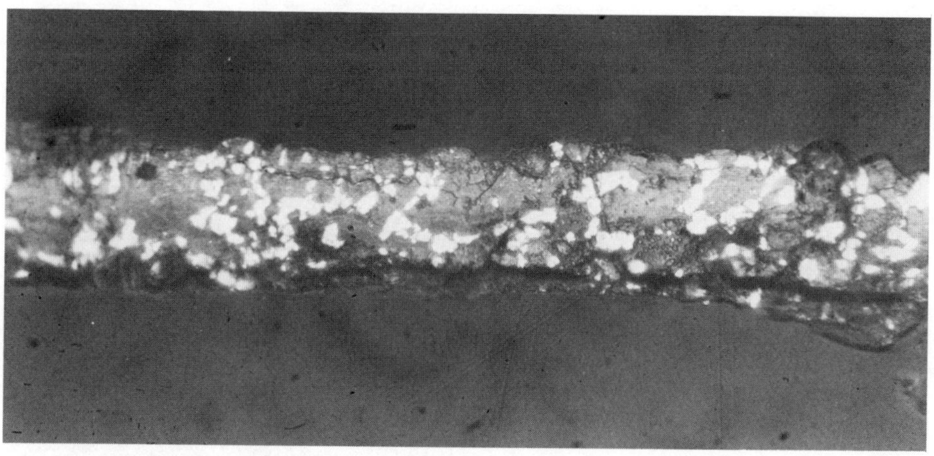

Fig. 30. Microstructure of black oxide, arrow 2, in Fig. 28. Depth is 0.002 in. Original Mag. 500X. (Reduced approximately 25 percent for reproduction).

REFERENCES

1. R.D. Barer and B.F. Peters, *Why Metals Fail*, Gordon and Breach Science Publishers, N.Y., p. 43–48 (1970).
2. Ibid. p. 57–66, also in *Corrosion*, Vol. XI, No. 4, p. 148 (April 1955).
3. Copson, H.R. "The Influence of Corrosion on the Cracking of Pressure Vessels", *Welding Journal,* pp. 75s–91s (Feb. 1953); *Corrosion*, 10, pp. 124–139 (1954).
4. Uhlig, H.H., Ed., *The Corrosion Handbook,* John Wiley and Son, p. 529 (1948).
5. R.D. Barer and B.F. Peters, *Why Metals Fail,* Gordon and Breach Science Publishers, N.Y., p. 50–57 (1970); also in "Marine Boiler Tube Failures – Three Case Histories", *Trans. of the Institute of Marine Engineers,* Canadian Division Supplement, No. 6, p. 154 (Dec. 1961).

FAILURE ANALYSES OF SURGICAL IMPLANTS FROM THE HUMAN BODY CAN IMPROVE PRODUCT AND PERFORMANCE RELIABILITY*

R.J. Gray**

INTRODUCTION

A relatively high measure of success is experienced today in the use of metal implants in the body. This success can be attributed to advanced metallurgical technology coupled with improved medical and surgical techniques. Ludwigson [1,2] has followed the history of metal implants and assigned several development periods: ancient times to 1875 when pure metals, copper, silver, and gold were used; from 1875 to 1925 when surgery and implantation became more successful; and finally the relatively current period from 1925 when more satisfactory results have been experienced with metals and alloys appropriately developed for such service. There are approximately 2,000,000 metal implants used in the United States today if we consider each pin and screw separately. The number of hip implants, the major joint replacement, is 50,000 per year in the United States alone [3].

At one time, implants and prosthetic devices were considered reserved primarily for the elderly. However, the increased interest in motorcycle, bicycle, and skate-board riding has "enrolled" many youths in the "implant and prosthesis club".

The number of implants and prostheses that fail during the expected service period is quite low——probably under 7%. For those unfortunate patients who experience failure of an implant, however, the odds have jumped to 100%. By careful diagnostic studies of failed implants, major contributing causes of failure often can be pointed out with a possibility of some remedial results; however, the root origin is not easily established [4]. Almost without exception, the failure of an implant cannot be attributed to a single cause.

The body is a hostile environment for some metals. "The environment about a bone implant includes fluid, tissue, and natural bone in close relation. It is a dynamic system of rapidly flowing fluids and electrolytes in living and moving tendon, muscle, and bone" [5]. The fluids which comprise about 70% of the body weight are saline solutions similar in nature and concentration to sea water. Cooperative efforts by the materials and medical/surgical disciplines can result in even greater improvements in the future in the durability and safety of these fixation devices.

The study of a failed implant usually begins with a critical low power microscopic review of the failure. This survey may show that a fissure originated in a gouge or a sharp corner of an implant which could have been the weak link in an otherwise success-

*Research sponsored by Union Carbide Corporation under contract with Energy Research and Development Administration.
**Oak Ridge National Laboratory, Oak Ridge, Tennessee USA.

ful service life. The survey may continue in depth to include magnification optical microscopy, scanning electron microscopy, and other sophisticated analytical procedures.

In this presentation, we are displaying some examples of implants that failed during service to the body. As much as possible, the cause(s) of failure will be pointed out.

THE ROLE OF AN IMPLANT IN THE BODY

The purpose of installing a fixation device over or through a bone fracture is to provide an aid to the body during its process of self-repair. An *implant* is considered a relatively temporary fixation which can be removed once the bone can accommodate its structure responsibility. A *prosthesis* is a permanent fixation which has replaced structural bone and is intended to remain in place for the duration of the patient's life. Some examples of various types of implants are presented in Fig. 1.

It is possible for an implant or a prosthesis to fail in the body just as metallic devices fail in any application that imposes excessive stress conditions, through steady or cyclic loading. Failure is even more apt to occur for any metallic device if stimulated by a corrosive environment; the body definitely provides such an envelopment.

CAUSES OF IMPLANT FAILURE

Basically, implants can be related to the six potential factors presented in Table 1. In reporting these factors, the order is not meant to represent any particular importance, priority, or statistic in the formation of failures. For example, the failure might have started due to one factor, but the continuation to failure might have involved several of the other factors. It must be pointed out that the first two factors listed relate primarily to the role of the body or the patient. The other four factors are related to the implant and its entry into the body.

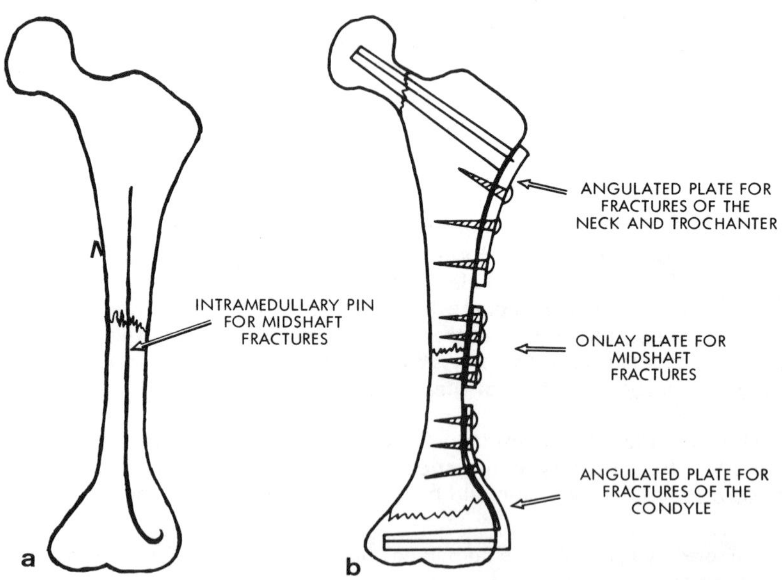

Fig. 1. Schematics of typical fixation devices for the femur.

Table 1. Potential Factors Influencing Implant Failure

- Corrosive Action of Body Fluids
- Inept Mobility of Patient
- Improper Selection of Materials
- Improper Installation of Implant
- Improper Fabrication
- Improper Design of Implant

Each factor will be described in detail and, where possible, examples will be presented. In some instances direct relationship with other factors will be presented also. Such information provides supportive evidence of the synergistic relationship of the causes of failure.

Corrosive Action of Body Fluids

In a suitable environment, there is a natural tendency for a metal or an alloy to revert to the lower form by oxidation through corrosion [6]. These reversion tendencies of metal placed in the body are aided by the internal environment of large and variable stresses and highly corrosive fluids and tissues. Stainless steel, for example, must rely on the presence of a closely adherent oxide surface layer to produce resistance to corrosion. The integrity of this oxide film depends on the continuous presence of oxygen in the environment. Differential oxygenation over the surface of an implant in the body can occur involving simple situations of contact areas. For example, the contact surfaces between a screw head and a plate provide areas with a low oxygen environment, and the reaction of metal with environmental oxygen is shifted in favor of the breakdown of the protective oxide film. In usual practice, a stainless steel implant is given a routine sterilization and passivation treatment which builds up an outer protective oxide layer which usually remains stable for the expected duration of service. If the passive surface is ruptured, as might be the case if the elastic limits of the material and film are exceeded, some form of corrosion will always occur. If the protective layer has been rubbed by another piece of metal such as a screw or an adjoining implant during the service in the body, corrosion may begin. It is well established that stainless steels have a passive protective oxide film that is much more noble in sea water than stainless steel that has become active through the loss of this protective layer [7].

An example of the extent of corrosion that can occur within the body can be seen in the study of the two fractured surfaces of a Jewett nail that broke while in service. This implant is used to bridge a femoral neck, or trochanteric fracture, as shown in Fig. 2. The attending surgeon reported that after an unknown service period the trifinned nail fractured as shown in Fig. 3. The patient remained mobile for an "extended" period——a remarkable feat because of the probable pain——before the implant was removed. The trifinned nail had been bent from its original straight position during the service period. A photomacrograph of the two fractures is shown in Fig. 4. All evidence of the normal, jagged fracture surface has been removed. Some areas of the two fractured surfaces are as smooth as an electrolytically polished surface obtained in the laboratory. Of course some erosion of the two fractured surfaces must have occurred jointly with the corrosive removal of metal from the fracture.

Another example of corrosion was found during the examination of two intramedullary pins (see Fig 1) retrieved from the tibia. A radiograph of the pins in position in the bone is shown in Fig. 5. It is apparent that the pins were crossed in bridging the tibia fracture. Examination of the contact areas revealed crevice corrosion as shown in Fig. 6. Further information on the latter study is available [8].

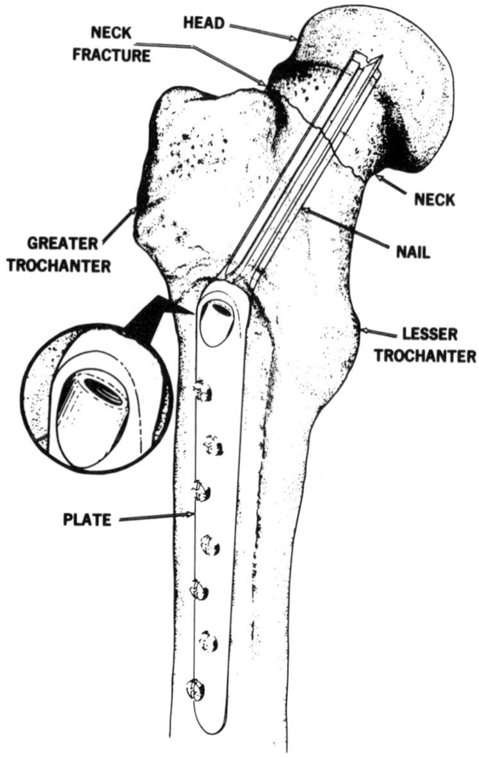

Fig. 2. Jewett hip nail-plate for bridging neck fracture of femur.

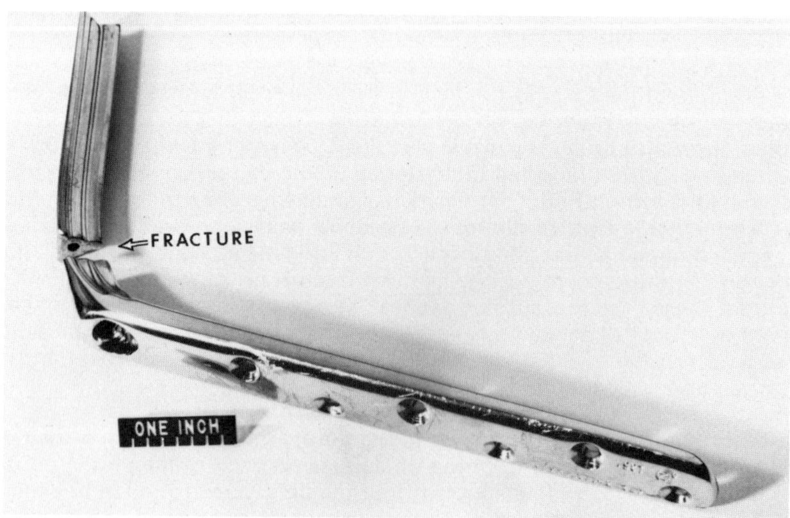

Fig. 3. Fractured Jewett nail-plate. Fracture has occurred through trifinned nail.

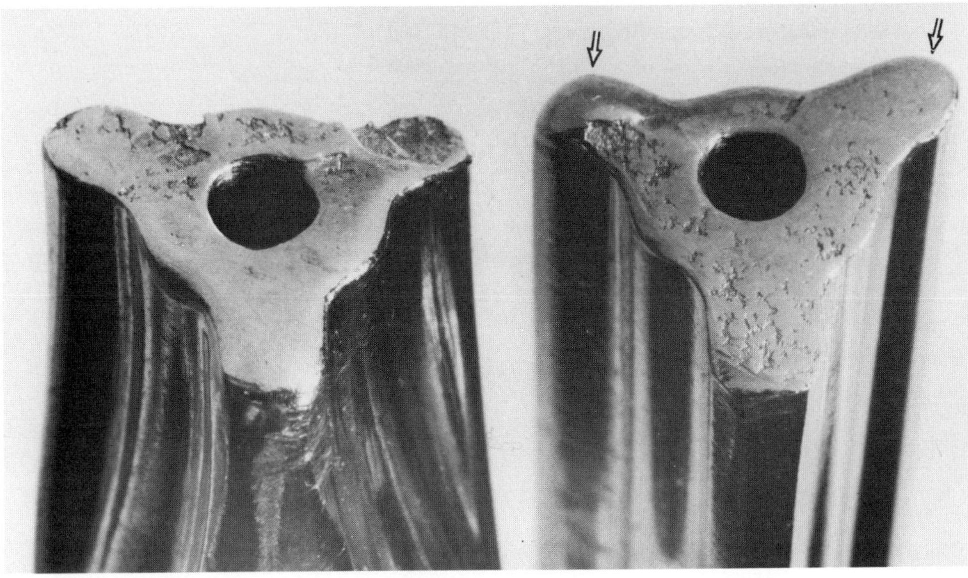

Fig. 4. Erosion-corrosion surfaces of fractured Jewett nail. Note rounded fin fracture (arrows) due to erosive action of the fractured surfaces against each other as well as the corrosion of body fluids.

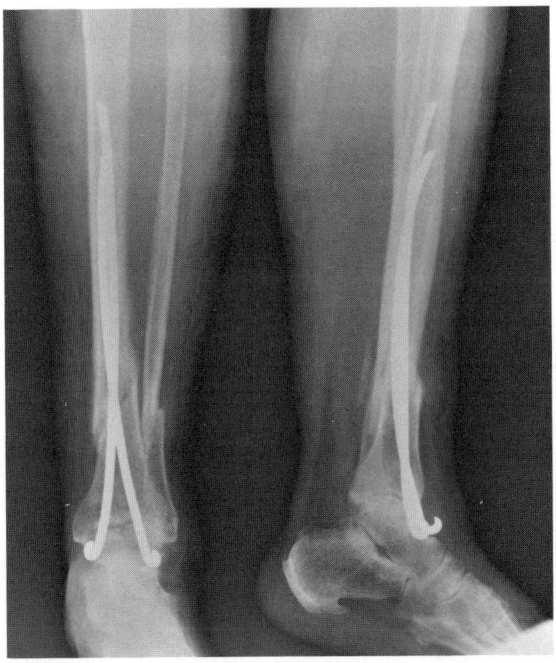

Fig. 5. Two radiographic views of the same fractures of tibia and fibula. Two intramedullary pins have been inserted in tibia.

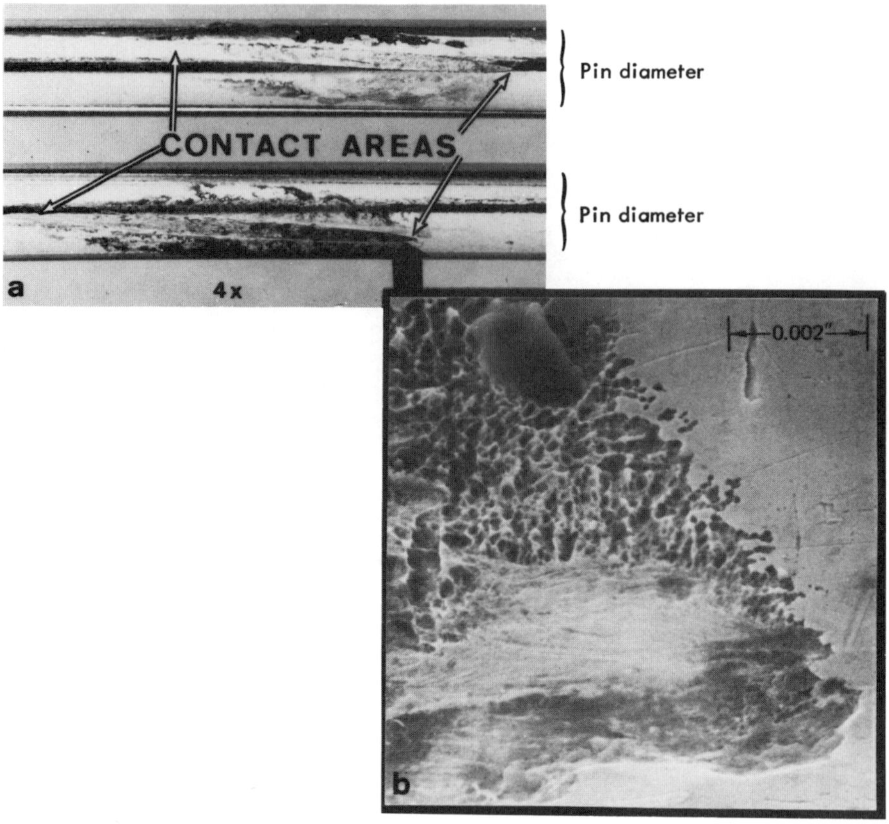

Fig. 6. Crevice type corrosion of two pins after being in contact while in tibia. (a) Contact areas of pin. (b) Close-up of corroded area.

Inept Mobility of the Patient

Very little can be pointed out specifically about the role of the active patient in the cause of a failure. The purpose of an implant is to restore some measure of mobility to the patient. The perfect orthopaedic implant material that can temporarily replace healthy bone on a one-to-one basis probably will never be developed. The natural, healthy bone sustains an orderly self-repair and ability to rearrange itself along imposed stress lines unlike any man-made implant. Quite likely some patients consider the installation of an implant an open invitation to return to the same active life as before the bone damage occurred. However, the patient trauma from a failed implant undoubtedly is as great a shock as the original bone failure.

An excerpt of an orthopaedist's account of a patient's experience in having an implant fail is as follows: "the patient was fully ambulatory without crutches when the patient was seen in the surgeon's office following pain in his thigh. He (the patient) heard a snap, the leg gave way and he buckled to the floor". In another failure case, a Jewett nail-plate (see Fig. 2) was used to bridge a subtrochanteric fracture [9]. The report describing the post-operative activity of the patient was that "he was bearing full body weight on the leg two weeks after surgery".

In almost every case, the surgeon desires the patient to gain post-operative mobility as soon as possible with the probable thought that pain restrictions will keep hyperactivity in check. With high level thresholds of pain and unrestricted motivations, some patients may have experienced implant failure that would not have occurred if a more subtle form of activity and reduced stress on the implant had been the case; particularly so in the early post-operation period.

Improper Selection of Materials

Fortunately much improvement has been made in the area of material selection. Commercial development of higher quality implant materials has made great progress in the past few years. The general knowledge of the surgeon on materials has made great strides. Some medical schools, such as the University of Alabama, include in the curriculum classes on metallic implants and their applications. The classes at Alabama are instructed by a metallurgist who has specialized in biomaterials and bioengineering. With such indoctrinations, the surgeon becomes aware of some very basic procedures. For example, it is mandatory that he use installation tools made of the same composition as the implant. This practice avoids mechanical transfer of dissimilar materials during inserting of the device which might lead to possible galvanic corrosion when later exposed to the body fluids. While the well-informed surgeon appreciates the need for such care, comments still have been expressed by surgeons, however, that this need for installation tools of the same composition as the implant was merely a sales gimmick by the manufacturer.

The most popular material in use since 1930 has been type 316L vacuum-melted stainless steel. Other materials, including cobalt-chromium-molybdenum alloys, sometimes referred to as Vitallium, began to be used about 1940 in both cast and wrought form. More recently, unalloyed titanium, and alloyed titanium with a 6 percent aluminum and 4 percent vanadium have come into use. Ceramics have played only a minor role in the United States as structural implants; however, alumina has been used to a relatively large extent in pediatrics as well as in complete hip prostheses in France.

We can become quite concerned about our potential personal involvement in the selection requirements for an implant. This presentation is concentrated on failures. Such a pointed presentation might be misconstrued by the reader to mean that any use of implants is a "Russian Roulette" experience. The low percentage of implant failures was presented earlier so the opposite is more the truth, and to help fortify this point, a photograph of a very old implant that functioned as desired is shown in Fig. 7. These Sherman-type plates were removed from the femur of a 60 year old man after 50 years of service. As compared with current onlay-type plates, these Sherman plates and the attachment screws are quite small for a femoral fracture. A major fabrication error also is apparent in the stamping of the company name on one plate and stamping the identification of the steel "vanadium" (vanadium steel was the principal material for implants at that time) on the other plate. So here are examples of past selections of materials that by today's standards definitely would not be approved; however, the plates served their intended purpose for 50 years.

Improper Installation of the Implant

The installation of an implant in a patient has been compared by an orthopaedic surgeon to "fixing a watch in a bottle of ink". Undoubtedly, the body fluids and tissues mask the repair operation considerably. Damage imposed on the implant during installation may be only microscopic, and it may not occur to the surgeon that the implant was altered in any way. In some instances, the possibility of small damage to the implant by the surgeon may not warrant, in his judgment, an extended period in the operating

Fig. 7. Sherman onlay type plates. Removed from femur after 50 years of service. Note stamping of company name pointed out with arrow.

room for the patient to institute a replacement. An example is described. A 21-year old soldier was wounded in the hip in the Republic of Vietnam. The wound was treated without the application of an implant. A surgeon examined the patient 21 months later, and in the interim, infection had developed. Also, corrective surgery was essential to restore the normal angle of the femoral head and neck relative to the bone shaft (see Fig. 8). A Jewett nail-plate identical to the one shown in Fig. 1b and Fig. 2 was inserted in the femur. In this patient, the fracture was located in the subtrochanteric region. In the normal sequence of surgery a Jewett nail-plate is inserted into the neck and head of the femur with an impactor-extractor as seen in Fig. 9. The impactor-extractor is screwed into the threaded cannula of the nail (see insert, Fig. 2) then used as a driver to position the implant. During the insertion of the nail, the surgeon found the head of the femur to be inordinately hard; a condition probably related to some vascular necrosis from the original wound and infection. In this instance, the threaded tip of the impactor-extractor broke off in the nail during the insertion. In the judgement of the surgeon, the nail was in place, the operation completed, and the patient should not be subjected to removal of the implant for a replacement. In four weeks the implant failed as shown in a radiograph (Fig. 10) thus requiring additional surgery. The implant was removed and replaced with a stronger Jewett nail-plate.

The retrieved Jewett nail-plate is shown in Fig. 11. The fracture had occurred across the chamfered opening of the cannula. The broken tip of the impactor-extractor can be seen. The plate-end of the fractured implant was sectioned along the dashed lines shown in Fig. 11 and examined along the chamfer with a scanning electron microscope. The results are shown in Fig. 12. The design and subsequent fabrication of this implant results in a chamfer ridge that is heavily stressed, particularly so if the bone fracture is directly under it. Most implant manufacturers of this style of implant do not recommend this device for subtrochanteric fracture. In this case, the ridge was gouged by the fractured edge of the impactor-extractor and the implant fracture can be seen to have propagated along this gouge. We have stated that failures seldom are restricted to single causes. The fracture surface of this implant shows evidence of corrosion as shown in Fig. 13. There is some possibility that the apparent intergranular corrosion could be the remains of dried body fluid not completely removed from the fracture surface. Evidence of intergranular corrosion, however, is fortified by the similarity of the corrosion-etched grain size (Fig. 14a), and the etched grain size of a metallographically prepared specimen (Fig. 14b) cut from the Jewett nail-plate. Both photomicrographs (Figs. 14a and 14b) were made from the same specimen and are located approxi-

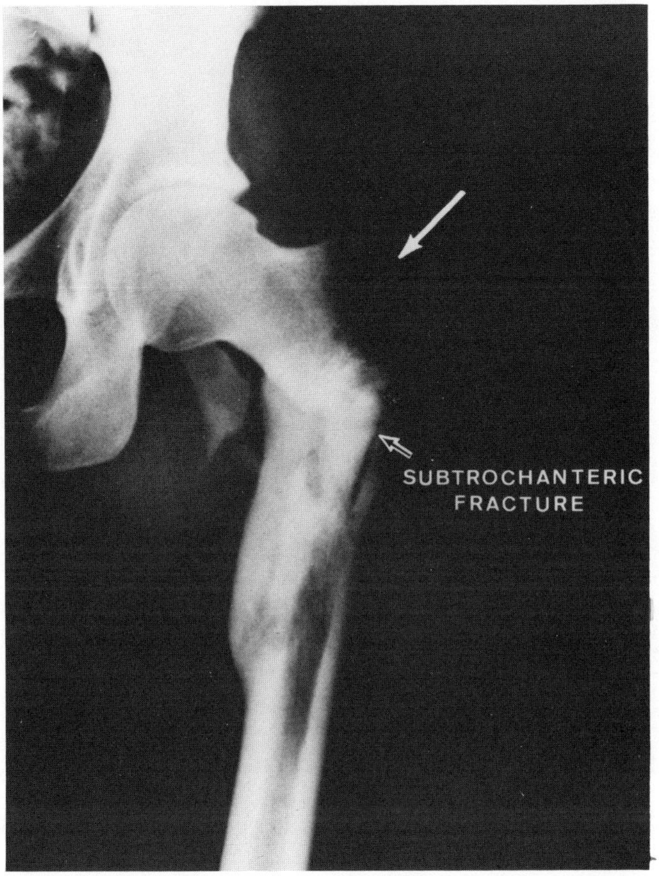

Fig. 8. Radiograph of proximal end of femur 21 months following initial injury and treatment to correct subtrochanteric fracture. Arrow points out loss of greater trochanteric bone area. Note varus angulation of bone shaft versus head neck-area. Correction required re-fracture and re-alignment with 135° Jewett nail-plate.

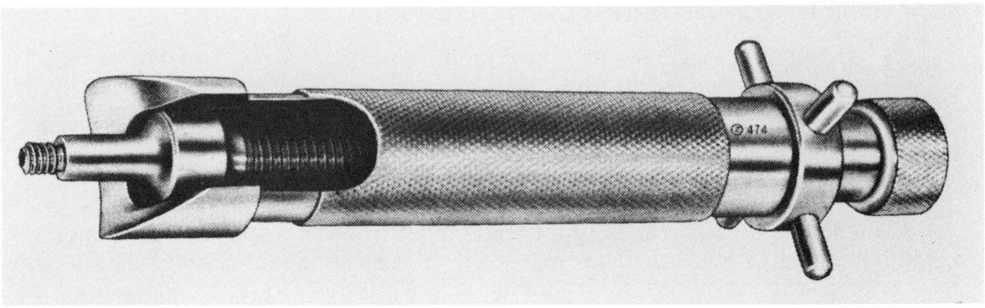

Fig. 9. Impactor-extractor tool for inserting or removing Jewett nail from femur. The threaded tip engages the threaded cannula of the nail for positioning or removing the fixation device.

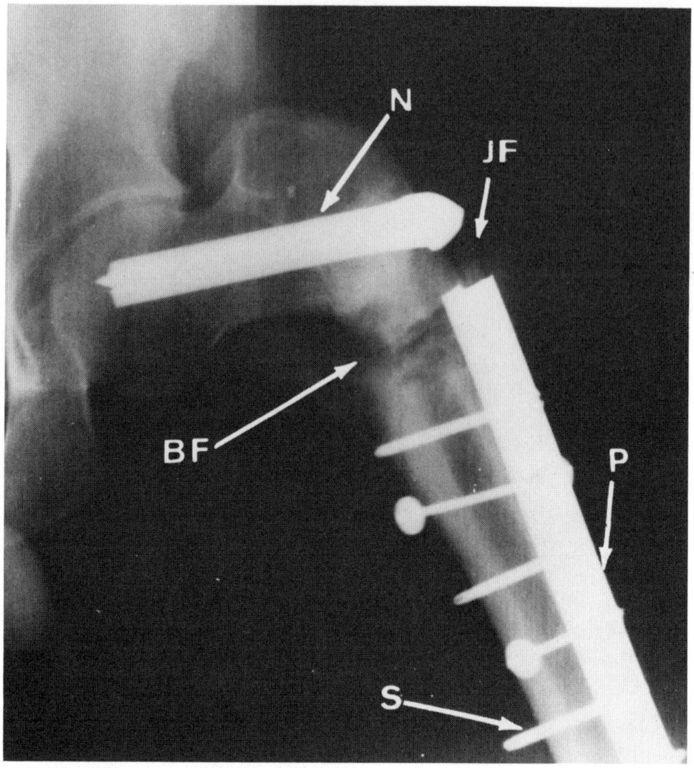

Fig. 10. Radiograph of failed Jewett nail plate in the femur. N – Trifinned nail, JF – Jewett nail-plate fracture, BF – Bone fracture (subtrochanteric region), P – Plate, S – Screw.

mately 1/16 in. apart. Such close areas of comparison would have the same grain size. It has been reported [10] that following injury and inflammation, there can be a pH drop of the tissue and fluids in the injured area to 5.3 to 5.6––a decided acidic build-up compared to the normal slightly alkaline pH of 7.4. Such an increase in the acidic environment might accelerate the corrosion. More detailed information on this particular study of this failed Jewett nail has been reported [9].

Improper Fabrication of Implant

In the fabrication of implants, high quality control must be exercised in the type of finish on the fixation device. The surfaces must be free of machining marks that could serve as stress raisers where a fissure and subsequent fracture could occur.

Intramedullary pins are used quite frequently to bridge a fracture of the femur, tibia, and humerus (see Fig. 1a). The cross sectional shape of an intramedullary fixation device may be a solid round rod, solid square rod, solid fluted rod, or a hollow cloverleaf. The choice of the type of rod selected is usually based on the training and experience of the surgeon and the service the implant must perform. The surgeon must depend on the quality control of the manufacturer to provide a properly fabricated implant. Very few surgeons have sufficient engineering/materials experience to evaluate the fabrication of the implant he plans to use.

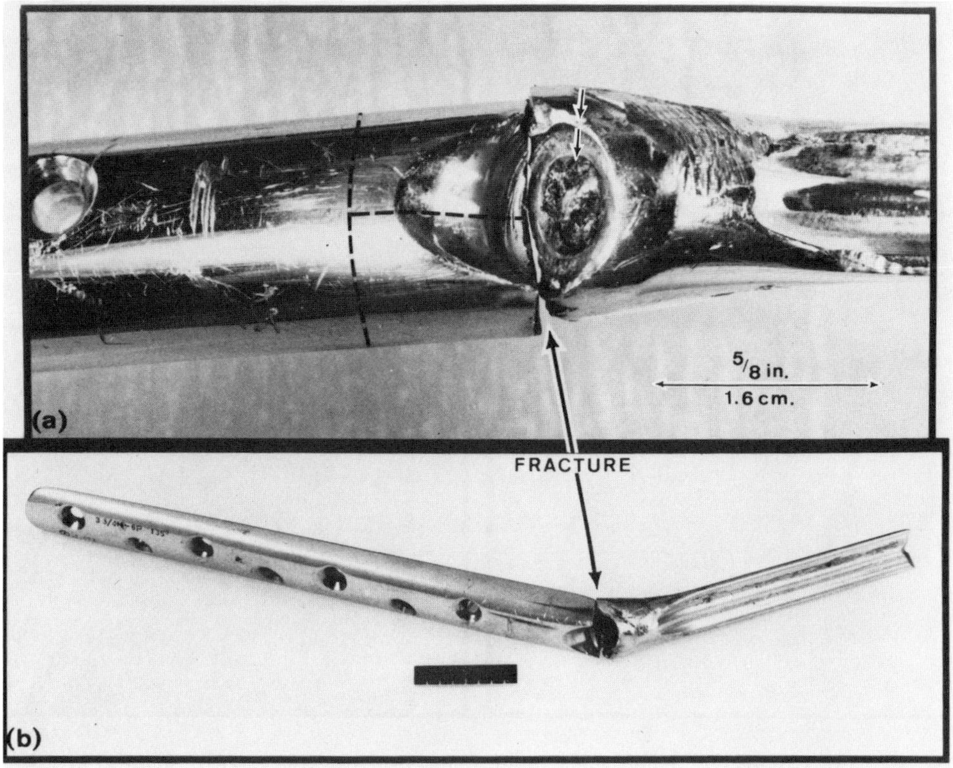

Fig. 11. Retrieved Jewett nail-plate after failure occurred during service. (a) Failure developed in chamfer of cannula. (b) Dashed lines indicate cutting guides. Note broken tip of impactor-extractor (double arrows) in tri-finned nail.

In one particular case, a pair of pins 3/15 in. (4.7 mm) x 12½ in. (31.6 cm) fractured after a service period of six months in the femur. A radiograph of the fractured pins in the fractured femur is shown in Fig. 15. A causal review of the surfaces of the pins in the vicinity of the fractures, as would be the opportunity of the surgeon, showed nothing that would have contributed to failure. A post-failure microscopic study of the pin surface, however, revealed circumferential grinding gouges which undoubtedly were produced during fabrication——probably by centerless grinding. These gouges, shown in Fig. 16, could become severe stress raisers in any cyclic bending action. Since the pins serve as internal structural supports across the femoral fracture, normal leg mobility in walking would generate low-cycle fatigue and cause bending of the pins. Such bending action would be restricted almost entirely to where the pins bridged the fractured bone. The decrease in ductility due to cold work induced hardness at the root of the gouge, and the subsequent notch effect offer a surface condition conducive to fatigue failure. Close examination shows fissures originating in many of the gouges and some examples of combinations of fissures could be found. The combination of several microscopic fractures undoubtedly led to the ultimate failure [8].

As will be shown later in describing another failure, improper fabrication may be closely allied with improper design, so much so, that both factors appear to contribute equally to the fracture.

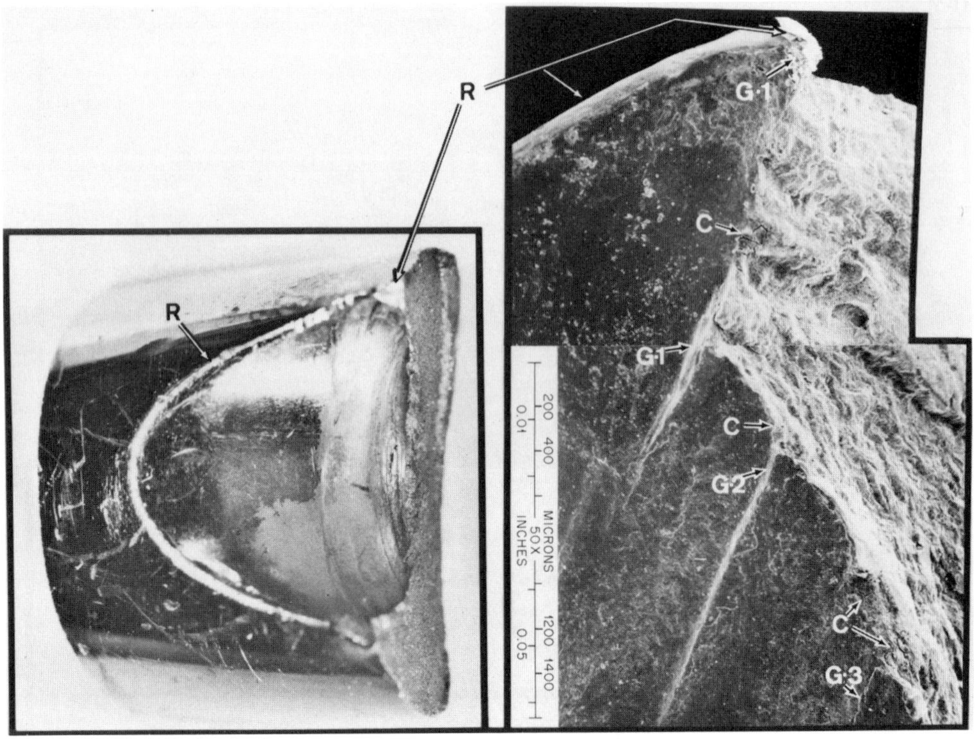

Fig. 12. Close-up view of failed Jewett nail-plate. (a) Note ridge (R) of chamfer as seen in optical photomicrograph. (b) Gouge (G–1) can be seen (SEM photomicrograph) on chamfer ridge where failure originated. Other gouges (G–2 and G–3) can be seen. Possible corrosion (C) areas are identified.

Improper Design of Implant

The limitations of current technology and design criteria may impose built-in characteristics in some implants that are prone to failure in service unless installed with extreme care. As shown previously, many bone implants are secured in place with self-tapping screws. The profile of a screw with the crest and root demonstrates a reduced cross section from thread root to thread root where stress build-up can occur. Some typical self-tapping screws that show failure are shown in Fig. 17. The screw identified as (A) was broken as it was being removed from the bone. The torque applied to the screw exceeded the ultimate strength of the material. Screw (B) failed in service. A longitudinal examination of screw (B) area (a) exhibited fissures in every thread root. These fissures display a pattern typical of stress corrosion cracking as shown in Fig. 18. Applications of screws to secure a Jewett nail-plate in a femur have been demonstrated in this report (see Figs. 1 and 2). If the screws secure the plate tightly to the bone shaft, there is an excellent chance of obtaining the desired service life from the screws. If, however, the nail had to be positioned into the femoral neck and head so space develops between the plate and bone as in Fig. 19, the possibility of screw failure is increased. If a screw head is placed in tension as demonstrated in the sketch shown in Fig. 20, a high tensile stress is concentrated in the thread roots. This tension on the screw coupled with the bending of the screw during the mobility of the patient, all within an environ-

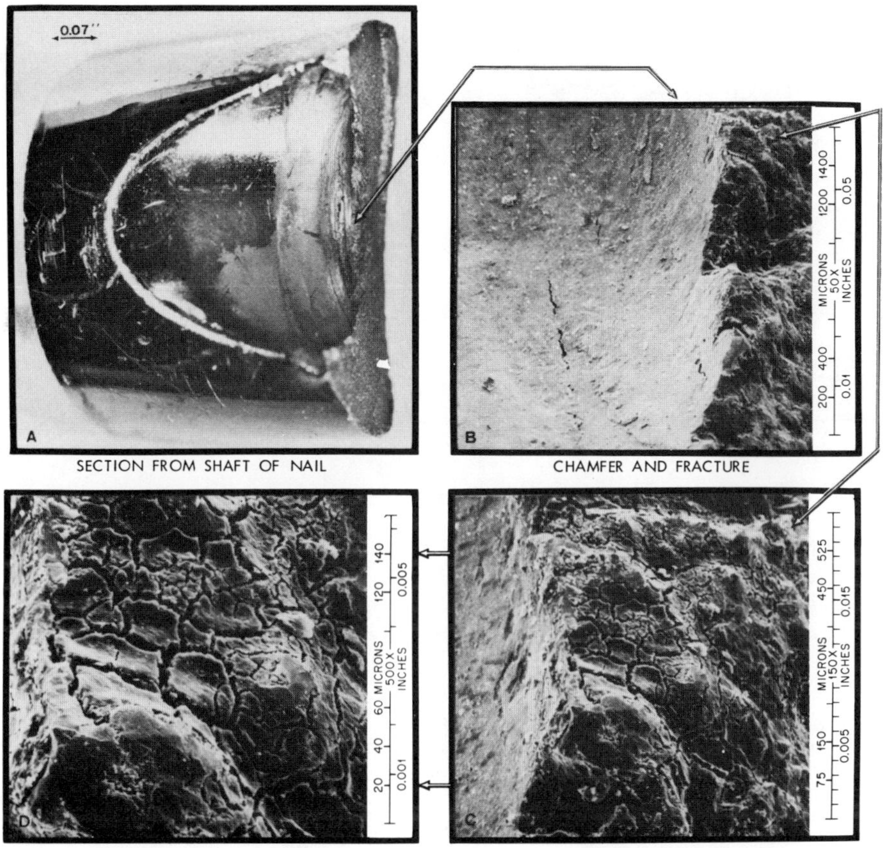

Fig. 13. Corrosion etched microstructure in fracture of failed Jewett nail-plate. (A) Photomacrograph of failed area. (B) (C) (D) SEM photomicrographs of corrosion etched areas on fracture.

ment of corrosive body fluids, could offer a strong probability of screw failure. It is possible that the screw shown in Fig. 17b failed as the result of just such a service history.

An example of another type of implant involving another style of screw is shown in Fig. 21. This implant is designed to bridge a neck fracture of the femur similar to the fixation device shown in Figs. 2 and 16. This implant however, is not a trifinned nail bridging the fracture, but a lag screw. The lag screw has a splined, sliding shaft which can telescope during the callus growth development of the fractured bone. A radiograph of a sliding screw plate in position in the femur is shown in Fig. 22. A two months "post-op" radiograph, Fig. 23, shows the same device is very sound and even some evidence of natural erasement of the bone fracture can be seen. After one year of service, however, a fracture has occurred between the threaded end and the sliding shaft as shown in the radiograph of Fig. 24. Repair required removal of the sliding shaft and the plate and a second sliding screw plate had to be positioned in the femur. The removal of the original broken screw end was not possible without serious damage to the femur so it remained in place. A radiograph which was made after the second sliding shaft was installed, is shown in Fig. 25. Close comparison of the profiles of the remains of the frac-

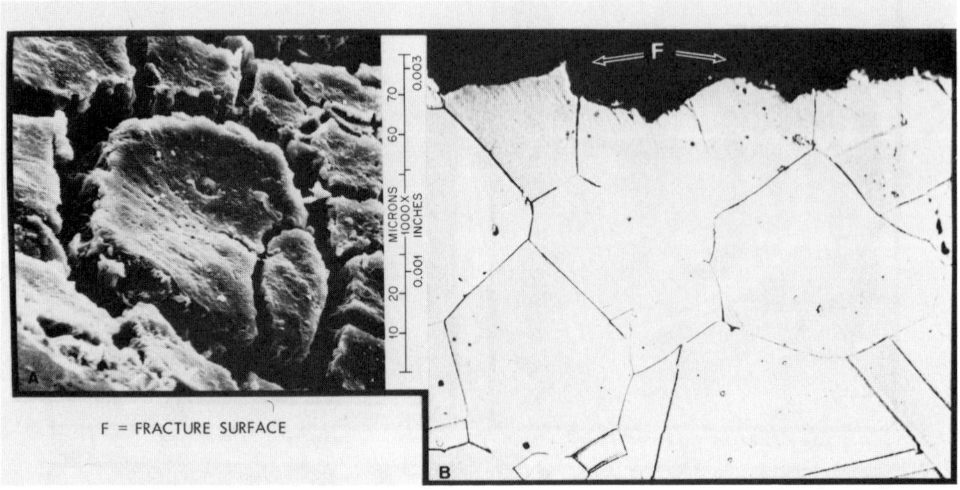

Fig. 14. Microstructure of failed Jewett nail-plate. Comparison is made of corrosion etched fracture surface (A) as revealed by SEM and the microstructure (B) of a metallographically prepared specimen obtained by optical microscopy.

tured shaft and an unfailed sliding screw plate is shown in Fig. 26. The fracture can be seen to have occurred in the forward notch of the last thread root. The first five thread roots were supported by the thread crests being imbedded internally in the femoral head. The last thread root where failure occurred probably was in line or close to the femoral neck fracture and would have been subjected to bending forces during patient mobility. A sketch demonstrating this movement is shown in Fig. 27. Examination of the thread root with the scanning electron microscope where the fracture occurred showed a number of matching gouges as shown in Fig. 28. Some gouges contained fissures which developed undoubtedly by low cycle fatigue bending during patient mobility. One or more of these fissures plus the sharp forward angle and reduced cross-section of the thread root and the low cycle fatigue movement in this thread root could have contributed to the failure in the location shown in Fig. 26.

The inherent design problems associated with screws have been presented. A related problem also exists with implants containing screw holes. The location of screw holes in a plate reduces the cross section and weakens the implant in that area. An example of failure through a screw hole is shown in Fig. 29. This implant is designed to bridge a fracture in the condyle of the femur, see Fig. 1b. The failure originated in the hole due to the reduced cross section and localized high stresses. In observing the fracture with a low power microscope, the fracture could be seen to have propagated in nearly a straight line on both sides of the hole (see Fig. 30). On one side of the hole, the straight line fracture continued to the edge. On the other side of the hole, the fracture continued in a straight line then assumed a more irregular pattern. Fractures in metals and alloys normally do not propagate in a straight line unless guided in some manner. Close examination of the plate surface showed numerous fine grooves assumed to be machining grooves (Fig. 30) that probably played a major role in initiating a fissure in one of the machining grooves that cut into the screw hole. The fissure propagated along the groove until a critical crack size developed that led to failure.

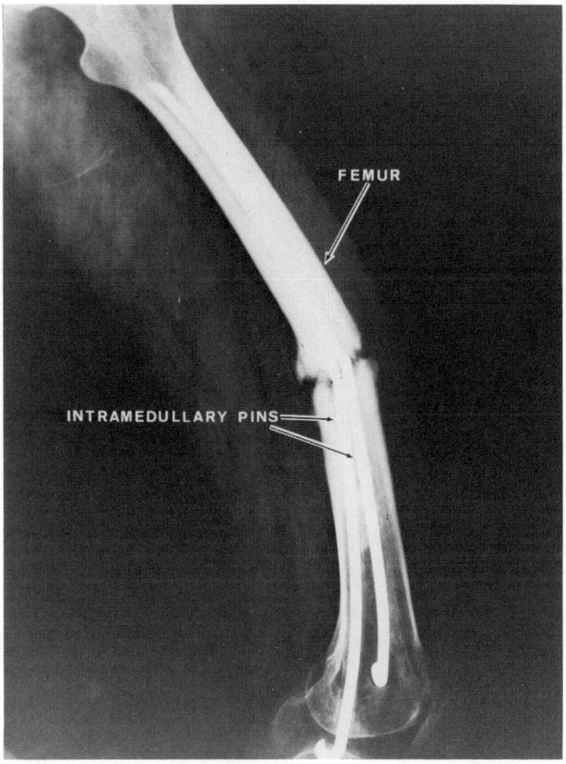

Fig. 15. Radiograph of fractured pins that fractured while in service. The pins had been inserted into the intramedullary cavity of the femur to bridge a midshaft fracture.

The Eggers plate (Fig. 31) is an onlay-type implant (see Fig. 1b) that can be attached to the shaft (diaphysis) of the femur, humerus, or tibia to bridge a fracture. Attachment is with self-tapping screws through screw holes in the plate. As described previously, the presence of the screw holes decreases the strength potential considerably. One screw hole reduced the cross section sufficiently for stresses in that area to exceed the elastic limit of the material.

Today, continuous strength plates are available with an increase in the plate thickness around the hole. Such logical improvements should offer improved performance.

DISCUSSION

The use of metallic implants and prostheses has given the surgeon a means for bone fixation to allow the patient to become active. Six causes for implant and prostheses failure have been presented. In most instances, a failure cannot be attributed to just one cause. Post failure studies, however, can point out some probable causes that, when made known, can be corrected, with subsequent probable reduced reoccurrence of these failures. The role of the implant is intended to assume normal bone structural responsibilities. Impeccable procedures must be followed during the manufacture of fixation devices as well as during the installation by the surgeon.

Fig. 16. Surface of intramedullary pin showing grinding gouges where fissures originated. The major fracture is to the right of this field. Arrows 1 and 2 point out fissures which combined at "A" to form a microcrack. Arrows 3 and 4 point out other fissures originating in grooves.

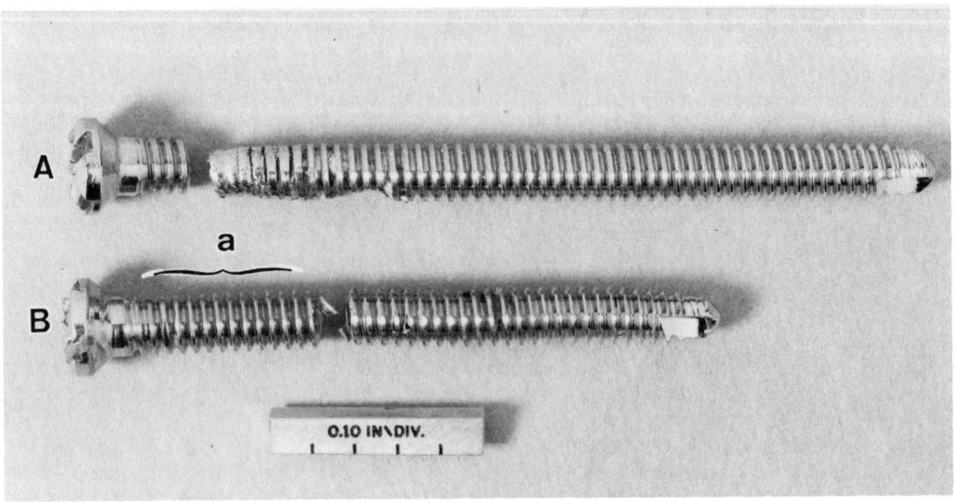

Fig. 17. Fractured self-tapping screws. (A) Failed during extraction. (B) Failed while in service. Thread roots in area (a) were examined for presence of incipient fractures.

Fig. 18. Longitudinal median view of 11th thread root from screw head showing typical stress corrosion fissure. Note the minor fissures in thread root.

We discussed one implant failure with a surgeon after the metallographic investigation revealed that the probable cause of failure was due to his improper installation procedures. The response was, "I can see now what I did wrong and I won't do it again". Implants that failed in service due to improper fabrication and design have been discussed with surgeons. In some instances, actual defects could be pointed out. Our response has been, "We were thinking about buying some of these devices——now I think we'll wait until that problem is corrected". Part of the reason for discussing these failures with surgeons is to erase some attitudes that "some implants are just expected to fail now and then and have to be replaced as a natural occurrence". Some surgeons do not understand the mechanics of failure, nor do they appreciate potential origins of failures.

We have pointed out that the patient has a responsibility in the prevention of implant failures. Excessive demands on the implant during patient mobility may exceed the stress limitations designed for the implant. The fixation device cannot be considered a one-to-one replacement for a healthy bone.

The technology in the materials and surgical fields is making great progress today. The future material for implants could be a biocompatible ceramic with a metallic core [12]. Another possibility for hip prosthesis is a metal alloy stem with an aluminum oxide (Al_2O_3) ball and a polyethylene socket [13]. High density graphite [14] is known to be compatible with the body. Considerable experience has been gained in its use as heart valves. As a bone implant, the high density graphite has great potential. An experimental implant designed for the femur of a dog is shown in Fig. 32. Vitreous carbon implants serve as base supports in the mandible or maxillary bones for a stainless steel post which in turn is the foundation for a crown. About 5,000 Americans are using such implants.

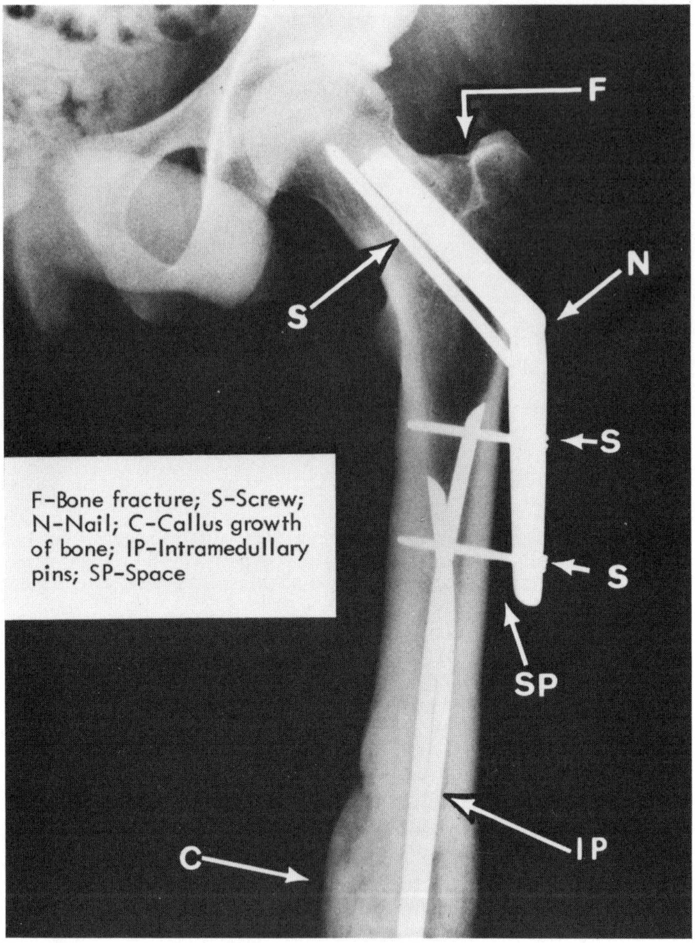

Fig. 19. Trifinned nail-plate for fixation of neck fracture of femur. The screws could be in tension after the plate was tightened against the femoral shaft. Some bending of the screws could occur during normal body movement. Note the ends of two intramedullary pins inserted into the femur from the distal end.

CONCLUSION

A general increase in the awareness of responsibilities by the manufacturer, surgeon, and patient is needed. Unbiased reporting of the causes of implant failures is one channel of information that is available today. Through such reporting, the performance reliability of implants and prostheses should improve.

ACKNOWLEDGEMENT

The author is very grateful to the contributions of Paul Spray, M.D., and Lewis Zirkle, M.D. in this study. Helpful review comments were provided by R. Crouse and J. Leitnaker. B. Auxier prepared the manuscript.

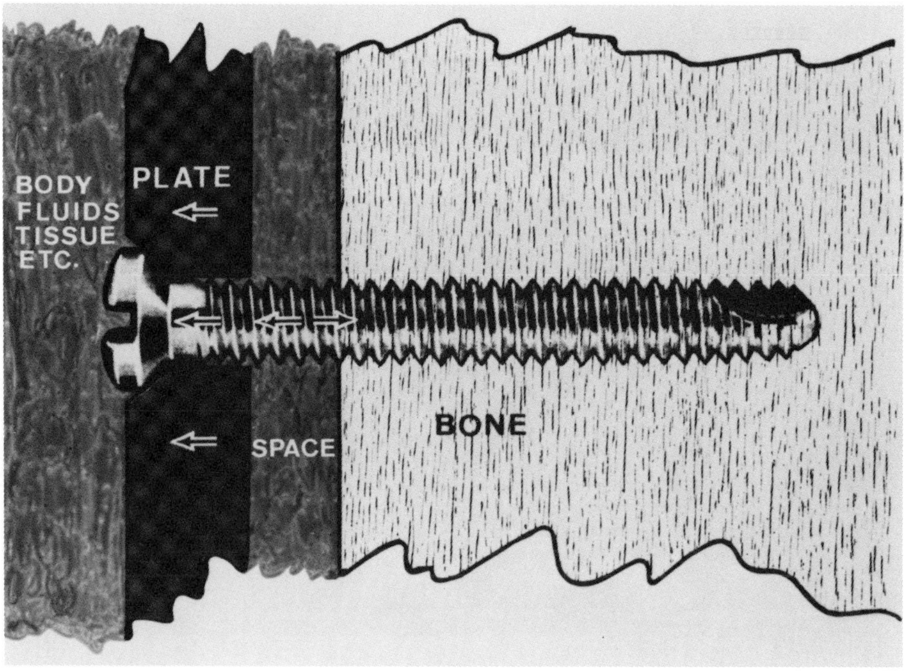

Fig. 20. Sketch showing screw in tension while holding plate against bone. A proposed origin for the tensile forces on the attachment screws that could lead to failure.

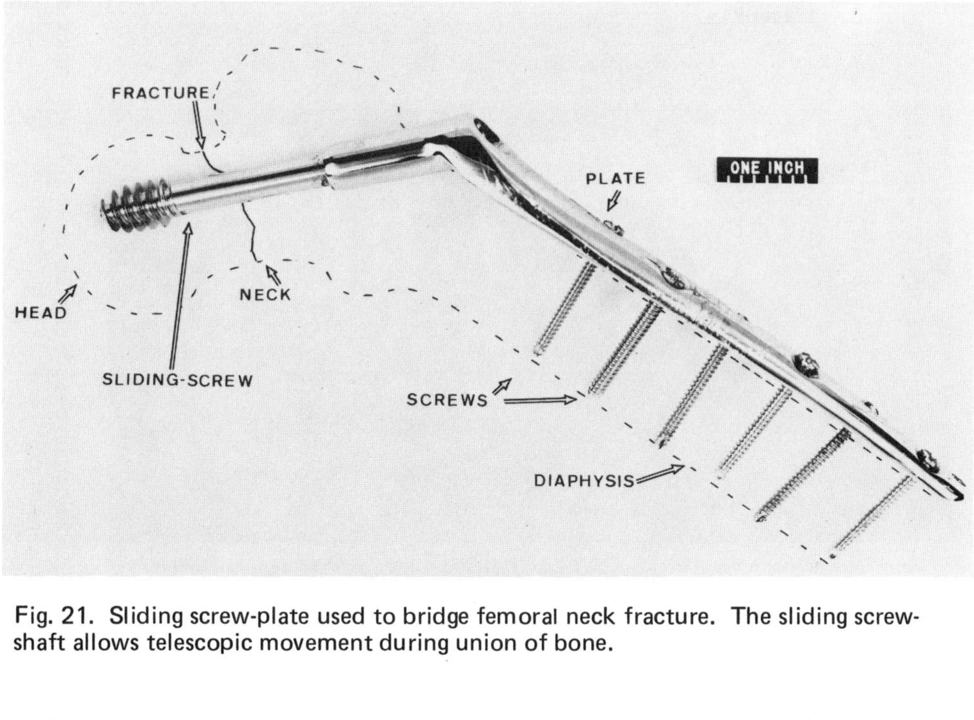

Fig. 21. Sliding screw-plate used to bridge femoral neck fracture. The sliding screw-shaft allows telescopic movement during union of bone.

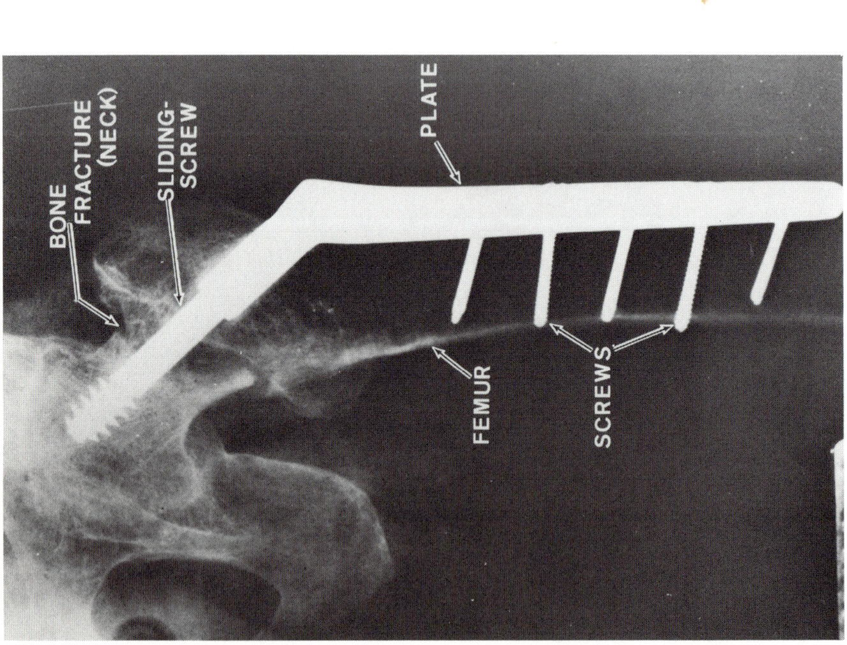

Fig. 22. Radiograph of sliding screw-plate in femur immediately following insertion.

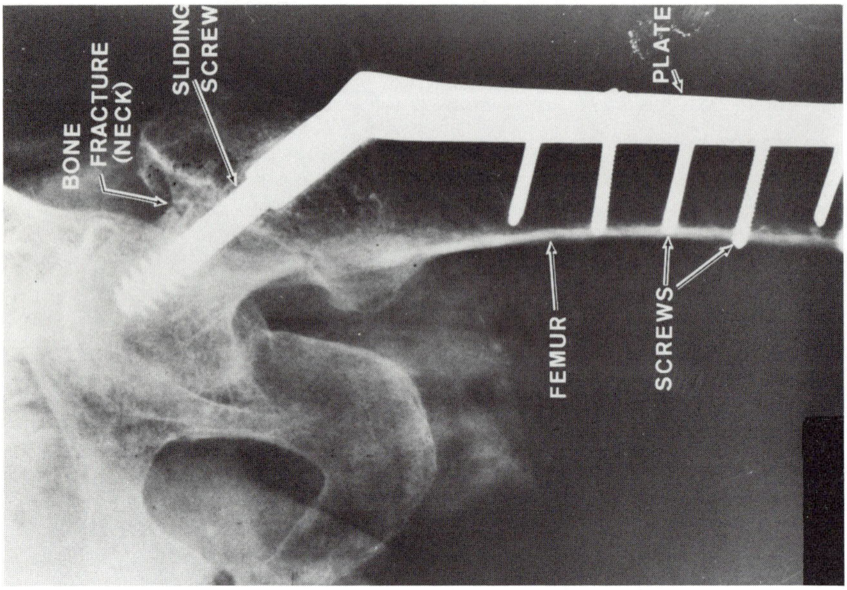

Fig. 23. Radiograph of sliding screw-plate two months following insertion in femur.

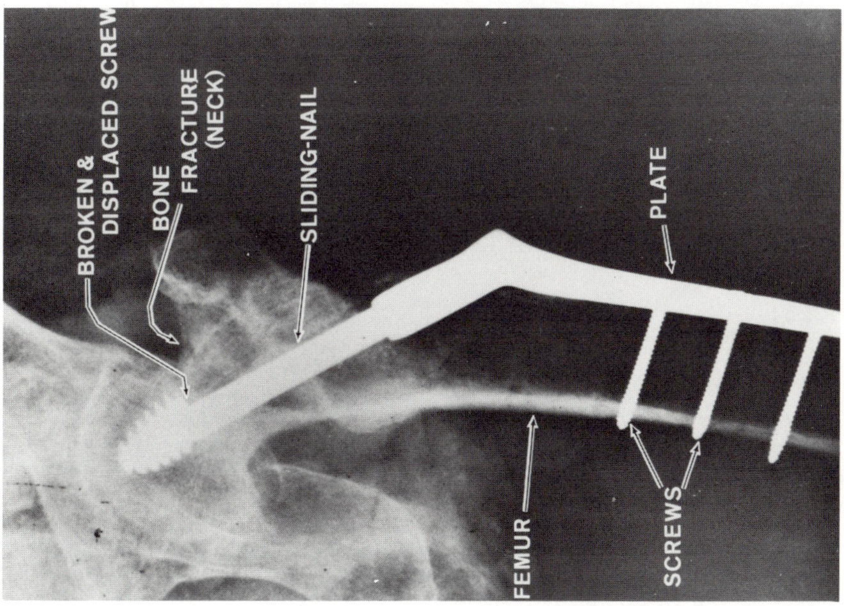

Fig. 24. Radiograph of sliding screw-plate one year after insertion in femur. Note fracture of screw shaft between threaded end and shaft.

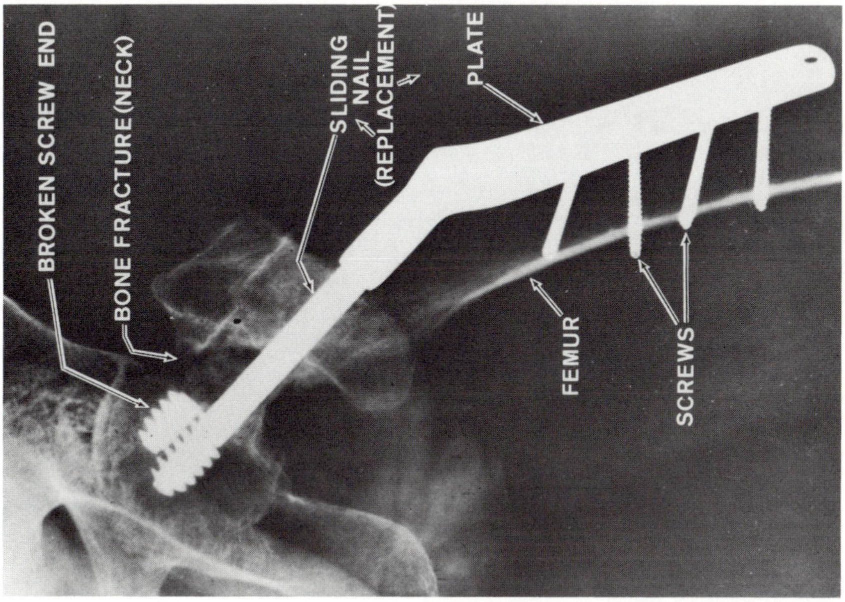

Fig. 25. Radiograph of proximal end of femur after second sliding screw-plate was inserted. Note original broken screw in femoral head.

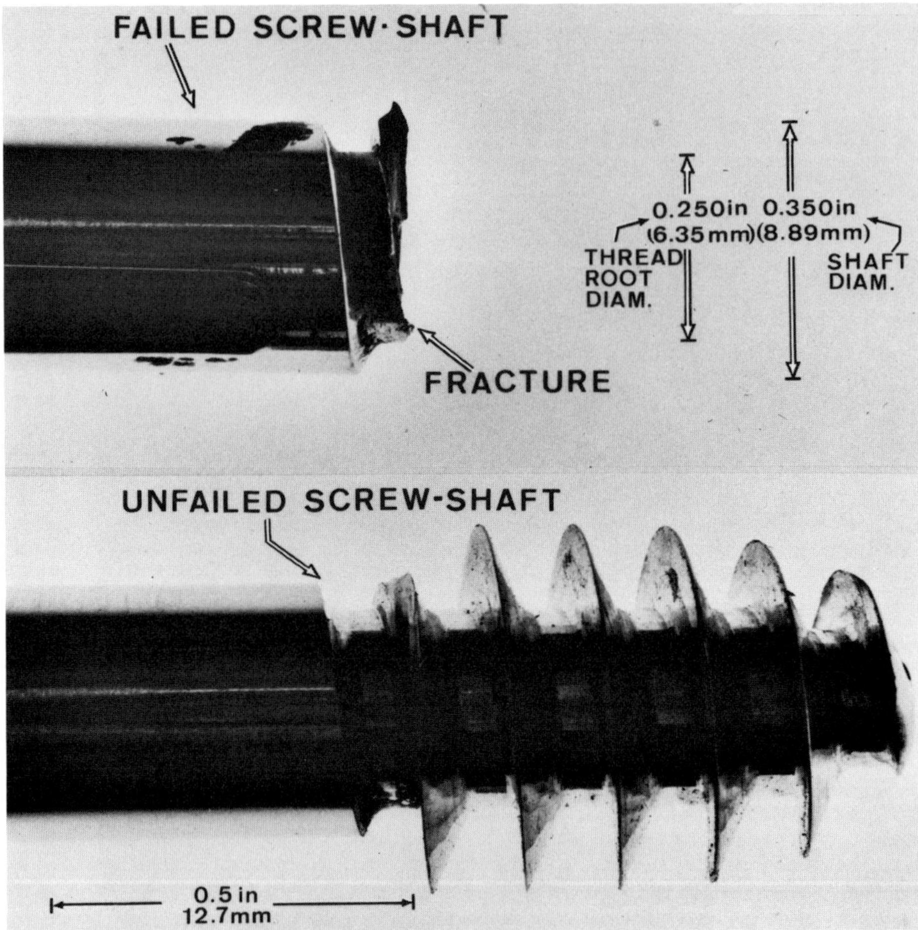

Fig. 26. Comparison of failed and unfailed sliding screw-shafts. Failure occurred in forward notch of last thread root. The diameter of the lag screw was reduced 29% at the thread root.

REFERENCES

1. Ludwigson, D.C., "Today's Prosthetic", *J. Metals,* Vol. 16, pp. 226–231 (March 1964).
2. Ludwigson, D.C., "Requirements for Metallic Surgical Implants and Prosthetic Devices", *Metals Quarterly* (Amer. Soc. Met.), pp. 1–6 (August 1965).
3. Fraker, Anna C., and Ruff, A.W., "Metallic Surgical Implants: State of the Art", *J. of Metals,* pp. 22–28 (May 1977).
4. White, William E. and Le May, Iain, "Optical and Electron Fractographic Studies of Fracture in Orthopaedic Implants", *Microstructural Science,* Vol. 3, Part B, (New York, NY: American Elsevier Publishing Co.), pp. 911–930 (1975).
5. Hulbert, S.F., Cooke, F.W., Klawitter, J.J.: "Investigation Into the Potential of Ceramic Materials as Permanently Implantable Skeletal Prostheses", *Biomaterials, Bioengineering Applied to Materials for Hard and Soft Tissue Replacement*, ed. A.L. Bement, Battelle Seattle Research Center, Seattle, Washington, (1971).

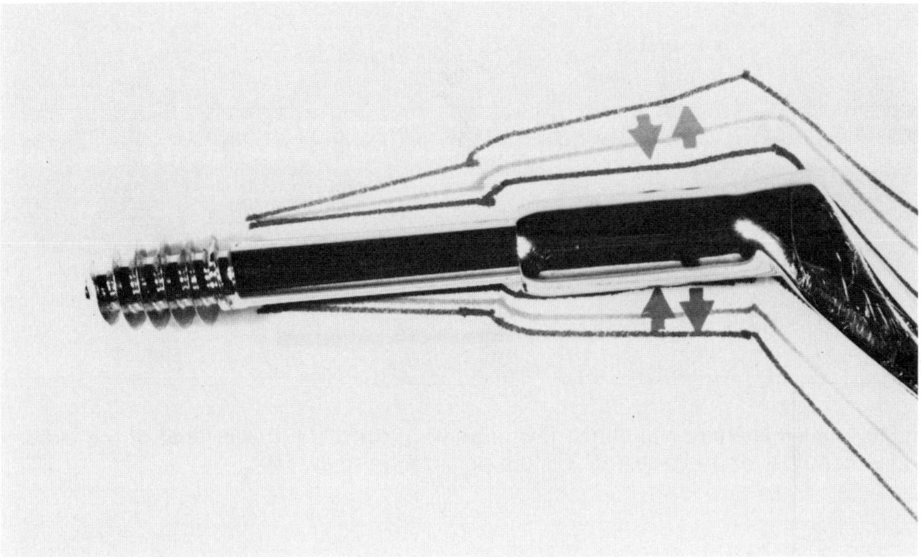

Fig. 27. Sketch demonstrating movement of screw shaft in femur. The first five threads were probably well supported in the femoral head as shown in Fig. 21. Patient mobility would impose severe bending on the last thread root.

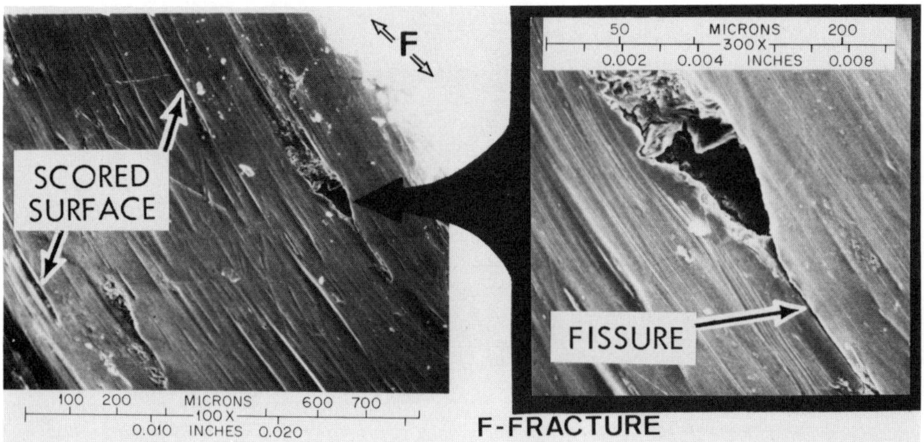

Fig. 28. Scanning electron micrographs of screw root showing machining gouges and fissures originating in gouges.

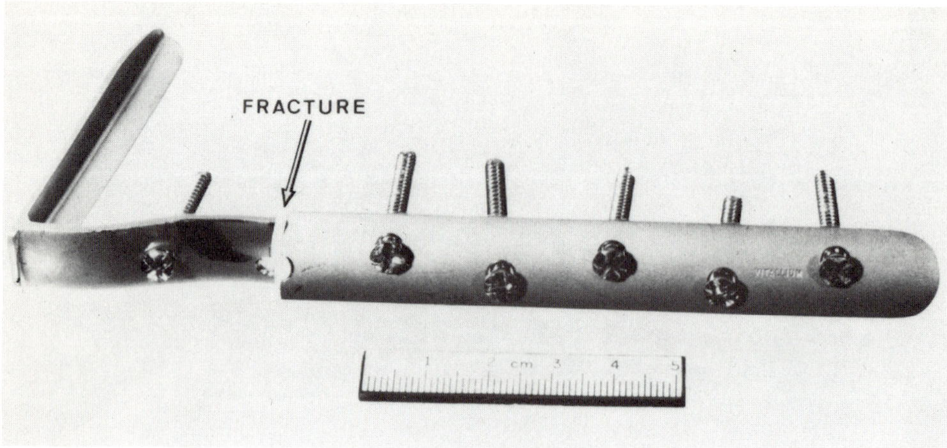

Fig. 29. Failed Neufeld nail-plate. For fixation of the distal (lower) end of the femur in supercondylar or T–fractures. Failure occurred in screw hole.

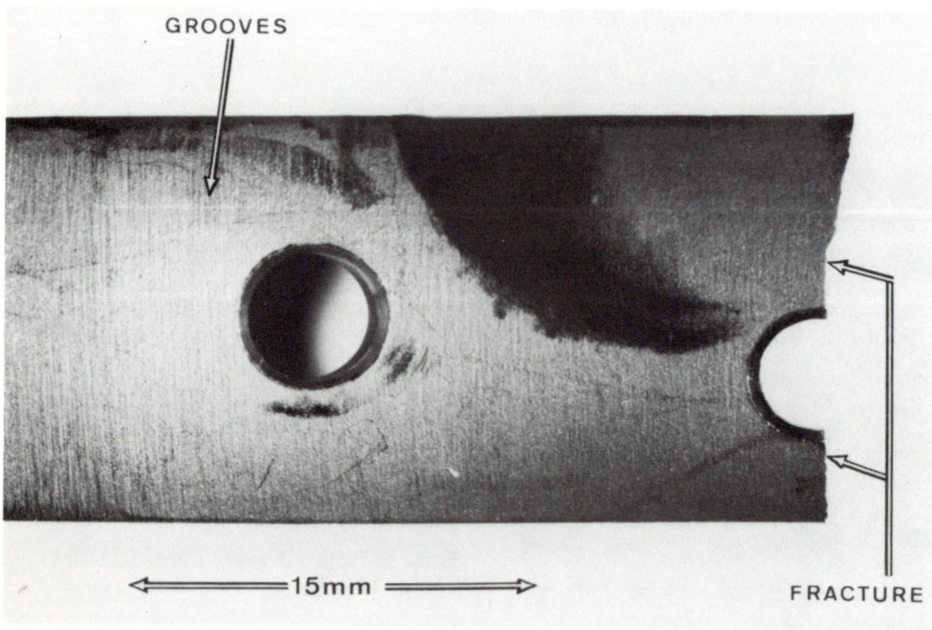

Fig. 30. Machining grooves on surface of failed Neufeld Condyle nail-plate. Note straight fracture pattern near hole. Grooves could have played major role in crack initiation and propagation.

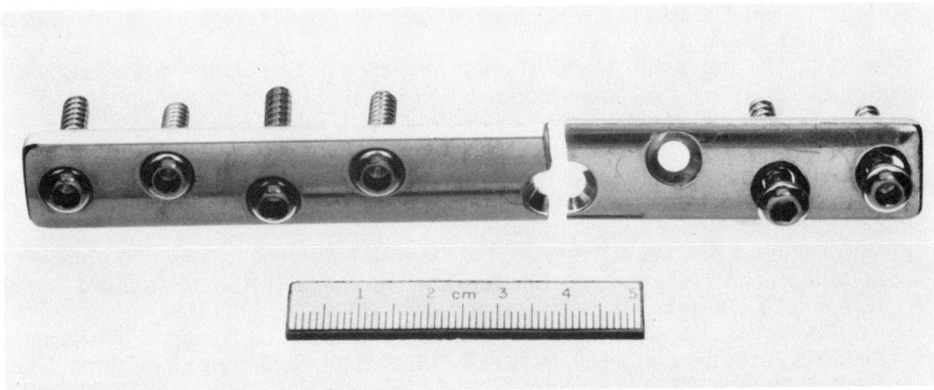

Fig. 31. Failed Eggers onlay type plate. Fracture occurred through screw hole.

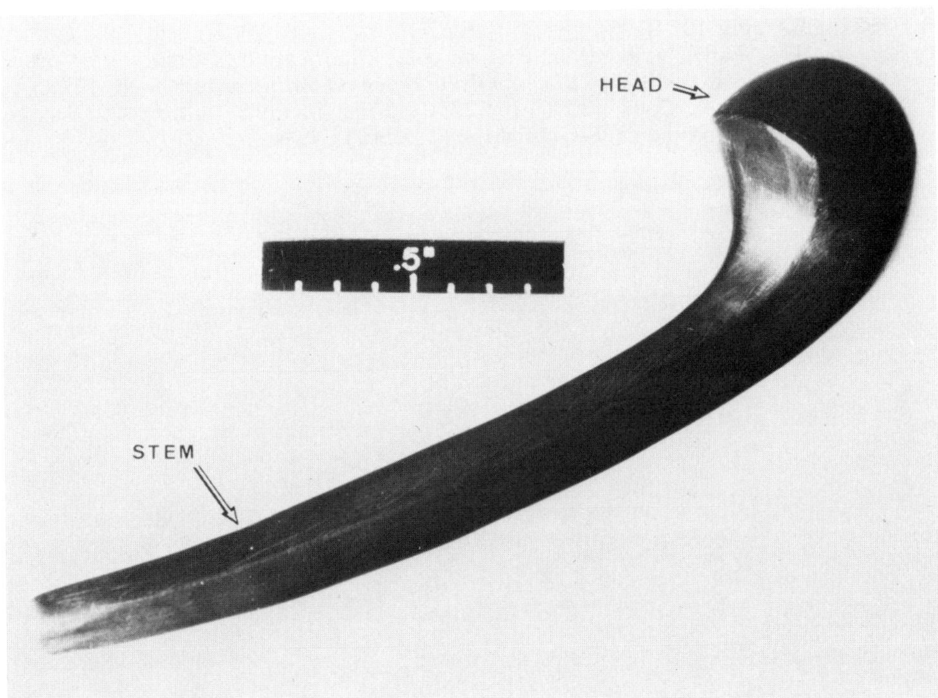

Fig. 32. High density graphite prosthesis for femur of dog.

6. Laing, P.G., "Available Metals", *Metal and Engineering in Bone and Joint Surgery,* (Baltimore: Williams and Wilkins Co.) (1959).
7. Stewart, D. and Tullock, D.S., *Principles of Corrosion and Protection – A Monograph,* (London: MacMillan & Co.).
8. Gray, R.J., Metallographic Examination of Retrieved Intramedullary Bone Pins and Bone Screws from the Human Body, *J. Biomed, Water, Res.* Symposium No. 5 (Part 1), John Wiley & Sons, Inc. (New York, NY), pp. 27–38 (1974).
9. Gray, R.J. and Zirkle, Lewis G. Jr., "Metallographic Examination of a Failed Jewett Nail-plate from a Human Femur", *Microstructural Science,* Volume 4, ed., E.W. Filer, J.M. Hoegfeldt and James McCall, American Elsevier Publishing Co., Inc., pp. 179–189, New York, NY (1976).
10. Ferguson, Albert B., "Metal Behavior, The Corrosion Process", *Metals and Engineering in Bone and Joint Surgery,* ed. Charles O. Bechtol, Albert B. Ferguson, and Patrick G. Laing; (Baltimore: The Williams and Wilkins Co.), p. 39 (1959).
11. Harth, George H., "Metal Implants for Orthopedic and Dental Surgery", Metals and Ceramics Information Center, MCIC 74–18, Battelle Columbus Laboratories, Columbus, Ohio, p. 43 (1974).
12. Hulbert, S.F., Klawitter, J.J., Leonard, R.B., "Compatibility of Bioceramics with the Physiological Environment", *Ceramics in Severe Environments,* ed. W.W. Kriegel and H. Palmour III, Materials Science Research, Vol. 5, Plenum Press, New York (1971).
13. Semlitsch, M., Lehman, M., Weber, H., "New Prospects for a Prolonged Functional Life–Span of Artificial Hip Joints by Using the Material Combination Polyethylene Aluminum Oxide Ceramic/Metal", *Jr. Biomed. Materials Res.,* Vol. II, No. 4, pp. 537–552 (July 1977).
14. Robbins, J.M., Eatherly, W.P., and Rossen, D.E., "Fabrication of Graphite for Use as a Skeletal Prosthesis", *11th Biennial Conference on Carbons,* June 4–8, 1973, Gatlinburg, Tenn., Conf. 730601 p. 123; National Technical Information Service, U.S. Dept. of Commerce, Springfield, VA. 22151 USA.

FAILURE ANALYSES IN ELECTRIC POWER SYSTEMS – A SELECTIVE REVIEW OF EPRI–FUNDED RESEARCH

Floyd E. Gelhaus*

INTRODUCTION

A number of EPRI failure analyses of components and systems utilized in electric power generating stations have involved a single dedicated-cost-center contractor (Table 1, Project 1). The work so performed on EPRI's behalf ranges from sending personnel to the site of a failure event for a first-hand evaluation, to completing specific metallurgical and analytical studies in support of larger efforts contracted elsewhere or in provision of Institute staff with the needed technical background data upon which future related work scopes are founded. Also, these activities typically initiate within the framework of separate vendor- and utility-supported failure analysis projects, a number of which have provided the nucleus for subsequent jointly-funded EPRI programs.

The following technical areas are a selected group of efforts which represent these various initiation paths for EPRI failure analysis activities:

- Intergranular Stress Corrosion Cracking (IGSCC) in BWR 304SS Piping Systems
- Stress Corrosion Cracking (SCC) in Zircaloy Fuel Rod Cladding
- U–Tube Steam Generator Denting Phenomena
- Near-Bore Flaws in Steam Turbine Rotors

Table 1 lists several specific EPRI projects which involve the metallurgy of failure analysis. The list is neither intended to be all-encompassing nor to deliberately omit any particular activity. Rather, the discussion which follows is intended to provide the reader with an appreciation of the types of research funded by EPRI and, by emphasizing the metallurgical data, to highlight some of the salient findings of this research to-date. References 1–5 provide certain details for the topics discussed here as well as offering a broader view of other related research efforts which include failure analyses as the key engineering discipline.

Table 1
Selected EPRI Projects†

<u>Failure Analysis and Failure Prevention in Electric Power Systems</u>.

Failure Analysis and Failure Prevention in Electric Power Systems: This ongoing project is directed toward more accurately defining the reliability of components and subsystems in power plants and reducing the frequency and/or severity of malfunctions that result in costly, extended outages. Tasks include: (1) root-cause failure diagnostics; (2) development of improved probabilistic failure prediction methods; (3) specific application of Task (2) failure prediction methods to pressure vessel reliability, weld HAZ sensitization quantification, condenser, steam generator, and feedwater systems malfunction analysis. The contractor is Failure Analysis Associates.

*Electric Power Research Institute (EPRI), Palo Alto, California 94303, USA.
†From K.B. Andrews, "Research and Development Projects", May 5, 1977 (EPRI Special Document).

IGSCC in BWR 304SS Piping Systems.

Corrosion Studies in Support of Nuclear System Reliability: The primary objective of this program is to determine and quantify the ranges of compositions and environments that produce damage to structural materials used in nuclear power plants. The program will emphasize the chemical-mechanical reliability of materials and those aspects of corrosion processes that may lead to costly equipment failures. Contractors are Ohio State University, San Diego State University, and Battelle, Columbus Laboratories.

Corrosion Studies of Nuclear Piping in Boiling-Water Reactor (BWR) Environments: This 2-year program is an investigation of the problem of intergranular stress corrosion cracking in BWR nuclear piping. Program emphasis is on evaluating the causes of cracks and finding near-term solutions to the problem. General Electric Company and Argonne National Laboratory are the contractors.

Stress Corrosion Cracking Investigation of Boiling-Water Reactor (BWR) Piping Remedies: This is a 2-year project designed to: (1) identify and confirm to higher assurance levels than now available, the conservative factors related to cracking in weldments of austenitic stainless steel piping; (2) demonstrate that recommended field remedies have a statistically determinable probability of being immune to cracking in weldments for the lifetime of the plant; and (3) further evaluate practical applications of highly discriminating acoustic emission monitoring techniques believed to apply to this type of stress corrosion cracking. Testing will be on full-size pipe segments. The contractor is General Electric Company.

Qualification of Alternate Boiling Water Reactor (BWR) Piping Material: The project objective is to provide a piping material alternative to the standard Type 304 stainless steel used in BWR piping systems. The material should have a high assurance of reliable performance for plant design lifetime and a substantial margin of tolerance for abuse or atypical operating conditions. The project will also provide a demonstration of predictive capabilities on crack propagation. General Electric Company and the Energy Research and Development Administration are cosponsors. General Electric Company is the contractor.

Basic Studies on the Variabilities of Fabrication-Related Sensitization Phenomena in Stainless Steels: Stress corrosion cracking in stainless steel piping has had a major economic impact on nuclear power plants. The objective of this research is to develop a quantitative understanding of the kinetics of sensitization reactions as influenced by material and fabrication variables and how this relates to grain boundary corrosion. The resulting effects of local environmental chemistries will be assessed for conditions relevant to power plant operation. General Electric Company is the contractor.

SCC in Zircaloy Fuel Rod Cladding.

EPRI/NASA Cooperative Program on Stress Corrosion Cracking of Zircaloys: This 30-month program is directed toward characterizing the nature of and prescribing remedies for fission-product-induced stress corrosion cracking of Zircaloy-clad fuel elements. The factors controlling crack initiation and crack propagation will be investigated. The EPRI/NASA jointly funded effort utilizes NASA-Ames and Stanford Research Institute facilities.

EPRI/Combustion Engineering (CE) Cooperative Program on Pressurized Water Reactor Fuel Performance Evaluation: In this 7-year project, EPRI will contribute to ongoing CE test programs to develop nuclear fuel performance data. Tests include: (1) evaluation of the effect of specific fuel rod design parameters on fuel performance using test fuel rods in 14 x 14 carrier assemblies; (2) evaluation of statistically significant numbers of precharacterized, new, standard design fuel rods in standard CE 16 x 16 fuel assemblies; and (3) detailed hot cell examination of irradiated fuel rods from Maine Yankee Core 1. It is expected that test results will explain anomalous behavior and furnish benchmark data for fuel performance models and for better statistical verification design basis. Item (3) tests have been concluded; see Final Report, Task C, No. NP218 (Project No. RP586–1), November 1976.

Determination and Microscopic Study of Incipient Defects in Irradiated Power Reactor Fuel Rods: The purpose of this 20-month project is to find incipient defects in fuel rods and correlate them with the power reactor duty cycle history which produces cladding strains. The study will characterize such defects in order to determine the rate of crack initiation and growth in BWR and PWR fuel rods. A few failed rods in the reactor core have a significant impact on plant output, so that a strong incentive exists to reduce both fuel failures and maneuvering restrictions. The contractor team includes Westinghouse Electric Corporation, General Electric Company, ASEA–Atom (Sweden), Battelle-Columbus, and Argonne National Laboratory.

U–Tube Steam Generator Denting Phenomena.

Optimization of the Metallurgical Variables to Improve the Stress Corrosion Resistance of Inconel 600: This 30-month project is directed toward improvement in Inconel 600 steam generator tubing. The objective is to demonstrate improved performance in this widely used alloy by optimizing thermo-mechanical treatments and by controlling minor alloy constituents. The expected result is to permit practical field application by all vendors of crack-and-corrosion resistant versions of Inconel 600. Westinghouse Electric Corporation is the contractor.

Steam Generator Model Boiler Program: This 2-year program is a study of materials and chemistry parameters, associated with pressurized water reactor steam generators, by use of two types of model boilers. This test apparatus will permit use of special instrumentation and also allow simulation of fouling conditions in steam generators caused by sludge accumulations. The information produced is expected to be valuable in helping to reduce or avoid the significant loss in plant availability, maintenance effort, and man-rem due to corrosion damage in steam generators. Combustion Engineering is the contractor.

Electromagnetic Transducers for Ultrasonic Inspection of Steam Generator Tubing: The primary objective of this project is to provide an evaluation of the feasibility of electromagnetic-acoustic concepts for generating ultrasonic waves to inspect pressurized water reactor steam generator tubing. A secondary objective is evaluation of this concept for inspecting fossil fuel boiler tubes. This project is limited to a proof of principle demonstration only. Steam generator tube failures are the second greatest cause of forced outage of pressurized water reactors. Thus, there is a need to develop alternative inspection methods. The contractor is Rockwell International Science Center.

Deposition of Salts from Steam: The aim is to learn which factors can and must be controlled to maintain corrosion-safe environments for the steam turbine internals. This project will provide data on the range of corrosive conditions on turbine surfaces relative to different boiler and feedwater conditions under both steady state and cycling loads for U-tube recirculation and once-through steam generators in PWRs. This information will be compiled in a guide to plant operation that will suggest ways to optimize the tradeoff between corrosion damage and rigid chemical controls at all phases of runup, loading, and running of steam turbines. Contractors are General Electric Company and Babcock and Wilcox.

AC Probe for Monitoring Corrosion of PWR Steam Generators: The objective is to test a laboratory electrochemical device under realistic conditions similar to the regions in which damaging corrosion occurs in a steam generator of a pressurized water reactor. Conditions of boiling heat transfer will be provided, using model boilers, for a primary feasibility test for this device. Brookhaven National Laboratory is the contractor.

Near-bore Flaws in Steam Turbine Rotors.

A Program to Increase the Reliability of Steam Turbine Rotors: The objective of this 24-month program is to increase the reliability of steam turbine rotors by advancing in-service nondestructive evaluation techniques and interpreting the significance of their results in an analytical lifetime prediction system. Primary program emphasis is placed on ensuring the integrity of the forged turbine rotor spindle. Contractors are Southwest Research Institute; Battelle, Columbus Laboratories; and Westinghouse Electric Corporation.

IGSCC IN BWR 304SS PIPING SYSTEMS

Background

As described by Reference 6, "The first of a series of small hairline cracks in Type 304 stainless steel piping was discovered in September, 1974. The cracking, associated with the heat-affected zones of pipe welds, was identified as intergranular stress corrosion cracking (IGSCC). Subsequent to this discovery, a thorough inspection campaign of stainless steel piping in all BWR plants was initiated by the then Atomic Energy Commission (AEC). Results of these inspections identified similar cracking occurrences in several BWR plants. To date, some 82 separate incidences have been recorded. In all of these cases, the cracking was isolated to main recirculation valve bypass lines and to the core spray piping systems. Only seven incidences resulted in through wall cracking behavior (leaking); the remainder were detected with ultrasonic inspection techniques at various depths into the wall thickness. None of these incidences involved safety issues although cracking constituted a major economic concern.

A number of major activities were initiated in response to this concern. The General Electric Company established an internal investigative task force and in July, 1975, issued a comprehensive two-volume report describing their findings[7]. The Nuclear Regulatory Commission (NRC) formed a special Pipe Cracking Study Group during the same general time frame and subsequently described their findings in a summary report issued in October, 1975[8]. Each of these studies made effective use of information available at that time; however, as with all evolving technologies certain needed information was unavailable or incomplete. The studies did not qualify remedial action, although the GE effort did recommend certain actions such as removing and capping bypass lines, replacing stainless steel core spray lines with carbon steel piping, and rerouting control rod drive return lines.

A third major activity was initiated by EPRI early in 1975. This activity concerned a program of complementary research activities dealing with the generation of needed information and qualification of remedial action[9,10]."

Description

Figures 2 and 3 show the general configurations for the core spray and recirculation systems, respectively [7].

The systems descriptions given in Reference 7, Vol. 1, indicate that "the core spray system is one of multiple, redundant systems designed to maintain a supply of cooling water to the core under conditions during postulated breaks in the main coolant lines. Each core spray system is made up of pumps, valves, piping, spray headers, and associated controls. Each plant is provided with two completely independent systems. . . . . The systems are designed to go on after the reactor pressure has decreased to approximately 20% of its normal value. Above this point, other systems, notably the high pressure coolant injection system, supply water to the core. The main recirculation line is the primary path for the coolant being circulated through the core by the main recirculation pumps. There are two of these loops in each plant . . . . . The main valve in the discharge line is bypassed by a smaller four-inch diameter line which also contains a valve. This line—— the recirculation bypass line——is shown in Fig. 2.

In BWR–2's, BWR–3's, and BWR–4's, the recirculation bypass line is provided mainly to equalize the pressure across the main discharge valve so that this valve can be opened and closed with zero differential across its seat."

Analyses Results

The analyses confirm that stress, material sensitization, and the environment which contacts the 304SS are the three synergistic causal factors in this IGSCC phenomenon.

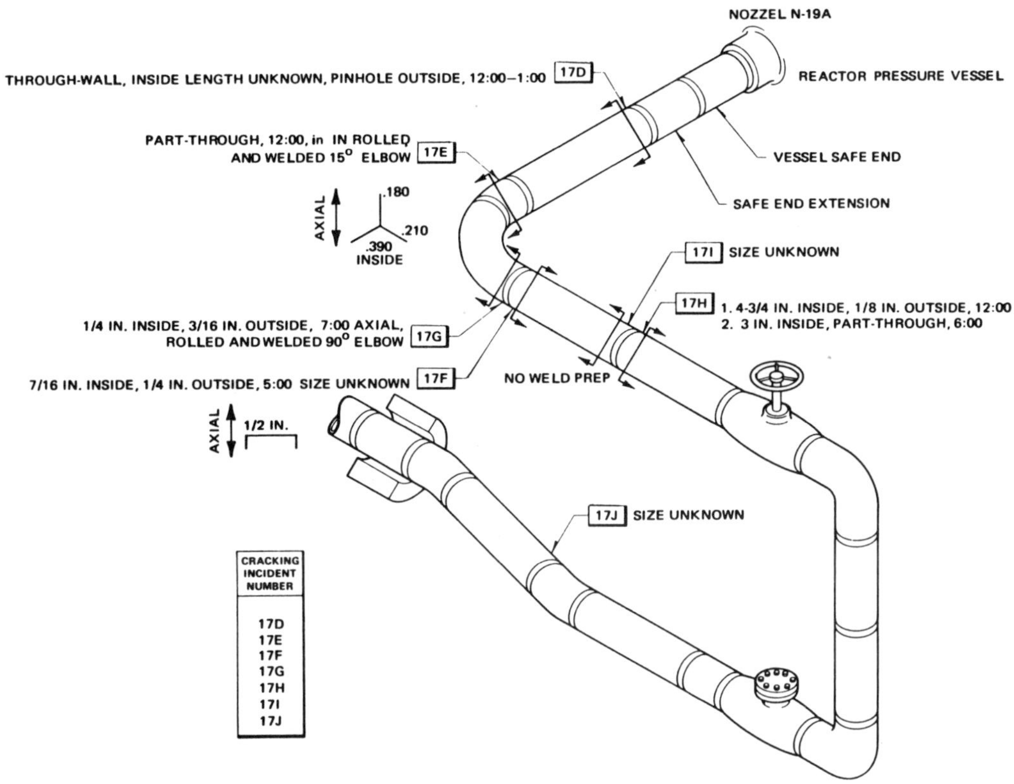

Fig. 1. Dresden 2, Core Spray System, Loop A.

The weld HAZ sensitization, typically producing only a mild susceptibility to IGSCC cracking in 304SS, is important here because of its presence in the zone of weld-shrinkage residual stress. Furthermore, the geometries shown in Figs. 1 and 2 apparently allow stress pile-up at specific locations, with thermal, mass, pressure, and fit-up stresses adding to the welding residual stress.

Fig. 3 shows the cross section dimensions of a typical circumferential crack in a four-inch diameter bypass line; the maximum crack depth shown is 0.31 inches, and the length-to-depth ratio is (typically) about 6 to 1.

The cracking mode is intergranular, with the cracks initiating on the inner surface of the pipe. Figure 4 clearly shows this intergranular attack, as well as providing an example of finding multiple cracks at various distances from the weld.

As pointed out by the authors of Reference 7, "Crack shape and orientation must be determined in three dimensions. If one viewed only Fig. 5, it might be concluded that the crack shown initiates in the inner weld bead. A different conclusion would, however, be drawn if Fig. 6 were the only one seen. Both are photos of the same crack. The correct conclusion can only be drawn when one is told that Fig. 5 is a section taken from the far right-hand end, and Fig. 6 is from the center near the through-wall portion of the very same crack. With all photos in hand, one can say that the crack did not initiate in, but stopped when it reached the weld metal.

The crack seen in Fig. 5 did not immediately stop when it reached the weld metal, as is the normal observation. The key to this difference is ferrite content. Figure 7 shows the weld prepared with a colloidal magnetic dispersive etch. This etch shows relative

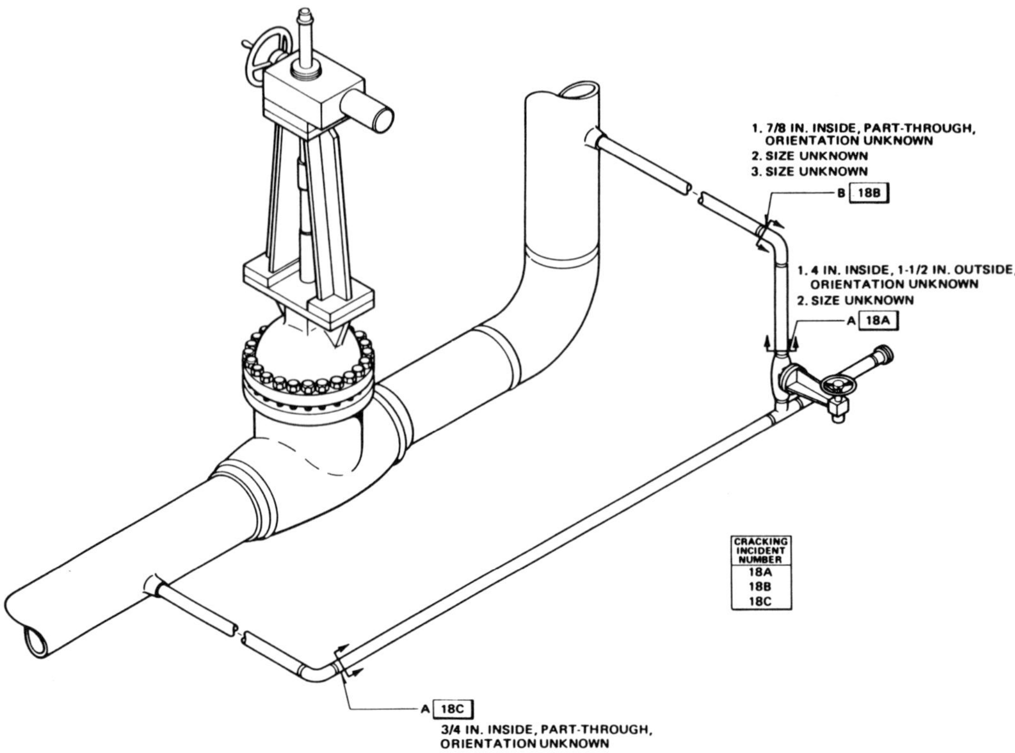

Fig. 2. Millstone Point 1, Valve Bypass, Recirculation System, Loops A and B.

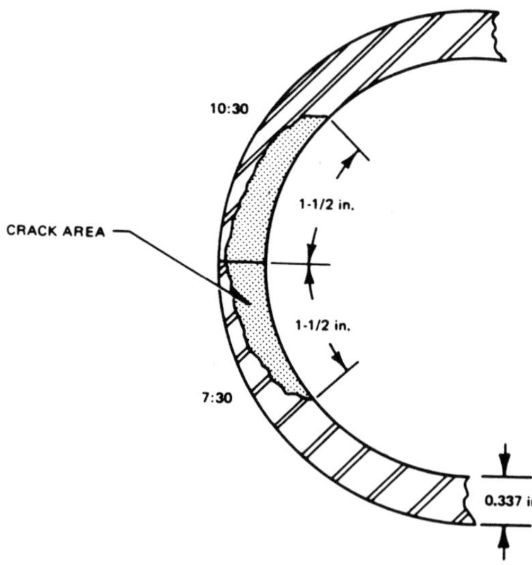

Fig. 3. Schematic drawing showing extent of cracking quad cities 2 Loop B.

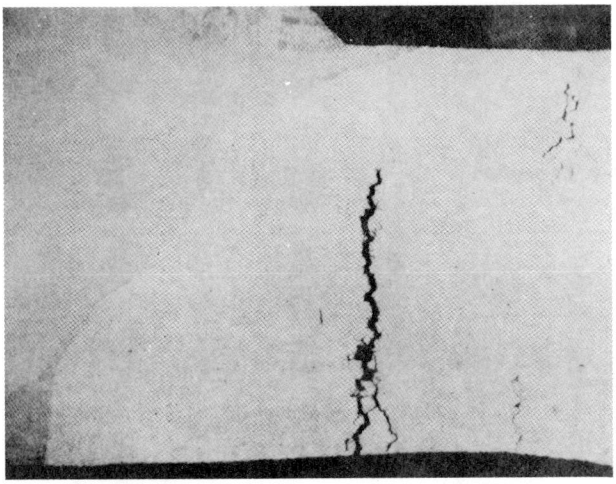

Fig. 4. Dresden 2 Bypass. Original Mag. 8X. (Reduced approximately 20 percent for reproduction).

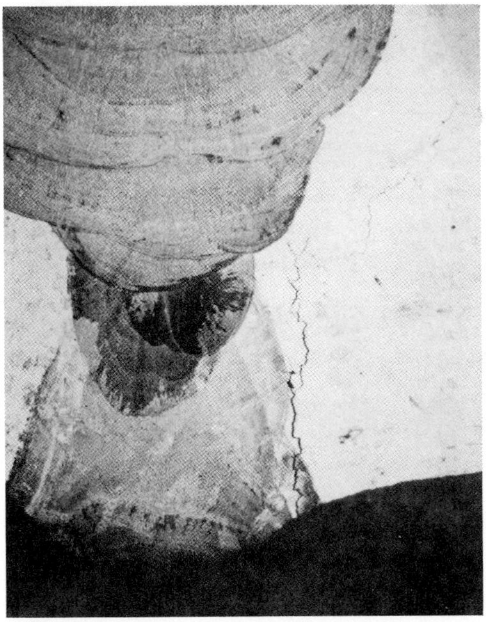

Fig. 5. Dresden 2 Core Spray at End of Crack. Original Mag. 8X. (Reduced approximately 20 percent for reproduction).

ferrite contents pictorially. Dark regions have a higher ferrite content than lighter regions. Portions of the root pass are barely discernible from the base metal. A quantitative ferrite reading of the root pass is not available."

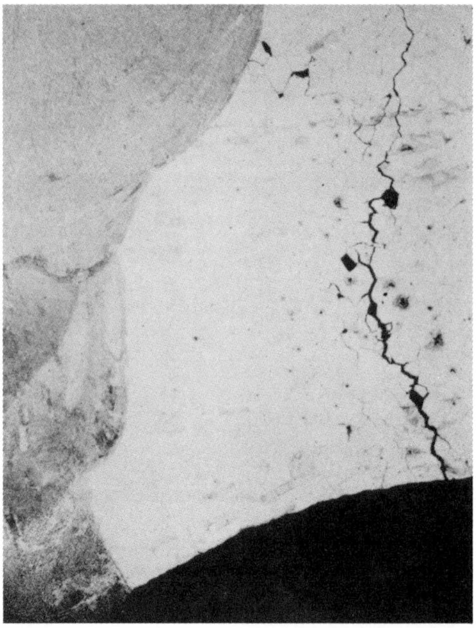

Fig. 6. Dresden 2 Core Spray, Middle of Crack. Original Mag. 7X. (Reduced approximately 20 percent for reproduction).

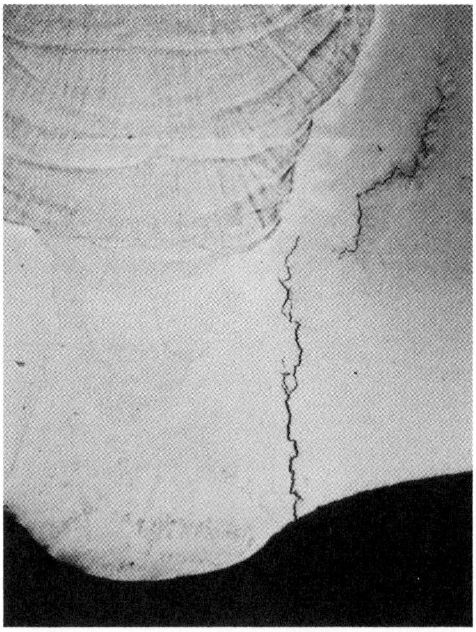

Fig. 7. Colloidal Magnetic Dispersion Etch of Dresden 2 Core Spray, Same View as Fig. 5. Original Mag. 8X. (Reduced approximately 20 percent for reproduction).

Summary

Intergranular stress corrosion cracking (IGSCC) has occurred in stainless steel piping of Boiling Water Reactors. Although the incidences have not involved safety issues, cracking has constituted a significant economic concern. A major research program has been established by the Electric Power Research Institute to address this problem. Quantitative information dealing with the three principal variables (stress, material sensitization, environment) has been generated as a result of the program and is being applied to understand and eliminate the cracking behavior. A realistic working model has been established. Although the phenomenological understanding is yet incomplete, the following quantification of factors controlling IGSCC is sufficient to proceed with qualifying remedial action and an alternate material [6].

Residual Stresses For any given azimuthal position across the weld on the inside pipe surface, the value of stress is a maximum at approximately the center of the weldment and decreases rapidly with axial distance from the weld center line (i.e., bell-shaped). The profile of maximum axial stress also exhibits a variation in magnitude around the circumference of the pipe. The maximum axial stress is a function of pipe diameter for Type 304SS schedule 80 pipe and this maximum stress decreases for increasing pipe diameter. Surface weld preparation such as grinding can dominate residual stresses on the inside surface.

Sensitization A variety of tests performed on different heats of Type 304 stainless steel pipes of various diameters has indicated that heat-to-heat variability is great (i.e., for the same nominal specification, some heats are very susceptible while some are very resistant to IGSCC). The additional observation that these heat-to-heat considerations can overshadow sensitization effects due to welding, has major impact on the weighting of the relative importance of the various parameters effecting the IGSCC process.

Environmental Effects In-plant measurements of water chemistry and electrochemical potential have been effective guides towards understanding expected plant environmental conditions and are being used to assure proper test conditions for laboratory qualification pipe testing of remedies. Oxygen and hydrogen peroxide levels were monitored as a function of temperature during steady state and transient operation (startup and shutdown). For steady-state operation, oxygen concentration varied from 0.1 to 0.3 ppm with no detectable peroxide. During shutdown the reactor system was normally opened to the environment once the water temperature was reduced below 100°C. The oxygen concentration increased rapidly to several ppm, and the peroxide concentration built up to 1 to 3 ppm. During normal BWR startups the oxygen and peroxide concentrations often remain significantly above the normal operating level until a temperature of at least 150°C is reached.

Concurrent with chemistry measurements during startup transients at one site, the electrochemical open circuit potential behavior (corrosion potential) of stainless steel was measured. Excellent correlation between the chemistry and electrochemistry measurements was obtained. For example, large potential and chemistry (oxygen and peroxide) changes were simultaneously monitored for control rod movements. Laboratory studies confirmed a strong electrochemical potential dependency on oxygen and peroxide levels, a strong temperature dependency, and initial evidence of limiting conditions for IGSCC in stainless steel.

SCC IN ZIRCALOY FUEL ROD CLADDING

Background

As part of its Fuel Performance Program, EPRI has initiated a number of projects to study stress corrosion cracking failure of the Zircaloy tubes that clad the uranium dioxide fuel pellets. (See Table 1). As indicated in Reference 11, "observed corrosion behavior in Zircaloy cladding can be broadly classified into hydriding effects and fission-product-assisted stress corrosion cracking (SCC). Primary hydriding, the designated cause of numerous early life defects in light water reactors (LWRs), has essentially been eliminated in current fuel rod designs by improved drying procedures in the manufacture of the fuel. Failures that follow reactor power changes have been given the generic title of pellet-clad interaction (PCI) failures and are generally attributed to SCC. Most investigators now accept the view that the release of embrittling fission product species (likely iodine or cesium) and an applied stress are prerequisites for power ramp induced cladding failures. However, the relative contributions of chemistry and stress are yet to be determined.

Both laboratory and reactor test data support the view that the SCC phenomenon can be divided into three stages: (1) crack initiation at the ID surface; (2) initial crack propagation; and (3) final crack propagation. Crack initiation requires breakdown of the protective oxide surface film on the cladding ID surface, either by chemical means if the film is thick enough, or by combined chemical/mechanical means if the film is thick. The mechanism must also produce a crack that deepens more rapidly than it widens; otherwise, the crack will become a pit or groove. Crack propagation requires that some mechanism permit transfer of the embrittling species to the vicinity of the sharpened crack tip. Also, the stress state must be such that a given initial increment of strain must cause crack advance by corrosion if it is to result in an equally large new increment of strain just ahead of the new crack tip position."

The power-plant-operational consequences of this phenomenon are fairly substantial. Plant availability has in the past been decreased due to shutdowns specifically to replace partial and entire core loadings and due to extended refueling outages when extensive non-destructive examination of the assemblies is required to permit the removal of failed fuel along with the depleted assemblies. Plant capacity factor has been impacted both by derating decisions (examples include clad flattening––densification––and fuel rod bowing) of the Nuclear Regulatory Commission, and by decisions of utilities to place lower-than-regulatory limits on in-plant and ex-plant radioactivity release levels. Further negative capacity factor impact results from limiting certain of the power-change-maneuvering flexibilities designed into the nuclear steam supply system in order to limit the occurrence of pellet-clad interaction failures.

The financial impact of these phenomena has also been substantial. The fuel suppliers have experienced costs in excess of contingency funds when reload batches expand in number to account for failed fuel; the utilities, paying the consequential costs, must provide replacement power when nuclear plant output decreases. Replacing the output of a 1000 MWe nuclear station can cost from $125,000 to $300,000 per day, depending upon the local price and availability of coal or oil. The cumulative effect of operation at less-than-maximum power because of the maneuvering limitations has been as much as $3–7 million per refueling cycle for some boiling water reactors (BWRs).

Description

The fuel for nearly all water-cooled commercial reactors consists of uranium oxide pellets (sintered ceramic, in cylindrical shape) clad in zirconium alloy tubes. (See Fig. 8). The fuel pellet o.d. and cladding i.d. are sized so that at the time of fabrication a clearance gap exists between these surfaces. This clearance does not persist throughout the irradiation lifetime of the fuel rod. Pellet-clad interaction occurs when a fuel pellet expands during power increases and stresses the cladding tube.

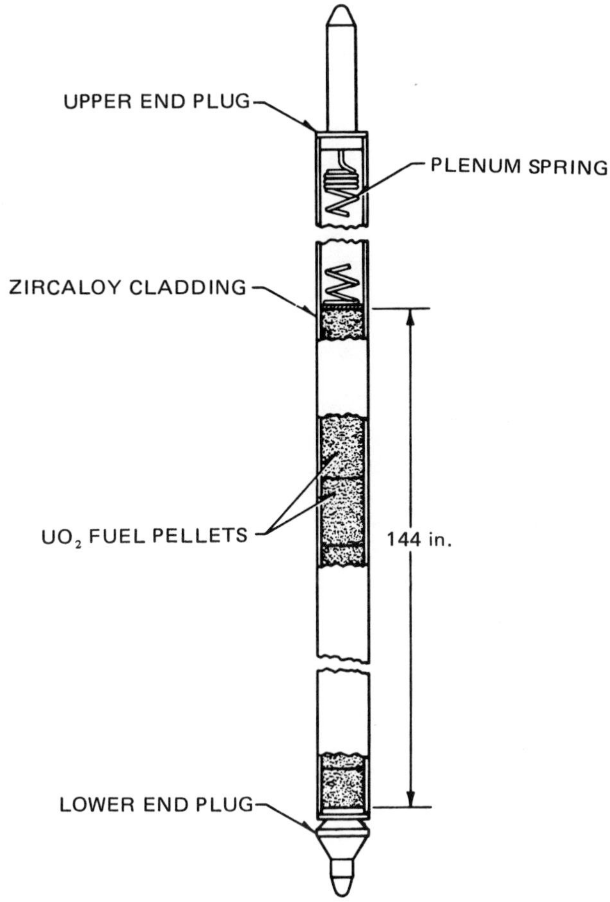

Fig. 8. Cross-section of typical LWR cylindrical fuel rod geometry.

These stresses are chemically augmented by fission products deposited on the cladding. These include chemical species known to embrittle Zircaloy in laboratory tests, such as elemental iodine. Based on thermodynamic arguments, iodine in the fuel rod should be combined with the more abundant fission product cesium to form nonaggressive CsI. However, other research has previously shown that radiolysis can liberate iodine from CsI. The local deposits of CsI on the clad might therefore serve as the source for embrittling the Zircaloy tube. The abundant fission product cesium is a common constituent of most cladding deposits.

In out-of-reactor laboratory tests, Zircaloy discs have been loaded in compression with an indenter in an iodine atmosphere at 300°C. The loaded area was characterized in detail by SEM. Information on the progress of crack development below the surface was obtained by removing material by argon-ion milling and re-examining the newly-exposed surface with the SEM. The results indicate that extensive cracking of the protective oxide layer occurred mechanically. Each crack in the Zircaloy substrate was associated with a surface oxide crack. The far smaller density of substrate cracks suggest that only very specific locations in the metal are susceptible to rapid localized attack (cracking) by iodine. The results indicate that these susceptible locations are associated with second-phase particles containing aluminum, perhaps as a zirconium-aluminum oxide (See Fig. 9).

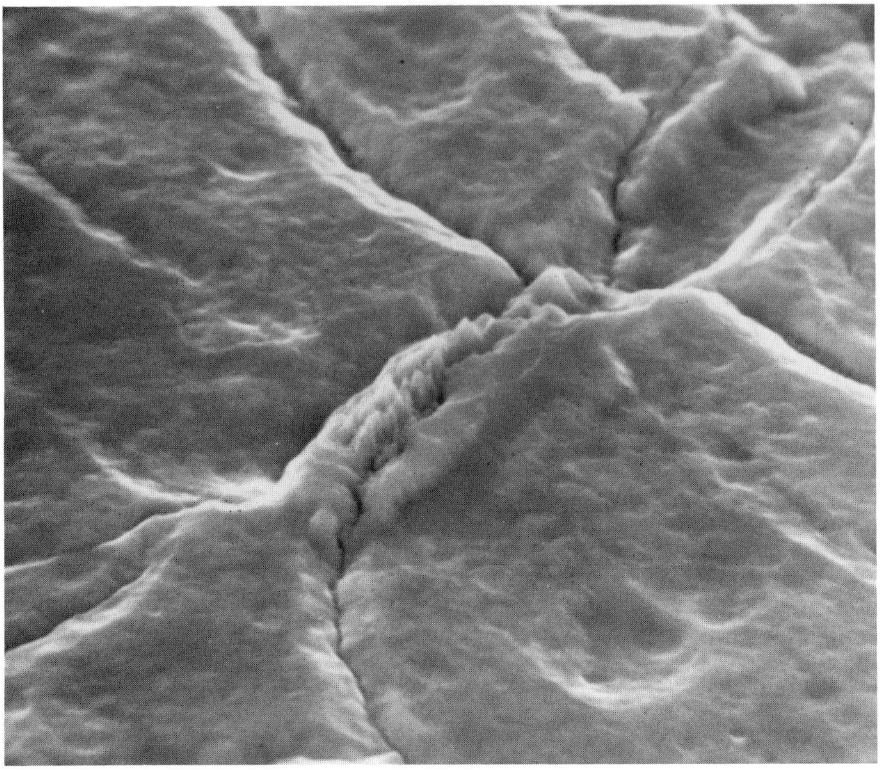

Fig. 9. Zircaloy substrate crack initiated by exposure to iodine and an applied stress at 360°C.

Analyses Results

As noted by the Reference 12 authors, the results from three related EPRI projects have led to the conclusion that PCI defects in LWR fuel are predominantly caused by SCC of Zircaloy. In cooperation with fuel vendors and utilities, EPRI has initiated projects to identify incipient defects in fuel rods that have been subjected to a power reactor duty cycle likely to produce PCI, and to conduct a detailed macro and microexamination of these defects with the objective of elucidating the rate controlling fracture mechanism. Freely utilizing References 12 and 13, a portion of the experimental results and associated analysis that justified this position are summarized below.

To date only one such examination program has been completed. Fuel rods from Maine Yankee Core 1 have been destructively examined in an attempt to determine the primary cause of cladding penetration that resulted in coolant iodine activity increases during Cycle 1[13]. This fuel was non-pressurized and contained UO_2 pellets that densified in-reactor. The defected fuel rods were virtually all located in high power positions in Batch B assemblies, which were the highest power assemblies in the core during the period of interest. Based on the coolant iodine activity history, defects started to occur after an average core burnup of 5,000 MWd/MTU (423 GJ/KgU) and were mainly associated with two power change events.

A total of seven cladding cracks, varying in depth to 65% of the wall thickness, were located in three intact fuel rods coincident with the assembly maximum power position.

The procedure used to locate these incipient cracks included a combination of non-destructive measurements, namely profilometry, gamma scanning (specific Cs137 isotope) and coil eddy current. These techniques were used in conjunction with the detailed knowledge of fuel rod operating histories, to locate fuel rod segments in which the probability of incipient cracks was high.

The typical appearance of the incipient cracks is illustrated in Fig. 10, and one of these cracks is shown at higher magnification in Fig. 11. These pictures are taken at a single surface, reached after several incremental grinding and polishing steps undertaken to map out the depth and length of the cracks. The maximum wall penetration of one crack is 65% and that of the other 45%. As sketched in Fig. 12, the axial length of the deeper crack is in the range 0.125 to 0.188 in., while that of the other crack is approximately 0.063 in. The cracks were tight (small opening compared to length in the direction of the cladding thickness) and difficult to detect in the as-polished condition; therefore, etching was necessary not only to characterize their appearance but to reveal their presence.

Since the characterization of Zircaloy fracture patterns can provide evidence of the cause of fracture, selected specimens were removed from their mounts and the crack was opened for fractographic examination by SEM. Low power examination of the fracture surfaces showed four separate zones; these zones are described in the photograph shown in Fig. 13.

As may be seen from Fig. 14, within the second zone, i.e., the original crack from the oxidized portion to the crack tip, shows transgranular cleavage with river patterns and some fluting. Stereo-pair photomicrographs verify the transgranular cleavage and the complete absence of ductile dimpling. This is the same type of fracture reported by others, and in the Reference [14] review of SCC in Zircaloy the authors note that "the only features that are unique indicators of the occurrence of SCC in the Zircaloys are the flat cleavage planes. No other conditions (e.g., fracture at low temperature, low ductility tensile failure after irradiation) give rise to cleavage in Zircaloys. Fractures always show ductile features when caused mechanically."

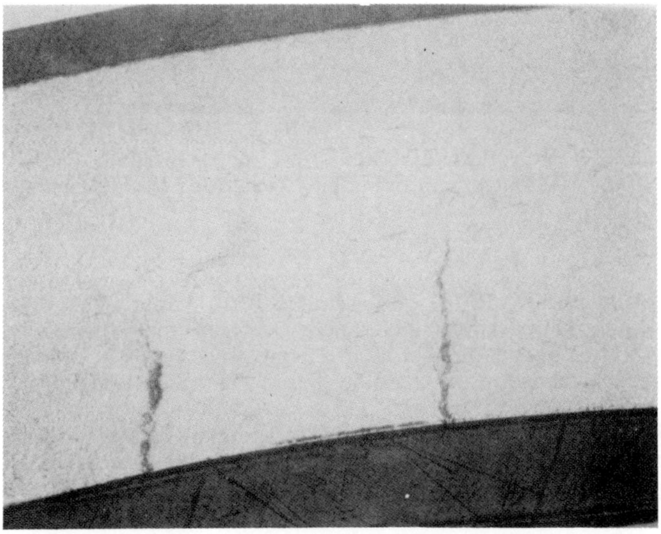

Fig. 10. Cracks at 75° and 80° from rod JBP—005 – 6th surface (etched). Maine Yankee, Core 1, Batch B Fuel Rod Cladding. Original Mag. 100X. (Reduced approximately 10 percent for reproduction).

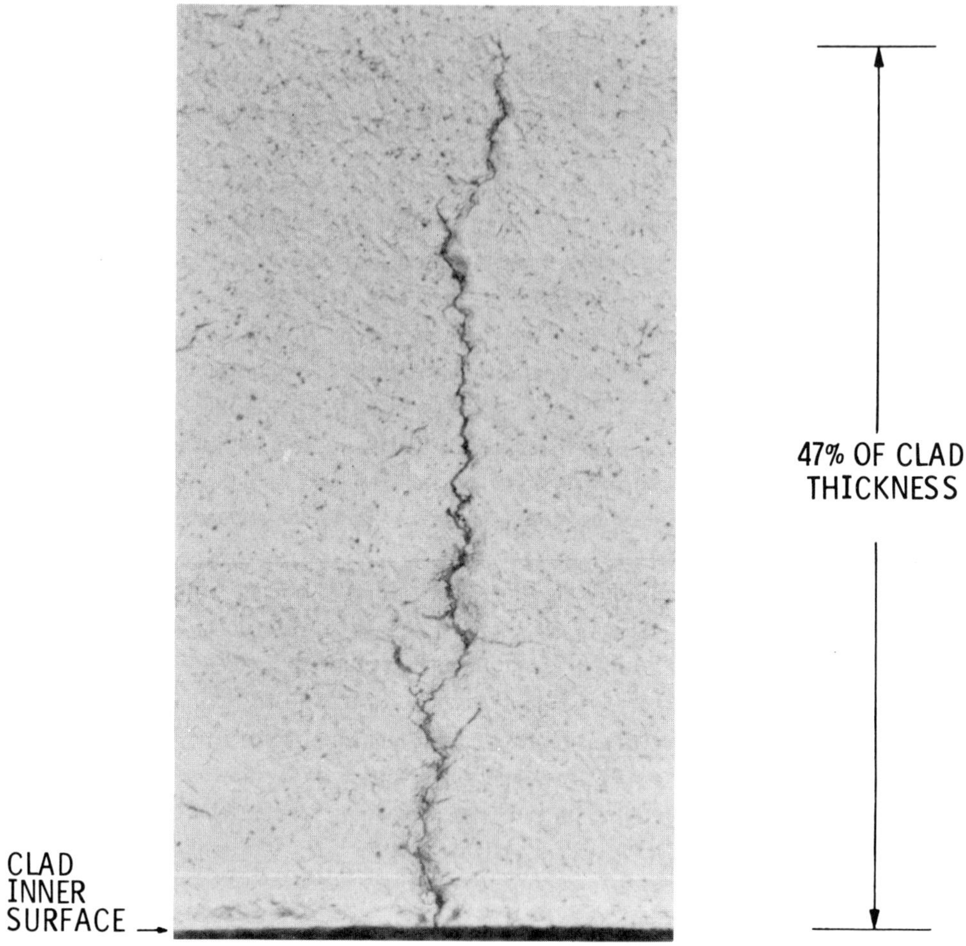

Fig. 11. Crack at 80° from Rod JBP–005 – 6th surface (etched). Refer to Fig. 10. Original Mag. 500X. (Reduced approximately 24 percent for reproduction).

Summary

Stress corrosion cracking (SCC) of Zircaloy is the principle mode of failure in PCI during normal fuel rod operations, including anticipated transients.

A simple analytical model of fracture in Zircaloy cladding[15] indicates that a critical stress must be sustained in the cladding wall in the presence of active fission product species.

A range of threshold stress is likely to exist, depending on the type and/or concentration of fission products. High threshold stresses will be required when fission product availability is limited (i.e. low power and/or burnups), but threshold stress will decrease significantly (by 60%) as fission products become available (i.e. high powers and/or burnups).

The shape and morphology of the cladding cracks should reflect the proportion of mechanical and chemical processes in crack propagation. For example, there should be more crack branching and more flutings on the fracture surface when high threshold

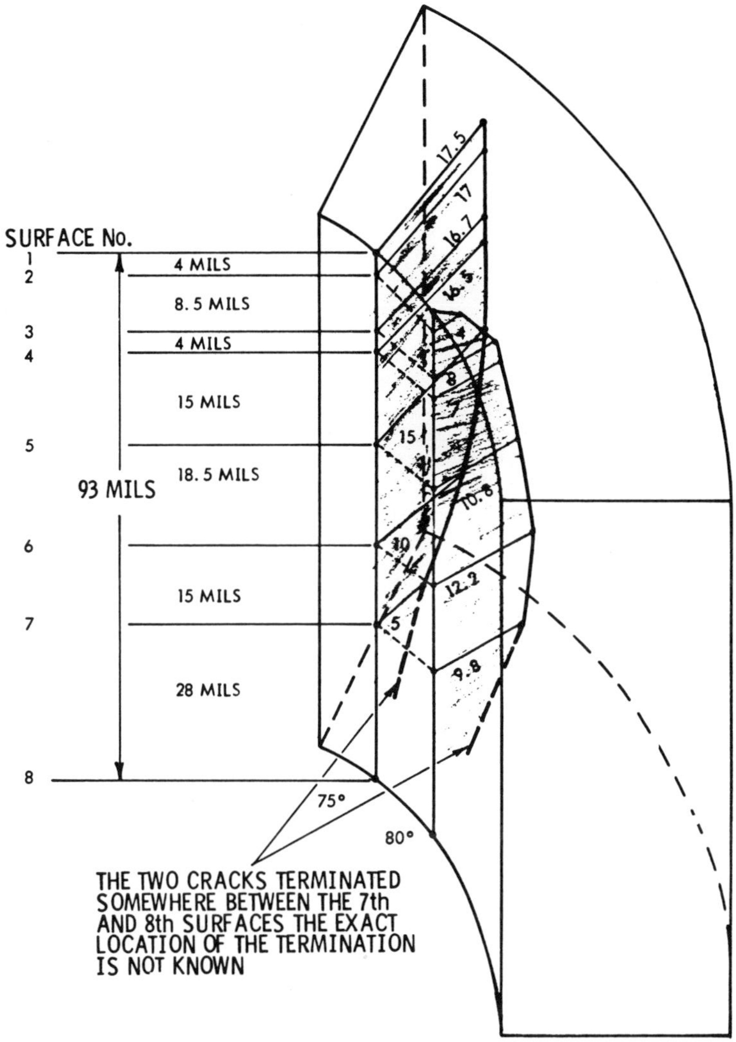

Fig. 12. Map of two cracks from Rod JBP—005 at 94-3/16 in. from bottom of rod. (Refer to Fig. 11).

stresses are required. If the threshold stress is low the crack is straighter and transgranular cleavage should predominate. The likely active fission product species are iodine vapor (released from CSI), liquid cadmium and liquid cesium, in that order. Cesium is most likely tied up in oxides or uranates and so its major contribution will be in fuel-clad bonding.

Crack formation (equivalent to very early stage propagation) is the key step in PCI failures. Laboratory data show that many small cracks (~10μm) form in the Zircaloy metal at artifacts such as pits, second phase particles or grain boundaries, once the oxide film cracks. The subsequent growth of these small cracks to say ~120μm, requires a critical stress to rupture the intervening group of grains, i.e., the ligament. The easier the chemical corrosion (cleavage), the smaller the group of grains and, hence, the lower the stress. Once the crack has formed to ~120μm, growth is fairly rapid and is predominantly transgranular cleavage.

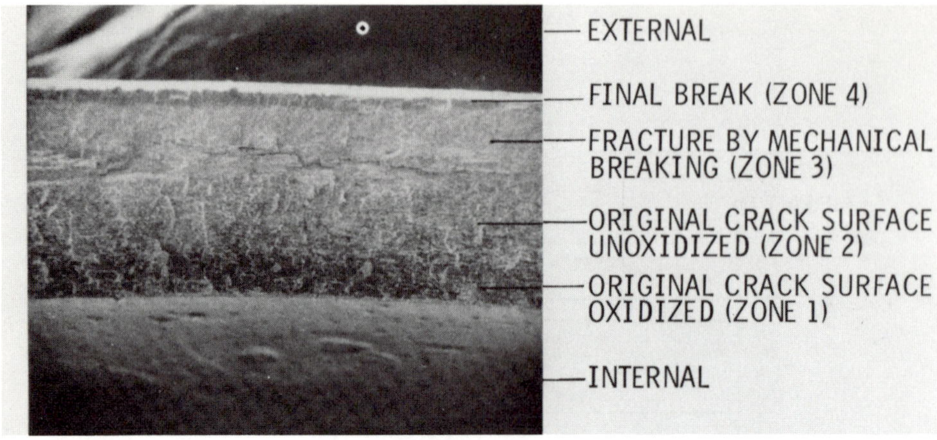

Fig. 13. Low magnification SEM-photograph of fracture. Maine Yankee, Core 1, Batch B, Fuel Rod Cladding. Original Mag. 55X. (Reduced approximately 25 percent for reproduction).

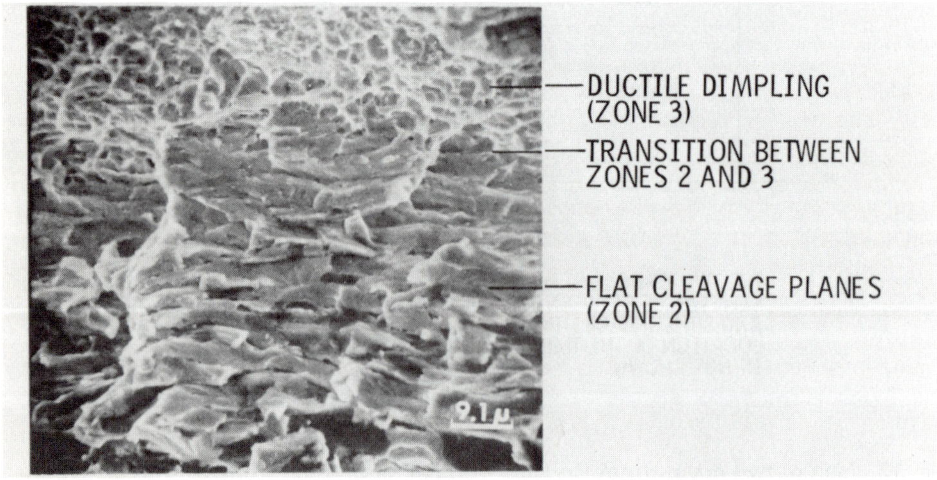

Fig. 14. Fracture surface at crack tip. Maine Yankee, Core 1, Batch B, Fuel Rod Cladding. Original Mag. 1100X. (Reduced approximately 25 percent for reproduction).

Crack propagation will slow and the crack can arrest if either the stress or the fission product environment is not sustained. For example, stress relaxation by creep will certainly affect crack growth. Slower propagation favors an intergranular path, and so intergranular regions on the crack front could be used as a basis for indicating whether PCI damage is cumulative.

Features of the Zircaloy cladding which may influence SCC include oxide film characteristics, local i.d. texture, yield stress (particularly as it is affected by irradiation), and i.d. surface features such as second phase particle distribution. Features of the fuel pellet which may influence SCC all relate to their influence on fission gas (product) release, e.g. porosity, grain size, O/M and cracking behavior. Of these, the first two are the easiest to manipulate to reduce fission gas release.

U–TUBE STEAM GENERATOR DENTING PHENOMENA

Background

In 1976, EPRI became aware of the phenomena associated with nuclear steam generator tube denting. (See Table 1). These generators are vertical shell, inverted U–tube heat exchangers employing over 3000 Inconel-600 tubes with either a three-quarters or seven-eighths inch o.d. and 0.050 in. nominal wall thickness[16]. The primary water inside these tubes is typically at 2000 psi pressure and at a temperature such that the tube wall operates near 525° F. At various elevations, this tube bundle passes through holes in three-quarter inch thick carbon steel plates. (See Fig. 15). The basic cause of the Inconel-600 tube denting is the acidic chloride attack of these support plates which causes rapid corrosion on the surfaces of the tube support plate holes through which the tubes pass. Since the corrosion product has a larger volume than the metal from which it forms, the tube is dented and the plate hole expanded.

A number of utilities with nuclear steam supply systems which utilize this design of steam generator formed a PWR Owners Group late in 1976 for the specific task of determining the best solutions to steam generator problems. EPRI has established a Steam Generator Project Office to manage the PWR Owners Group Steam Generator program. A broad and comprehensive technical program has been defined by the EPRI Project Office and at the present time final arrangements are being made to fund and implement the program.

Description

Figure 16 shows an example of a dent in the tube due to the formation of corrosion product in the tube-to-support-plate annulus. (Nominal 0.014 in. radial clearance). Numerous samples of this material have been subjected to chemical analyses, and it consists primarily of magnetite (Fe_3O_4) along with lesser amounts of other metals, chlorides and phosphates. Denting has occurred in operating units that have employed either phosphate or all-volatile-treatment for water chemistry control.

Denting has been noted at various tube-support plate intersections (hot and cold) leg and at all tube support plate elevations[17].

The formation of the corrosion product at dents also produces geometric distortions in the three-quarter-inch thick support plate itself. The flow circulation holes interspersed among the expanding (oxide growth) tube holes may become oval and, as depicted by Fig. 17, the divider lane flow slots pinch down in the center as the overall plate tries to expand to accommodate the high volume of oxide. In addition, as the flow slots close, the U-bends (especially the smallest radius bends) are bent producing strains in the tubing at the apex of the U-bend. Cracks have been observed both in the smallest radius U-bends and in the vicinity of dents away from the U-bends. Though there are a number of discrete (but related) phenomenon associated with denting, the cracking results at high strain (stress) locations caused by the denting oxidation.

Analyses Results

Sections of tubes and of tube support plates have been removed from a number of steam generators for analysis[18]. At the time of the writing of Reference 18, the maximum diametral restriction was 0.060 in. Wall thinning greater than 0.001 in. was not observed on the Inconel-600 tubing at dents. The annular gap between the tube and tube support plate was full of corrosion product which had two distinct layers. The layer next to the tube had a width equal to approximately the initial, nominal annular gap dimension and was phosphate-rich. The adjacent layer next to the support plate was an in-situ generated corrosion product of the carbon steel with the metallic element being almost pure iron. Chloride ion was a major impurity in these materials. Figures

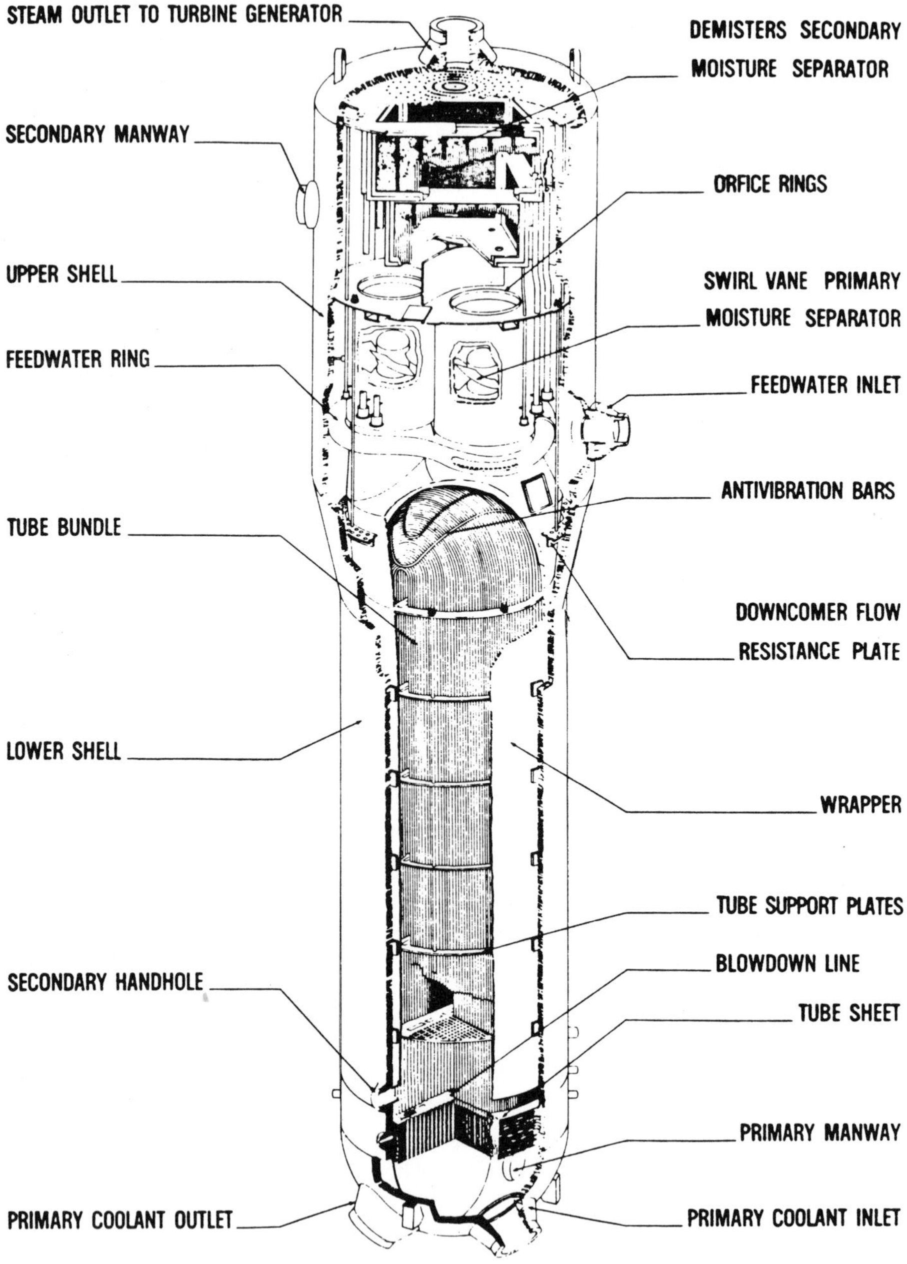

Fig. 15. Westinghouse steam generator

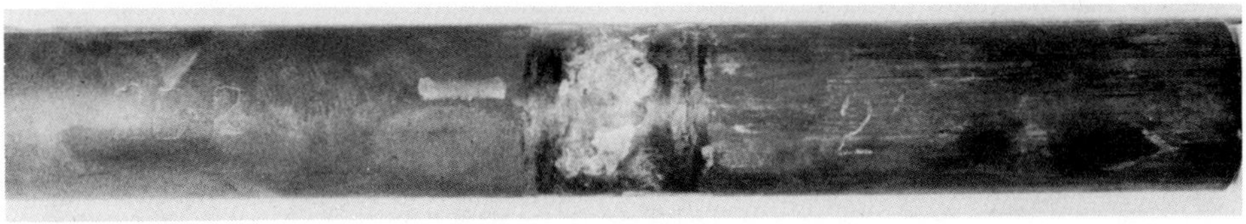

Fig. 16. Dent in the Inconel-600 Tube due to the formation of corrosion products in the clearance annulus between the tube and the support plate. Mag. 1X.

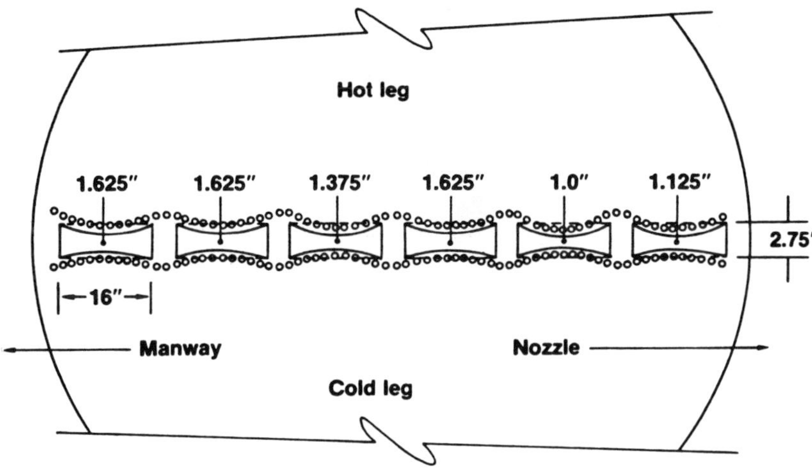

Fig. 17. Flow slot deformation

18 and 19 from Reference 18 show the dented area of the Inconel-600 tube and the extent of the corrosion product in the annulus. Intergranular stress corrosion cracks occur on the tube surfaces and an example of an i.d. initiated flaw is shown in Fig. 20. Since this tube had to be pulled through a number of support plates in removing it for examination, some of the distortion may not have been present in service. However, the high local values of strain or in-situ strain rate produced by this distortion apparently enhance the cracking process. Figure 21 shows at higher magnification the intergranular nature of the crack extension.

Summary

Failures of the Inconel-600 tubes in a number of nuclear steam generators has occurred due to intergranular stress corrosion cracking and is associated with a denting corrosion phenomena that produces locally high stress and strain. The accelerated carbon-steel-support-plate corrosion which causes this tube denting also produces distortion of the flow holes and slots in the support plates which in turn cause cracks in support plates and tight radius, U-bends.

A considerable and expanding joint vendor — EPRI research effort is underway which is aimed at the solutions to the denting phenomena problems. These projects include gathering operational and water chemistry data from the sites, performing eddy-current inspections and working to improve the NDE Systems, analyzing the constituents of

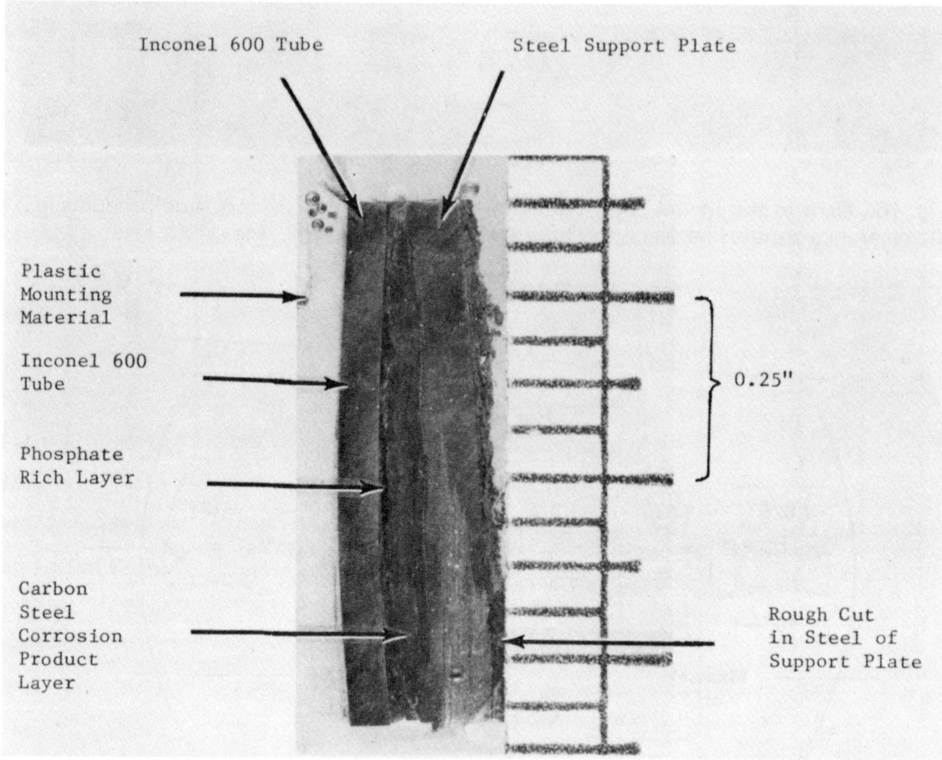

Fig. 18. Longitudinal cross sectional view of tube and tube support plate. Original Mag. 6X. (Reduced approximately 25 percent for reproduction).

corrosion product in order to evaluate the possibility of removal by chemical cleaning as well as to prevent/limit further formation of this porous annulus. In addition, analytical work is underway to evaluate the thermal-hydraulic and stress-strain conditions when denting prevails, and alternate materials are being evaluated and qualified as replacement candidates for both the Inconel-600 tubing and the carbon steel support plate.

NEAR–BORE FLAWS IN STEAM TURBINE ROTORS

Background

As discussed in detail in Reference 19, "on June 19, 1974 during a cold start of the TVA Gallatin No. 2 steam turbine, the combined intermediate pressure (IP) – low pressure (LP) rotor burst at approximately 3400 RPM subsequently fracturing into 30 major pieces. The rotor was forged from an air melted basic electric furnace heat of Cr-Mo-V steel similar in composition to current ASTM A470, Class 8".

In November, 1974, Electric Power Research Institute and Failure Analysis Associates [1–5] staff met with Tennessee Valley Authority and Westinghouse personnel to review the data associated with the Gallatin Unit 2 rotor burst. Based on this meeting and dis-

Fig. 19. Tube-support plate intersection 0.18" below top surface of top plate. Numbers denote SEM–EDAX point analyses. 1 = tube; 2–7 = annulus material.

cussions with other utilities, steam turbine vendors, and others technically active in the field, EPRI concluded that the phenomena involved were of generic concern to the utility industry. This led to a Request for Proposal in 1975 for the scope of work which was to become the Reliability of Steam Turbine Rotors project. (Table 1, Project 5).

The overall goal of this project is to develop a rotor lifetime prediction system to permit utilities to perform reliability analyses from which they can make run/retire decisions. These analyses can be performed using actual steam turbine operational and inspection data or on a parametric basis to further understand the sensitivity of the run/retire conclusion to key mechanistic assumptions and model approximations.

Description

The Gallatin No. 2 steam turbine operated at 2000 psig/1050°F/1050°F conditions, and was rated at 225 MW. During the unit's 106,000 hour lifetime, operation included five overspeed trip tests and cyclic behavior including 105 cold starts (rotor cooling time exceeds 72 hours) and 183 hot or warm starts (rotor cooling time is less than 72 hours).

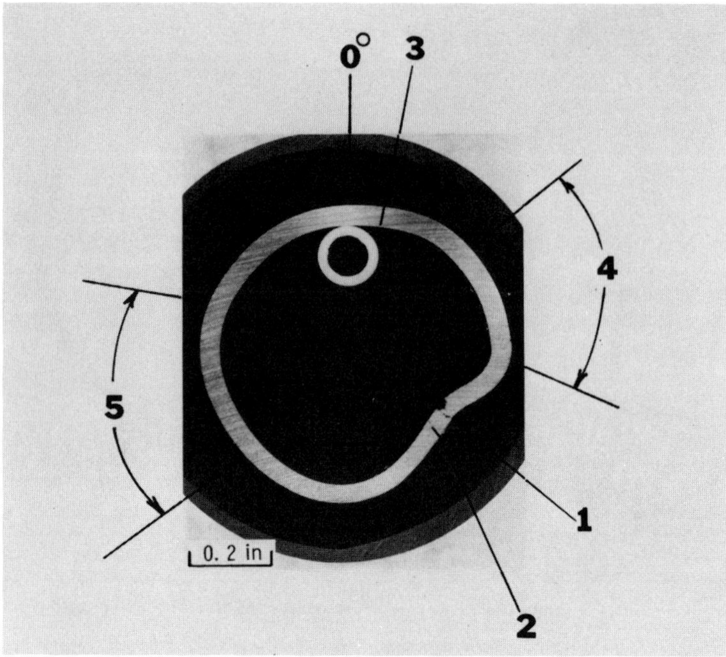

Fig. 20. Denting can result in a non-uniform circumferential tube profile and can initiate intergranular stress corrosion cracks in the Inconel-600 tube.

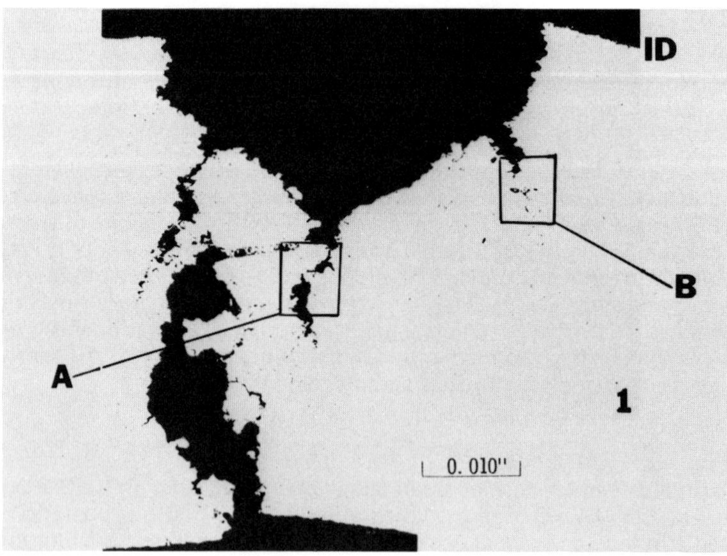

Fig. 21. Major crack; refer to Fig. 20. Original Mag. 100X. (Reduced approximately 45 percent for reproduction).

Reference 19 points out that "the burst IP—LP rotor was forged from a 1954 state-of-the-art Cr-Mo-V grade alloy steel most closely confirming to the modern ASTM A470, Grade 8 material but without the stress rupture, impact, and bore non-destructive examination demanded by modern practice. The rough machined forging weight was 56000 pounds and was manufactured from an air case 108 inch round exterior diameter corrugated ingot. Three separate basic electric furnace heats using standard two slag practice were utilized for the ingot. Vacuum degassing was not yet available at the time of manufacture. Forging practice, heat treatment, thermal stability, and machining practices were generally similar to current techniques. This rotor was austenitized at 1750°F which is a temperature not favoring the notch sensitive stress rupture problem known to occur in this vintage of the alloy, and data show notch sensitivity not to be a factor". Two event-originating flaws, nearly 180° apart circumferentially, were discovered by fractography and are associated with the surfaces of pieces A and B and pieces G and D that are shown schematically in Fig. 22. The major fracture plane is defined by the lines drawn from balance hole number 5, through the shaft, then out to the periphery at balance hole number 2. Figure 23 shows the crack morphology on the face of piece A, and the insert indicates details which will receive further discussion in the presentation of results.

Analyses Results

The EPRI pursuit of a quantitative understanding of the Gallatin rotor burst as part of the contract work scope[20] has led to several possible mechanisms, but none to date has been verified. Linkup between clustered inclusions by cyclic stress-rupture is concluded to be the critical process, with additional effects possibly resulting from microstructure and grain boundary segregation.

A complete analysis of the Gallatin No. 2 IP rotor has been presented by the manufacturer[19]. Subcritical crack growth took place in a 5½ by 3/8 inch region of banded segregation near the bore at the axial location where the sum of the centrifugal and transient thermal stresses reaches its maximum value.

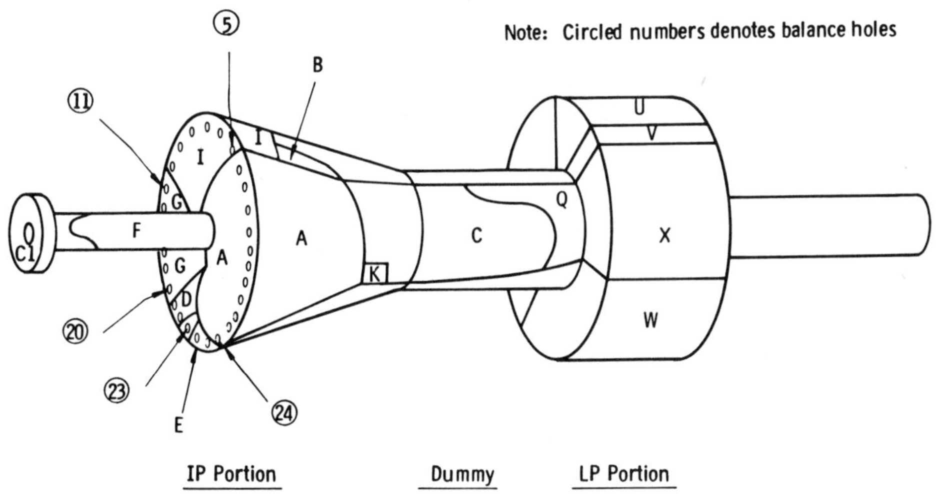

Fig. 22. Schematic of Gallatin No. 2 rotor showing fractured sections.

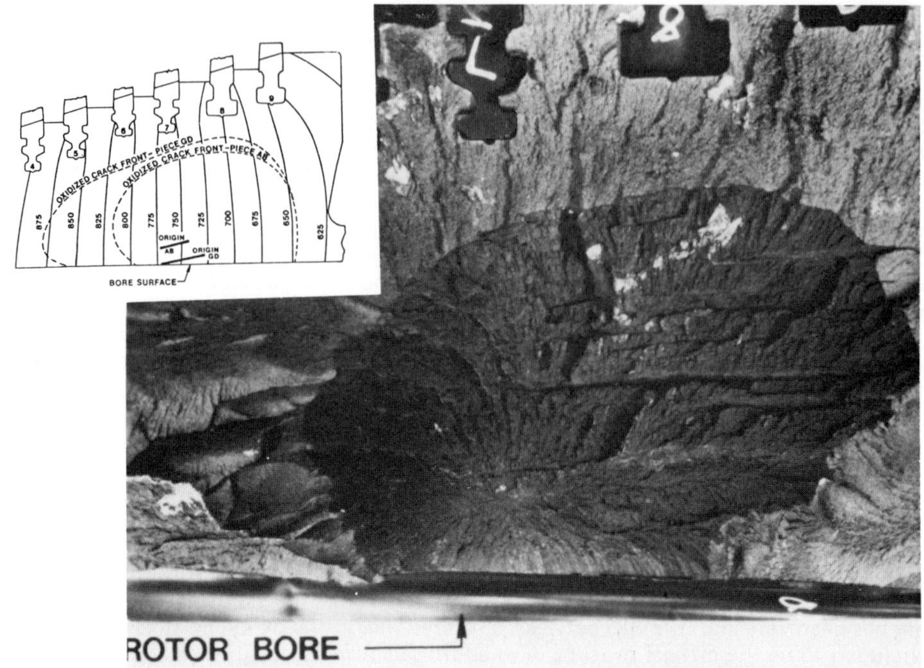

Fig. 23. Crack morphology and steady state temperature distribution in the IP rotor.

Crack initiation was attributed to a high density of plate-like manganese sulfide inclusions. Initial crack growth was intergranular, and a region of intergranular cavitation and inclusion-matrix decohesion was observed adjacent to the fracture surface, indicative of a significant stress rupture component of damage[19]. A typical example of inclusion decohesion and linkup is shown in cross-section in Fig. 24, and Fig. 25 shows a fractograph revealing the intergranular cracking.

Mechanical properties testing was conducted by the manufacturer[19] utilizing both specimen location and orientation as a test variable. "Tensile, impact, creep and fracture toughness (by both conventional ASTM E399–74 and instrumented precracked impact methods) testing revealed no gross deviations in expected properties although some bore tangential tensile and creep data was marginally lower than the other tests. Low cycle fatigue tests with and without hold time were also conducted." Further detailed characterization of the composition and microstructure of the rotor material was carried out under EPRI's RP502 program. Within the segregated bands in the vicinity of the fracture origin, the microstructure was composed of very fine ferrite grains and massive carbide particles, rather than the normal bainite with a distribution of fine grain-boundary carbides. This structure suggested that the center of the forging had been subjected to thermal lag during heat treatment and that the material in the bands might exhibit reduced stress-rupture properties[21].

As noted above, specimens of the Gallatin rotor were tested by the manufacturer in low-cycle fatigue and creep. Figure 26 summarizes the low-cycle fatigue results and indicates the Gallatin failure point for comparison.

With the exception of the interspersion test data described below, the maximum calculated strain range for the rotor is on the order of one tenth the range required for fracture in the total number of starts and stops in its operating history. In comparing the stress-rupture results with the failure point of Gallatin, the steady-state stress and

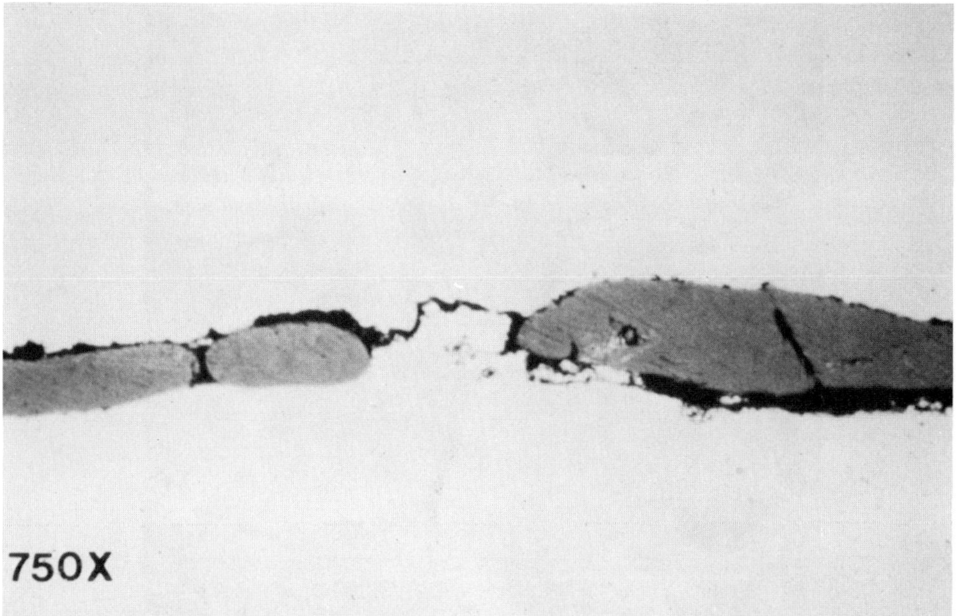

Fig. 24. Decohesion of MnS inclusions and inclusion linkup, typical of Gallatin No. 2 Rotor material near the initiation site. Original Mag. 750X. (Reduced approximately 48 percent for reproduction).

temperature were selected, since the transient thermal stress occurs at temperatures far below the creep range. The data indicate that failure by this mechanism would require a stress of 56 ksi rather than the 44 ksi calculated for the rotor. Extensive metallographic and fractographic analysis was performed on these specimens to verify the presence of the suspect microstructure. Unfortunately, although significant banding was confirmed, no reduction in mechanical properties has to date been attributed to this microstructure. Moreover, with the exception of one low-cycle fatigue test condition with a 23-hr. hold time at peak tensile strain (see Fig. 26), all the specimens tested have failed transgranularly, in contrast to the primarily intergranular fracture of the rotor[21].

Linear elastic fracture mechanics was applied to fatigue crack linkup between inclusions[22]. The small number of cycles experienced by Gallatin results in a negligible amount of fatigue crack extension unless the inclusions are very closely spaced.

The concentration of inelastic strain at defects and increases in the net steady-state stress have also been considered. In the case of a high volume fraction of inclusions or porosity, linkup by intense shear and accompanying void formation is known to be accelerated by plane strain conditions in monotonic ductile rupture; also stress rupture ductility is substantially reduced, but rupture lifetime only slightly so. No data are available on the effect of plane strain in cyclic creep or low-cycle fatigue with hold time.

The Metal Properties Council has funded a number of creep-low-cycle fatigue interspersion tests on material from a vacuum-degassed Cr-Mo-V rotor forging[23] and has made specimens available to EPRI for analysis. Typical LCF failure data are plotted in Fig. 26. From these data it is apparent that interrupting a creep test by one or more fully-reversed strain cycles reduces the rupture lifetime to about one third the static value and the low-cycle fatigue lifetime to between one tenth and one hundredth of the steady cyclic life at the high strain ranges tested to date. Extrapolation to lower strains of concern cannot be done at this time. These results are considered significant since the

Fig. 25. Intergranular crack, typical of those linking inclusion sites in the Gallatin No. 2 rotor. (See Fig. 24). Original Mag. 3000X. (Reduced approximately 50 percent for reproduction).

specimens are of normal bainitic microstructure and contain a very low density of inclusions. The SEM photographs in Fig. 27 show the microstructure of one of the MPC specimens. Vacancy accumulation at the grain boundaries is very evident, with some indications that only the grain boundary carbides maintain the cohesiveness along these surfaces. Considerable oxidation of the fracture surfaces makes the fracture mode difficult to ascertain with certainty; although transgranular behavior is predominant, selected areas hint that intergranular failure may play a role in initiating the failure.

Summary

Steam turbine rotor stresses are highest at the rotor bore surface, and the sum of the thermal plus centrifugal stress peaks during startup. Under "severe conditions", subcritical flaw growth can occur with the possibility that a brittle extension of the critical flaw can result in a rotor burst event.

In attempting to quantify the boundary between "severe conditions" and acceptable conditions, first the turbine manufacturer and then EPRI have probed many possibilities. The research results to-date clearly indicate that inclusion linkup is the dominant subcritical growth mechanism. However, the exact effect played by grain boundary denudation, possible ferrite grain boundary embrittlement, vacancy accumulation, or some combina-

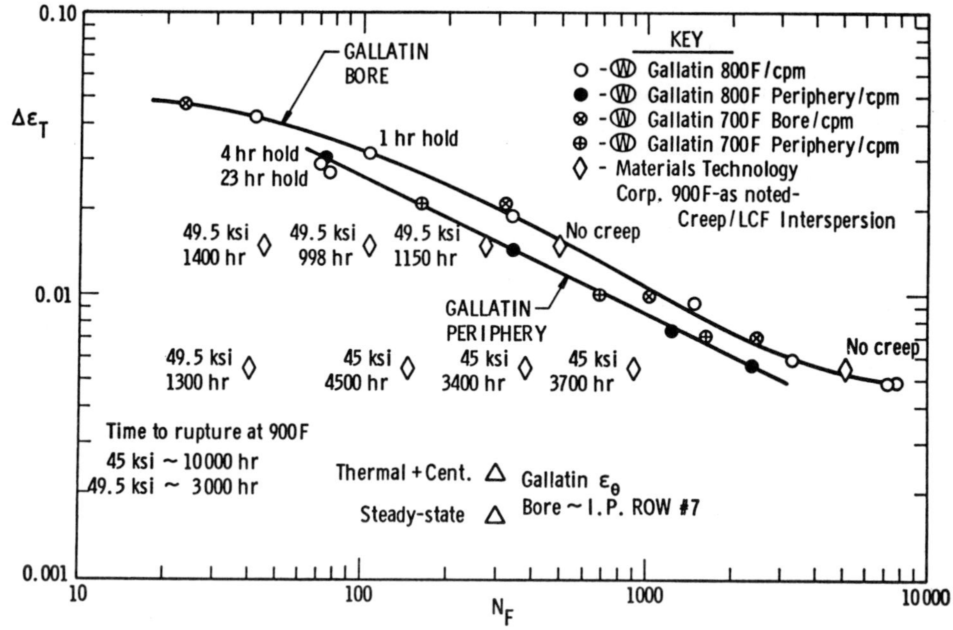

Fig. 26. Gallatin rotor and MPC creep interspersion test results.

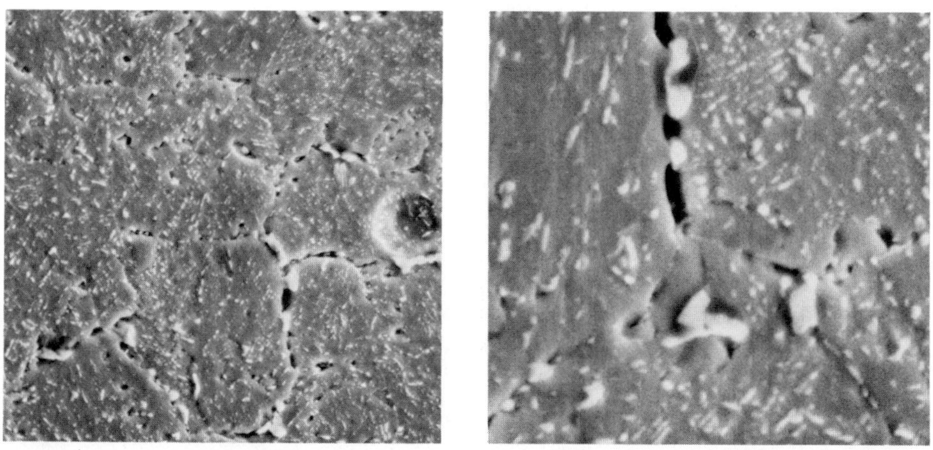

(A) (B)

Fig. 27. SEM photographs of a creep-low-cycle fatigue interspersion test specimen of vacuum degassed Cm-Mo-V steel. (Provided to the EPRI RP502 project by the Metal Properties Council.) Original Mag. (A) 2400X, (B) 6000X. (Reduced approximately 30 percent for reproduction).

ation of these phenomenon in lowering the inter-inclusion ligament's ability to withstand the thermal-mechanical duty without cracking has not been determined. Also in need of further research is the quantification of the area (or volume) fraction of inclusions which must be exceeded before linkup becomes a highly probable process. Developing the critical relationship between the defect as inferred by bore sonic non-destructive inspection results and the same defect as defined by destructive metallurgical examination is the key thrust of the current EPRI project work.

ACKNOWLEDGEMENTS

Dr. Charles A. Rau, Failure Analysis Associates, is warmly thanked for his detailed technical help. In compiling this paper, liberal use has been made of text written by others. Special recognition for such contributions go to Dr. Richard E. Smith, EPRI, for the IGSCC in BWR 304SS Piping Systems writeup, to Dr. J.T. Adrian Roberts, EPRI, for the SCC in Zircaloy Fuel Rod Cladding discussion, to Mr. Lou Martel, Technical Project Manager for the EPRI Steam Generator Project, and to Dr. Clifford H. Wells, as Technical Project Manager for EPRI contract RP502 on Reliability of Steam Turbine Rotors. General thanks and recognition is offered to the many persons who, as part of an EPRI contract team and also independently, have worked to provide the data which has herein been summarized.

REFERENCES

1. FAA Staff, "Failure Analysis and Failure Prevention in Electric Power Systems: First Annual Progress Report", (August, 1975); (EPRI 217–1–IR1).
2. P.M. Besuner, "An Engineering Fracture Mechanics Analysis of the Pilgrim I Nozzle-To-Pressure Vessel Weld Discontinuities", (October, 1975); (EPRI 217–1–TR6).
3. FAA Staff, "Failure Analysis and Failure Prevention in Electric Power Systems", Final Report for Project RP217–1, (November, 1976); (EPRI NP–280).
4. FAA Staff, "Failure Analysis and Failure Prevention in Electric Power Systems", Annual Report, Project RP700–1, (July, 1977).
5. I. Roman, *et al.,* "A Study of Crack Growth Under Operational Conditions in Steam Turbine Steel: Phase 1 Report", (July, 1976); (EPRI NP–325).
6. R.E. Smith, "Progress in Reducing Stress Corrosion Cracking in BWR Piping:, APC Paper No. 3, Session V, (April 18–20, 1977).
7. H.H. Klepfer, *et al.,* "Investigation of Causes of Cracking in Austenitic Stainless Steel Piping", Volumes 1 and 2, GENED 21000– 1 and 2, 75NED32, Class 1, (July, 1975).
8. NRC Pipe Cracking Study Group, "Technical Report, Investigation and Evaluation of Cracking in Austenitic Stainless Steel Piping of Boiling Water Reactor Plants", NUREG–75/067, (October, 1975).
9. P. Rao, "Microstructural Studies of BWR Pipe", Topical Report for EPRI Project RP449–2, NEDC–21229, (April, 1976).
10. First Semiannual Progress Report, EPRI RP701–1 – "Evaluation of Near-Term BWR Piping Remedies", NEDC–21463–1, (November, 1976).
11. J.T.A. Roberts, *et al.,* "Planning Support Document for the EPRI Light Water Reactor Fuel Performance Program", EPRI Special Report; (January 1977); (NP–370–SR).
12. J.T.A. Roberts, *et al.,* "On the Pellet-Cladding Interaction Phenomenon", ANS Winter Meeting; (November 1976). (To be published in *J.Nuc.Tech.,* August 1977).
13. N. Fuhrman and V. Pasupathi, "Joint CE/EPRI Fuel Performance Evaluation Program: Task C; Evaluation of Fuel Rod Performance in Maine Yankee Core I", Report CENPD–221, December 1975; and N. Fuhrman, *et al.,* ibid., Final Report for Task C of Project 586–1, (November 1976); (EPRI NP218).
14. B. Cox and J.C. Wood, "Iodine Induced Cracking of Zircaloy Fuel Cladding – A Review", Corrosion Problems in Energy Conversion and Generation, Ed. C.S. Tedmon, Jr., *Pub. The Electrochemical Society,* pp. 275–321 (1974).

15. E. Smith, "A Criterion for Failure of Zircaloy Cladding in a Water Reactor Fuel Rod Subjected to Power Ramps", Proceedings of the ANS Topical Meeting on Water Reactor Fuel Performance, St. Charles, Illinois, (May, 1977), (to be published).
16. W.D. Fletcher and D.D. Malinowski, "Operating Experience with Westinghouse Steam Generators", *Nuclear Technology,* 28, p. 356 (1976).
17. "Steam Generator Update 1976", Nuclear Energy Systems, Westinghouse Electric Corporation.
18. E.P. Morgan, *et al.,* "Examination of Denting and Characterization of Associated Materials in the Plate-Tube Intersections of Westinghouse Nuclear Steam Generators", Scientific Paper 76–7D2–SGEXM–P1, (September 27, 1976).
19. L.D. Kramer and D.D. Randolph (Part I Metallurgical Considerations), D.A. Weisz (Part II Mechanical Analysis), "Analysis of the Tennessee Valley Authority Gallatin No. 2 Unit Turbine Rotor Burst", ASME–MPC Symposium on Creep-Fatigue Interaction, ASME, New York, (1976).
20. C.H. Wells, "The EPRI Program to Increase the Reliability of Steam Turbine Rotors", Presented at the EPRI Workshop on Improved Turbine Availability, January 17–19, 1977; and "EPRI RP502, Reliability of Steam Turbine Rotors", Interim Report, (June 16, 1977).
21. G.A. Clarke, L.D. Kramer, and L.K. Tu, "RP502–4, Reliability of Steam Turbine Rotors", Semi-Annual Report No. 1, (August 19, 1976); and G.A. Clarke and L.D. Kramer, Semi-Annual Report No. 2, (April 19, 1977).
22. C.H. Wells and T.S. Cook, "EPRI RP502, Reliability of Steam Turbine Rotors Task I. Lifetime Prediction Analysis System", First Semi-Annual Progress Report, Southwest Research Institute, (July 1, 1976).
23. R.M. Curran and B.M. Wundt, "Continuation of a Study of Low-Cycle Fatigue and Creep Interaction in Steels at Elevated Temperatures", *1976 ASME–MPC Symposium on Creep-Fatigue Interaction,* ASME, New York, pp. 203–282 (1976).

AN ENGINEER'S OVERVIEW OF PRODUCT LIABILITY

Gordon W. Powell* and Salah Mahmoud**

The ramifications of product liability suits are diverse in nature and are of considerable economic importance. With regard to the latter, Senators J.C. Culver and G. Nelson[1] have reported recently in the *Congressional Record* that the number of product liability suits will rise from approximately 50,000 cases 10 years ago to 2,000,000 in 1980, the average award having increased from $11,000 per case to approximately $100,000 currently. The cost to the insurance industry is several billion dollars per year and, consequently, insurance premiums have increased to the point at which they have become a major business cost in some industries. Consumer awareness has been a powerful force behind the upsurge in product liability suits.

The objective of the discussion which follows is to examine the current situation in several areas pertinent to product liability. The overall topic is broad in scope and extremely complex. Hence, some of the discussion may border on being trivial from an expert's point of view, but nevertheless, the authors hope that the discussion will provide a good perspective of this interesting and vital field. The coverage will include commentary on the product liability trial and the effects which product liability suits have had on the manufacturing and insurance industries, education, technical personnel and the law.

SOME ASPECTS OF PRODUCT LIABILITY TRIALS

The expert witness is an important figure in a product liability suit. The courtroom demeanor and dress of the expert witness can exert a strong influence on the jurors. It is not uncommon to observe some of the members of a jury which is hearing a technical case dozing or at least showing signs of boredom. Consequently, although the jurors may not recall and comprehend all the technical facts of a case, they will definitely have a decided impression (correct or not) of the competence of each expert witness. Was the witness hesitant in his speech? Did he become flustered during cross examination? Was he forced to acknowledge that a statement he made or an opinion he had put forth was incorrect? Thus, it is quite possible that a witness who creates a good impression with a jury may simply be much more of an actor than an expert.

Again, relative to expert testimony, an attorney during cross examination may not realize that the line of questioning is leading into a critical area about which the witness may have to express some uncertainty. Often times, however, the attorney does not press his advantage because of his lack of technical training and a full understanding of the problem at hand. If the attorney has sufficient resources, he can retain a technical expert to advise him in the courtroom as the testimony evolves and, in cases involving large corporations, this is the procedure which may be followed. The immediate

*Department of Metallurgical Engineering, The Ohio State University, Columbus, Ohio 43210 USA.
**CTL Engineering, Inc., 2860 Fisher Rd., Columbus, Ohio 43204 USA.

availability of technical assistance in the courtroom puts a great deal of pressure on the expert witness who is giving testimony and, consequently, it is easier for the attorney to expose the *actor* or any weaknesses in the expert's testimony.

Although the expert witness may play a prominent role in a product liability suit, it is very often the case that the plaintiff is the most influential figure in a trial. The attorney for the plaintiff will make repeated references to the severity of his client's injuries and will eventually parade his maimed or disfigured client before the jury. Quite recently, one of the authors was conversing with an attorney who was representing a young and handsome paraplegic farmer injured in a chain accident. The attorney stated "I have the best metallurgist in the world on my side——my client". Attorneys are not loath to use their clients in this manner because the sympathy factor can readily overwhelm a reasonable defense position based upon contributory negligence or assumption of risk.

With regard to the matter of evidence, the defense is not obligated to provide evidence which is not requested specifically by the plaintiff's counsel. On the other hand, the plaintiff's attorney also may find that it is to his advantage to withhold evidence which his experts have uncovered. The following example of this kind of situation is based upon an actual experience of one of the authors. A mother died and her two children were severely burned in a trailer fire caused by a propane gas leak. The source of the fire was traced to the small furnace in the trailer. The plaintiff's attorney filed suits against the trailer manufacturer, the manufacturer of the furnace, and also the nationally-known manufacturer of the controller on the furnace. One of the plaintiff's experts found positive evidence that the gas had leaked from a loose gas-inlet fitting on the furnace controller; such being the case, the trailer manufacturer was liable. However, the attorney chose not to divulge this evidence because the trailer manufacturer was a small firm and did not have adequate assets or insurance coverage to merit the settlement which the attorney thought was justified. Subsequently, the experts for the controller manufacturer not being able to uncover the same evidence nor being able to show the controller was not defective because it had been partially destroyed by the fire, the company agreed to an out-of-court settlement of a few million dollars.

A product liability trial often appears to be composed of a sequence of random events, the presentation of technical evidence being dispersed between testimony relative to the injury-producing accident and the nature of the injuries. In order to provide for a more orderly and logical flow of evidence, Weinstein et al.[2] have proposed that a seriated trial be used for product liability cases. The initial segment of the trial would involve evidence pertinent only to the product and ultimately whether it was defective and unreasonably dangerous considered apart from the actual injury-producing event. The jury would then be asked to make the decision as to whether or not the product was defective and unreasonably dangerous. If an affirmative decision is reached, then the trial continues and evidence relative to the injury-producing event is presented on the basis of which the jury must determine if the defective product caused the accident. If an affirmative decision is reached on this matter, the trial then turns to the final question of the extent of liability of the manufacturer. The step-by-step methodology of the seriated trial insures that the jury will be properly informed of the design and use of the product and consequently will be able to make more knowledgeable decisions.

Some large corporations use their laboratory facilities to provide test data in support of their case and also use their technical personnel as expert witnesses during the trial. The authors believe it is far wiser for a company to retain an independent testing laboratory and independent technical experts to develop and support its case in court. The opinions of an expert witness who is employed by the company on trial cannot be regarded as unbiased and independent because he is obviously an advocate of his employer's cause. For the same reason, test data obtained by an independent laboratory are more convincing to a jury. Some additional comments on testing laboratories seem appropriate at this point because they are involved in a significant way in many of the various aspects of product liability. Firstly, a testing laboratory must establish a quality assurance program which is a quaranty of the validity of all test results. The essential features of such a program are the proper training of the laboratory personnel, strict adherence to standardized test methods and periodic calibration of test equipment. Secondly, in some situations, engineering analysis of test data may yield tenuous conclusions because either non-representative samples or an insufficient number of samples were furnished by the client. Consequently, the writing of an engineering report may require close attention to word selection and, in particular, the preference for moderate, conditional phraseology and the avoidance of superlatives and absolute phraseology. For example, in some cases, prudence may dictate the use of "the test results suggest that..." rather than "the test results prove that...;" the use of unwarranted wording may have dire consequences in a product liability suit. And finally, testing laboratories which become involved in product testing may be asked to certify or endorse a product on the basis of the test results. If the certification or endorsement is used by the client to induce the consumer to purchase the product, then the testing laboratory becomes a potential third-party defendant in a product liability suit.

And as a final matter for discussion, the question of fees paid to expert witnesses and engineering firms is worthy of comment. The contingent fee usually charged by counsel in a product liability case is one-third to one-half of the award after payment of the pretrial expenses. In a recent and typical case, the attorney received $120,000 of a $410,000 award. The fee paid to the engineering firm which established the technical aspects of the plaintiff's case was approximately $3,000. The relative remuneration of the counsel and the technical people appears at times to be distributed disproportionately. It is doubtful that it can ever be otherwise because, if a technical expert, for example, also was to be paid for his services on a contingent fee basis, then he would have to be regarded as an advocate and thus incapable of independent, unbiased testimony. With regard to this matter, it is noteworthy that the code of ethics approved by the Board of Directors of the Engineers' Council for Professional Development on October 1, 1974, states that "engineers shall not request, propose nor accept professional commissions on a contingent basis under circumstances under which their professional judgments may be compromised, or when a contingency provision is used as a device for promoting or securing a professional commission."

BURDENS IMPOSED BY PRODUCT LIABILITY ON INDUSTRY

The specter of product liability has put pressure on manufacturers to market products which are not only economically competitive but also safe for consumer use and even abuse. Consequently, the design and testing, manufacture, quality control, packaging, marketing and advertising of a new product must involve consideration of potential product liability. One very important component of a product liability control program within a corporation is the recording of all design meetings and calculations, test data on the materials used to manufacture the product and on the performance of the product, etc. In the event of a product liability claim, the corporation can use such records to demonstrate that its engineers made a conscious effort to design and manufacture a product that was safe and consumer proff; engineers must be consumer conscious. Another important component of a product liability control program is the rapid acqusition of information on in-the-field product performance and the transformation of this information into changes in the design or manufacture of the product. It is to be expected that trade associations will become important sources of information and assistance for their members on matters relevant to product liability. The National Electric Manufacturing Association has established a product liability program which provides its members with information on sources of expert witnesses, insurance and defense against suits. [3]

The instruction manual and the labeling of a product also must be subject to close scrutiny in any viable corporate product liability control program. The manufacturer has a duty to warn the consumers of any hazards associated with the use of a product and also to warn the consumer of dangers associated with unintended abnormal use of a product if such circumstances are foreseeable; here again, it is patently evident that an engineer who is writing an instruction manual for a product must be consumer conscious. As a simple example, if a manufacturer has determined that a vehicle which it manufactures is unsafe above a certain speed and foresees situations in which the consumer may exceed the speed, then there is a duty to warn of the danger. The warning must be presented in such a form as to attract the consumer's attention and make him aware of the potential danger. If the warning is on a package or a label or even an instruction manual, the size and coloring of the lettering can effectively draw the consumer's attention to the warning. Each of us has undoubtedly had the experience of being dissuaded from the purchase of a product by a warning on a label or package. Consequently, disagreements over the wording of a warning can arise between the technical and marketing personnel when there is a possibility that sales will be affected by the warning.

There are many other and diverse facets to the product liability problem faced by industry. For example, in past years, the relationship which existed between the public utilities and the manufacturers of power-generating equipment was a strong and friendly one. If a problem arose with a major piece of equipment, the costs of the repair or replacement of the equipment were simply passed along to the consumers in the form of higher rates. However, today the public utility commissions are more consumer conscious and, if it can be established that an equipment failure was caused by a design or manufacturing defect, the commissions are reluctant to pass the costs along to the consumers, thus effectively making the manufacturer liable. Consequently, product liability suits in this industry are not uncommon today. An example of more direct consumer action to forestall increases in public-utility rates caused by an industrial accident is provided by the following situation. A class-action suit has been initiated against a public utility to prevent the costs of the damages which resulted from the failure of a dam from being passed along to the consumer by an increase in rates. Although the suit may be successful, the costs for this catastrophe will be paid ultimately by the general consumer through an increase in the price of some product, insurance, etc.

And yet another facet to product liability is the ethical problem which confronts the production engineer when he knows that in order to meet a production quota faulty equipment or materials are leaving the plant. Such a situation is very real (production of faulty military weapons, engine blocks, etc.); other examples of situations which try an engineer's ethics have been reported by Peters. [4] What recourse is available to the engineer whose attempts to have production of a faulty product stopped or a faulty product recalled are unsuccessful? The threat of loss of his job may resolve the matter. On the other hand, should he retain a lawyer? Should he go public? It is an appropriate time for the professional engineering societies to address themselves to this issue. The societies might take some relatively passive action such as the formulation of a set of rules of conduct or provide active support and counsel to individuals who are faced with an ethical problem and who request help. The societies also should support any legislation which calls for the licensing of all practicing engineers and the cancellation of an engineer's license on grounds of unethical conduct or incompetence. The latter would be at least as important as the possible loss of a job when one is forced to consider the consequences of any course of action.

Reference has been made earlier to the rising cost of product liability insurance. Many smaller industrial concerns have chosen to carry only very limited coverage or none at all. Under these circumstances, a plaintiff who has a legitimate claim against such a company is faced with the strong possibility that he will be denied any recovery to which he is entitled. Senate Bill 527 which is co-authored by Senators J.C. Culver and G. Nelson is an attempt to alleviate the burden of rising product liability insurance premiums on small businesses; the Select Committee on Small Business began hearings on this bill on March 9 of this year (1977). The objective of this bill is "to furnish reinsurance for product liability insurers for small business concerns which would not otherwise be able to obtain product liability insurance on reasonable terms". As pointed out in a *Washington Post* editorial of February 24, 1977: "At best——if the small businessmen are right and there is a crisis——this can only be a band-aid. At worst——if they are wrong and the consumer advocates are right——it would put the government in the position of underwriting the carelessness and mistakes of those it is trying to help." The best and cheapest insurance is unquestionably a well-engineered, well-manufactured safe product.

THE PLIGHT OF THE INSURANCE COMPANIES

The principal reaction of the insurers of the manufacturers of consumer products, machinery, pressure vessels, etc. to the escalating costs associated with product liability suits has been simply to increase premiums. However, a significant change may be presaged by the fact that an insurer of boiler and pressure vessels has acquired a technically-based company with chemical and metallurgical capabilities to conduct failure analyses in its behalf; in one case alone, the failure analysis conducted by the subsidiary company resulted in a saving of 1.3 million dollars to the insurer. Other insurers may very well follow this example.

But the authors believe there is an alternate path open to the insurers to achieve the same capabilities and much more at minimal cost. Interested insurers could band together and establish what will be called here an "Engineering Institute" which is staffed by technical personnel with expertise in a wide variety of disciplines and also has the types of equipment used routinely in product liability investigations. The functions of this jointly-supported but independent institute would be

1) to determine technical liability when members (supporters) of the institute find themselves confronting one another in some product-liability-related matter, i.e., an arbitration function,

2) to perform failure analyses for its members,

3) to review inspection procedures, new plant and equipment and designs, etc., and

4) to provide input to governmental and industrial committees assigned the task of revising and/or formulating standards and specifications.

The institute could obviously perform other services, but, of those listed above, perhaps the arbitration function is the most important because of the savings in time and money to the members.

State and Federal legislation which would provide for reinsurance, establish limits on awards, etc. would benefit the insurance companies and may be forth coming in the future.

RECENT LEGISLATION

The National Commission on Product Safety which was established in 1967 concluded that "the imposition of liability on responsible parties for harm caused by their products was not a sufficient deterrent to the sale of unsafe products"[5] and therefore recommended the enactment of federal legislation to protect the consumer. This recommendation resulted in the Consumer Product Safety Act of 1972 which established a Consumer Product Safety Commission (and an Advisory Council) to protect the consumer from unreasonable risks of injury associated with manufactured products through the formulation of uniform safety standards. The Commission maintains an Injury Information Clearinghouse which collects data from over 100 hospital emergency rooms and communicates to each manufacturer any information which indicates that there may exist significant risks associated with the use of one of its products. Enforcement of the Consumer Product Safety Act can be initiated by governmental action or action originated by a private person.

A most noteworthy event has been the passage of the first state law related directly to product liability. This law was passed by the Utah State legislature and became effective on May 22, 1977. The Utah Product Liability Act provides that:

1) liability expires 6 years after the date of original purchase or 10 years after the date of manufacture;

2) alteration, modification or misuse of a product relieves the manufacturer of liability;

3) the defect must exist or the danger be inherent in the product at the time of manufacture or sale;

4) the amount of damages sought in an injury suit cannot be specified;

5) the existence of unreasonably dangerous defects must be established beyond normal use of a product; and

6) if a product was designed and manufactured according to a government standard, then conformity to the standard is a reasonable defense in a product liability suit.

There will certainly be a great deal of interest in the outcome of suits tried under this statute. Attorneys will undoubtedly welcome the additional avenue of defense provided by the conformity-to-a-standard provision of the law. However, standards are fluid and subject to change with the evolution of design and materials technology. Furthermore, although specifications are in principle unbiased documents, they are in fact industrial instruments because the membership of a committee charged with the formulation of a specification will more likely be composed predominantly of representatives of industry and/or persons with ties to industry due prior employment. Consequently, economic as well as technical considerations may determine the wording and content of a specification.

Representatives of the member states of the Council of Europe have prepared a document entitled "European Convention on Products Liability in Regard to Personal Injury and Death". This convention was opened to signature on January 27, 1977, and has been signed by Belgium, France and Luxembourg, one-third of the member states. Amongst a diversity of matters, the Convention provides for a reduction or disallowment of a claim due to contributory negligence, liability limitation periods and the amount of compensation to be paid under certain circumstances. It is worthy of note that an earlier draft [6] of this Convention contained an article (not included in the final draft) which stated "The producer shall be liable even if the article could not have been regarded as defective in the light of the scientific and technological development at the time when he put the article into circulation". This article would disallow a defense based upon conformity to a standard but which is permitted by the Utah Product Liability Act referred to earlier.

ENGINEERING EDUCATION

Some departments within a College of Engineering may offer a course with a title such as "Professional Aspects of Civil Engineering", for example. The course content may include material which pertains to the professional societies and codes of ethics, employment practices, the engineer-client relationship, etc. The material is not particularly exciting and, consequently, the existence of such a course may depend upon a single faculty member who believes the course is an important part of the students' education. However, the situation may change because there is a movement within the Engineers' Council for Professional Development (ECPD) to examine the amount of classroom time devoted to ethics and professionalism [7]. ECPD accredits engineering curricula and thus ECPD's concern over ethics and professionalism may presage a curriculum requirement on this topic for all engineering students.

At The Ohio State University, a course entitled "Legal Aspects of Engineering" is offered in the Department of Chemical Engineering. The course is unique in that the instructor, Professor T. Sweeney, has a law degree in addition to his engineering degrees. The course content includes discussions of the American legal system and basic legal concepts, patents, trade secrets, personal liability of engineers, expert testimony and many other topics.

With regard to the continuing education of the graduate engineer in the field of product liability, symposia on this topic are becoming more frequent. As recent examples, the Milwaukee School of Engineering sponsored a two-day (May 16,17, 1977(symposium entitled "Products Liability Exposure Control" and another symposium entitled "Product Liability Prevention" was held in Palo Alto, California, on March 16–18, 1977. A new journal, *Journal of Products Liability,* is being published by Pergamon Press and it should be a valuable source of current information on this field.

REFERENCES

1. Congressional Record–Senate, S1752–1753, January 31, 1977.
2. Weinstein, A.S., Twerski, A.D., Piehler, H.R. and Donaher, W.A., "Product Liability: An Interaction of Law and Technology", *Duquesne Law Review,* 12, No. 3, p. 425 (1974).
3. Schreiber, H.A., "The Role of Marketing in Product Liability Exposure Control", Products Liability Exposure Control Symposium, Milwaukee School of Engineering, May 16 and 17, 1977.
4. Peters, G.A., "The Engineer–Lawyer Interface: Abrasive Conflict or Harmonins Interaction?", *Proc. Product Liability Prevention (W),* p. 189 (1977).
5. Frumer, L.R., and Friedman, M.I., *Products Liability,* Matthew Bender, New York (1976).
6. "Proposal for a Council Directive relating to the approximation of the laws, regulations and administrative provisions of the Member States for defective products", *Official Journal of the European Communities,* No C241/9–No C241/12 (October 14, 1976).
7. Dean M.L. Smith, The Ohio State University, private communication.

AUTHOR INDEX

Z

SUBJECT INDEX